# GUIDE TO BUILDING CO

## FOR DOMESTIC BUILDINGS

# GUIDE TO BUILDING CONTROL
## FOR DOMESTIC BUILDINGS

Anthony Gwynne MRICS, MIFireE

**WILEY-BLACKWELL**

A John Wiley & Sons, Ltd., Publication

This edition first published 2013
© 2013 John Wiley & Sons, Ltd

Blackwell Publishing was acquired by John Wiley & Sons in February 2007. Blackwell's publishing programme has been merged with Wiley's global Scientific, Technical and Medical business to form Wiley-Blackwell.

Registered office:
John Wiley & Sons Ltd, The Atrium, Southern Gate, Chichester, West Sussex, PO19 8SQ, UK

Editorial offices:
9600 Garsington Road, Oxford, OX4 2DQ, UK
The Atrium, Southern Gate, Chichester, West Sussex, PO19 8SQ, UK
2121 State Avenue, Ames, Iowa 50014-8300, USA

For details of our global editorial offices, for customer services and for information about how to apply for permission to reuse the copyright material in this book please see our website at www.wiley.com/wiley-blackwell.

The right of the author to be identified as the author of this work has been asserted in accordance with the UK Copyright, Designs and Patents Act 1988.

All rights reserved. No part of this publication may be reproduced, stored in a retrieval system, or transmitted, in any form or by any means, electronic, mechanical, photocopying, recording or otherwise, except as permitted by the UK Copyright, Designs and Patents Act 1988, without the prior permission of the publisher.

Designations used by companies to distinguish their products are often claimed as trademarks. All brand names and product names used in this book are trade names, service marks, trademarks or registered trademarks of their respective owners. The publisher is not associated with any product or vendor mentioned in this book. This publication is designed to provide accurate and authoritative information in regard to the subject matter covered. It is sold on the understanding that the publisher is not engaged in rendering professional services. If professional advice or other expert assistance is required, the services of a competent professional should be sought.

All reasonable attempts have been made to contact the owners of copyrighted material used in this book. However, if you are the copyright owner of any source used in this book which is not credited, please notify the Publisher and this will be corrected in any subsequent reprints or new editions.

Library of Congress Cataloging-in-Publication Data
Gwynne, Anthony, author.
  Guide to building control: for domestic buildings / Anthony Gwynne.
      pages cm
  Includes bibliographical references and index.
  ISBN 978-0-470-65753-9 (pbk.)
  1. Standards, Engineering–Great Britain.  2. Building–Great Britain–Quality control.  I. Title.
  TH420.G89 2013
  690.02′1841–dc23
                                            2012031586

A catalogue record for this book is available from the British Library.

Wiley also publishes its books in a variety of electronic formats. Some content that appears in print may not be available in electronic books.

Set in 10/12.5 pt Times by Toppan Best-set Premedia Limited
Printed and bound in Malaysia by Vivar Printing Sdn Bhd

Cover image courtesy of Anthony Gwynne
Cover design by Workhaus

# Contents

| | |
|---|---|
| *About the Author* | xi |
| *Acknowledgements* | xiii |
| *Notes to the Reader* | xv |
| *Expected Changes to the Building Regulations to Come into Force in 2013* | xvii |

### Section 1  General information — **1.1**

| | |
|---|---|
| Introduction | 1.3 |
| The Building Act 1984 and the Building Regulations 2010 | 1.3 |
| Approved Documents | 1.4 |
| Other ways of satisfying the Building Regulations requirements | 1.5 |
| Technical and condensation risks | 1.5 |
| Timber-sizing tables independently calculated by GEOMEX for solid timber members | 1.6 |
| Engaging a property professional | 1.6 |
| Obtaining Building Regulations approval | 1.7 |
| Notices of stages of works | 1.9 |
| Exempt buildings and work | 1.9 |
| Preliminary works | 1.11 |
| Matters related to the Building Regulations | 1.14 |
| The Party Wall Act 1996 | 1.15 |
| Rights of Light | 1.17 |

### Section 2  Domestic extensions — **2.1**

| | |
|---|---|
| Part A: Structure | 2.9 |
| A1: Sub-structure | 2.9 |
| Foundations | 2.9 |
| Ground floors and sub-structure walls | 2.18 |
| A2: Superstructure | 2.28 |
| Minimum headroom heights | 2.28 |
| External walls | 2.28 |
| External cavity wall construction | 2.35 |
| Lateral restraint strapping of upper floors to walls | 2.49 |

| | |
|---|---:|
| Lateral restraint strapping of roofs to walls | 2.51 |
| Lateral restraint strapping of walls at ceiling level | 2.52 |
| A3: Separating walls and floors | 2.53 |
| Masonry party walls separating dwellings | 2.53 |
| Double-leaf timber-frame party walls separating dwellings | 2.54 |
| A4: Internal partitions | 2.56 |
| A5: Intermediate upper floor(s) | 2.56 |
| Floor joists | 2.56 |
| Trimming and trimmer joists | 2.58 |
| A6: Pitched roofs | 2.60 |
| Pitched roof coverings | 2.61 |
| Pitched roof structure | 2.62 |
| A7: Flat-roof construction | 2.70 |
| Option 1: Flat roof with 'cold deck' | 2.70 |
| Option 2: Flat roof with 'warm deck' | 2.72 |
| Option 3: Flat roof with inverted 'warm deck' (insulation on top of waterproof coverings) | 2.74 |
| Option 4: Flat roof with green roof on 'warm deck' (either intensive or extensive) | 2.74 |
| The design, workmanship and selection of materials for flat roofs | 2.75 |
| A8: Mortars, renders and gypsum plasters | 2.75 |
| Cement mortars and renders | 2.75 |
| Gypsum plasters | 2.77 |
| Part B: Fire safety and means of escape | 2.78 |
| Fire detection and fire alarm systems | 2.78 |
| Means of escape | 2.79 |
| Surface spread of flame: internal wall and ceiling linings including roof lights | 2.89 |
| Part C: Site preparation and resistance to contaminants and moisture | 2.92 |
| C1: Resistance to contaminants | 2.92 |
| Radon gas | 2.93 |
| Methane and other ground gas protection | 2.96 |
| C2: Resistance to moisture | 2.96 |
| Part D: Cavity wall filling with insulation | 2.98 |
| Part E: Resistance to the passage of sound | 2.98 |
| Part F: Ventilation | 2.99 |
| Purge (natural) ventilation | 2.99 |
| Mechanical extract ventilation and fresh air inlets for rooms without purge ventilation | 2.99 |
| Background ventilation | 2.100 |
| Intermittent mechanical extract ventilation | 2.100 |
| Part G: Sanitation, hot-water safety and water efficiency | 2.100 |
| Wholesome hot and cold water supply | 2.100 |
| Solar water heating | 2.102 |
| Electrical water heating | 2.102 |
| Insulation of pipework to prevent freezing | 2.102 |

| | |
|---|---:|
| Supply (Water Fittings) Regulations 1999 | 2.103 |
| Part H: Drainage and waste disposal | 2.103 |
| H1: Foul- and storm-water drainage | 2.103 |
| H2: Septic tanks, sewage treatment systems and cesspools | 2.113 |
| Septic tanks | 2.113 |
| Sewage treatment systems | 2.114 |
| Percolation tests | 2.116 |
| Cesspools | 2.118 |
| H3: Rainwater drainage and harvesting | 2.119 |
| H4: Building over or close to, and connections to, public sewers | 2.122 |
| Building over or close to a public sewer | 2.122 |
| Connections to public sewers | 2.123 |
| H5: Separate systems of drainage | 2.124 |
| H6: Solid waste storage | 2.124 |
| Part J: Combustion appliances and fuel storage systems | 2.124 |
| Solid fuel appliances up to 50 kW rated output | 2.125 |
| Appliances other than solid fuel | 2.133 |
| Fuel storage tanks | 2.134 |
| Renewable energy/micro regeneration installations | 2.134 |
| Part K: Protection from falling, collision and impact | 2.135 |
| Internal stairs, guarding and landings for changes in level of 600 mm or more | 2.135 |
| External stairs, guarding and landings for changes in level of 600 mm or more | 2.138 |
| Loft conversion stairs | 2.139 |
| Ramps | 2.140 |
| Part L: Conservation of fuel and power in existing dwellings | 2.141 |
| Areas of external windows, roof windows and doors | 2.141 |
| New thermal elements | 2.141 |
| Energy-efficient lighting | 2.143 |
| Insulation of pipework to prevent freezing | 2.143 |
| External walls, roofs, floors and swimming-pool basin | 2.143 |
| Renovation/upgrading of existing thermal elements | 2.144 |
| Part M: Access to and use of buildings for disabled | 2.145 |
| Part N: Safety glazing, opening and cleaning | 2.146 |
| Safety glass and glazing | 2.146 |
| Part P: Electrical safety | 2.147 |
| Electrical installations | 2.147 |
| External works – paths, private drives, patios and gardens | 2.149 |

## Section 3  New dwellings — 3.1

| | |
|---|---:|
| Parts A and L: Starting point | 3.4 |
| Conservation of fuel and power in new dwellings | 3.4 |
| Criterion 1 – Achieving the Target Emission Rate (TER) | 3.4 |

| | |
|---|---|
| Criterion 2 – Limits on design flexibility | 3.5 |
| Criterion 3 – Limiting the effects of solar heat gain | 3.5 |
| Criterion 4 – Calculation of the Dwelling Design Emission Rate (DER) | 3.6 |
| Criterion 5 – Provision for energy-efficient operation of the dwelling | 3.10 |
| Insulation guidance details for floors, walls and roofs | 3.10 |
| Guidance on the Code for Sustainable Homes for new dwellings | 3.12 |
| Guidance on PassivHaus | 3.18 |
| Part B: Fire safety and means of escape | 3.19 |
| Part C: Site preparation and resistance to contaminants and moisture | 3.20 |
| Part D: Cavity wall filling with insulation | 3.20 |
| Part E: Resistance to the passage of sound | 3.20 |
| Part F: Ventilation to new dwellings | 3.21 |
| Ventilation systems | 3.21 |
| Purge (natural) ventilation to habitable rooms: system 1 – new dwellings | 3.22 |
| Background ventilation: system 1 – new dwellings | 3.23 |
| Intermittent mechanical extract ventilation: system 1 – new dwellings | 3.24 |
| Part G: Sanitation, hot-water safety and water efficiency | 3.24 |
| Part H: Drainage and waste disposal | 3.25 |
| Part J: Combustion appliances and fuel storage systems | 3.26 |
| Part K: Protection from falling, collision and impact | 3.26 |
| Part L: Conservation of fuel and power | 3.26 |
| Part M: Access to and use of buildings for disabled | 3.26 |
| Guidance on Lifetime Homes Standard for new dwellings | 3.31 |
| Part N: Safety glazing, opening and cleaning | 3.32 |
| Part P: Electrical safety | 3.32 |

## Section 4   Domestic loft conversions    4.1

| | |
|---|---|
| Converting an existing loft space | 4.3 |
| Assessing the feasibility of your loft for conversion | 4.3 |
| Part A: Structure | 4.5 |
| A1: Inspection of the existing roof and building structure | 4.5 |
| A2: Alteration, modification and strengthening of the existing roof structure | 4.5 |
| A3: Roof conversion details | 4.5 |
| Upgrading existing external walls | 4.5 |
| Internal load-bearing timber stud walls | 4.7 |
| Part B: Fire safety and means of escape | 4.11 |
| Single-storey dwellings with loft conversion | 4.11 |
| Two-storey dwellings with loft conversion (or new third storey) | 4.13 |
| Part C: Site preparation and resistance to contaminants and moisture | 4.17 |
| Part D: Cavity wall filling with insulation | 4.17 |
| Part E: Resistance to the passage of sound | 4.17 |
| Part F: Ventilation | 4.17 |
| Part G: Sanitation, hot-water safety and water efficiency | 4.17 |
| Part H: Drainage and waste disposal | 4.17 |

| | |
|---|---|
| Part J: Combustion appliances and fuel storage systems | 4.17 |
| Part K: Protection from falling, collision and impact | 4.17 |
| Part L: Conservation of fuel and power in conversions | 4.17 |
| Part M: Access to and use of buildings for disabled | 4.18 |
| Part N: Safety glazing, opening and cleaning | 4.18 |
| Part P: Electrical safety | 4.18 |

## Section 5 Domestic garage and basement conversions into habitable rooms and conversion of barns and similar buildings into new dwellings — 5.1

| | |
|---|---|
| Assessing the feasibility of your building for conversion | 5.6 |
| Part A: Structure | 5.14 |
| A1: Underpinning works | 5.14 |
| Traditional underpinning | 5.14 |
| Sections through proposed underpinning (*not to scale*) | 5.15 |
| A2: Single-wall garage conversions (or similar buildings) into habitable rooms (typical details indicated in Figures 5.2–5.12) | 5.16 |
| Upgrading pitched roofs | 5.16 |
| Upgrading flat roofs | 5.19 |
| Infilling of garage door openings | 5.20 |
| Upgrading single-skin external walls | 5.21 |
| Upgrading garage ground floors (or similar) with upgraded enclosing single-skin walls | 5.26 |
| A3: Cavity wall garage conversions (or similar buildings) into habitable rooms (typical details indicated in Figures 5.16–5.26) | 5.29 |
| Upgrading pitched roof | 5.29 |
| Infilling of garage door opening | 5.31 |
| Upgrading external cavity walls | 5.33 |
| Upgrading garage ground floors (or similar) with enclosing upgraded cavity walls | 5.34 |
| A4: Basement conversions into habitable rooms | 5.38 |
| Existing basements and tanking systems | 5.38 |
| A5: Conversion of barns and similar buildings into new dwellings – Technical and practical guidance | 5.40 |
| Part A: Structure | 5.40 |
| Part B: Fire safety and means of escape | 5.51 |
| Part C: Site preparation and resistance to contaminants and moisture | 5.51 |
| Part D: Cavity wall filling with insulation | 5.51 |
| Part E: Resistance to the passage of sound | 5.51 |
| Performance standards | 5.51 |
| Part F: Ventilation to new dwellings | 5.52 |
| Part G: Sanitation, hot-water safety and water efficiency | 5.52 |
| Part H: Drainage and waste disposal | 5.53 |
| Part J: Combustion appliances and fuel storage systems | 5.53 |
| Part K: Protection from falling, collision and impact | 5.53 |
| Part L: Conservation of fuel and power in conversions | 5.53 |
| Energy Performance Certificate (EPC) | 5.53 |

| | |
|---|---|
| Part M: Access to and use of buildings for disabled | 5.53 |
| Part N: Safety glazing, opening and cleaning | 5.54 |
| Part P: Electrical safety | 5.54 |

## Section 6  Upgrading old buildings using lime and modern applications **6.1**

| | |
|---|---|
| Upgrading old buildings using lime and modern applications | 6.3 |
| Re-pointing and repair of existing buildings | 6.4 |
| Types of lime mortar, lime render/plaster and decorative finish suitable for breathable buildings | 6.10 |

*Index*      I.1

# About the Author

Anthony Gwynne, MRICS, MIFireE, is a Chartered Surveyor and Fire Engineer and has 35 years' experience in the construction industry. He co-manages a building control section and has been in building control for over 19 years. He has been responsible for overseeing the building control function of major developments, including commercial, industrial, healthcare, residential, housing developments, bespoke dwellings, extensions, conversions and works to heritage buildings.

From 1976 to 1977 he worked in Canada on construction projects and from 1977 to 1986 he was apprenticed as a banker mason and was responsible for conservation projects with CADW (Welsh historic monuments and buildings). Following further academic study, he was later with English Heritage as a professional and technical officer, responsible for historic monuments in the south of England. From 1986 to 1993 he was a Building Surveyor with a local authority, dealing with the repair and planned maintenance of buildings including contract procurement and contract administration.

# Acknowledgements

Special thanks to Trud, Craig and Gem.

I would also like to acknowledge and thank the following people and organisations for their contributions to the book:

**Hertfordshire Technical Forum for Building Control**
Extracts of thermal insulation values and tables taken from Technical Note 10: U-Values of Elements
Contact: Trevor Clements www.north-herts.gov.uk/gold_guide_tech_note_10_2010-3.pdf

**Sovereign Chemicals Ltd (Bostik)**
Guidance on tanking systems
Contact: Mark Gillen Mark.gillen@bostik.com; www.sovchem.co.uk

**Ty-Mawr ecological building products**
Breathable buildings and products
Contact: Joyce Gervis www.lime.org.uk

**Kingspan Insulation Ltd**
Insulation values and calculations
Contact: Peter Morgan technical@kingspaninsulation.co.uk; www.kingspaninsulation.co.uk

**Celotex Insulation**
Insulation values and calculations
Contact: technical@celotex.co.uk; tel: 01473 822093

**Knauf Insulation**
Insulation values and calculations
Contact: Chris Roughneen Chris.roughneen@knaufinsulation.com

**Nationwide Fire Sprinklers**
Guidance for domestic sprinklers and fire consultant
Contact: Keith Rhodes Keith.rhodes@nationwide-fire.co.uk; www.nationwidefiresprinklers.co.uk

**Geomex**
Span tables for solid timber members and structural consultant
Contact: Paul Smith Eur.Ing, DipHI, BEng, MSc, C.Eng, MICE, MCMI, MIHT, MCIOB www.geomex.co.uk

**Rockwool**
Insulation values and calculations
Contact: James Rees technical.solutions@rockwool.co.uk

**Lifetime Homes**
Lifetime Homes guidance
Contact: Chris Goodman cgoodman@habinteg.org.uk; www.habinteg.org.uk and www.lifetime-homes.org.uk

**Midland Energy Services Ltd** (trading as MES Energy Services)
Code for Sustainable Homes and PassivHaus guidance and
Sustainable Building Solutions
Contact: Alex Hole Alex.hole@mesenergyservices.co.uk; www.mesenergyservices.co.uk

**SureCav Ltd**
Cavity wall spacer system details
Contact: Charlie Ayres info@surecav.co.uk; www.surecav.com

**Cordek Ltd**
Clay heave product details
Contact: Alistair Seaton aseaton@cordek.com; www.cordek.com

**Liddell Associates** (Architects)
Plans layouts, sections and elevations
Contact: Chris McGonagle cm@liddellachitects.co.uk; www.liddellarchitects.co.uk

**Apex Architecture**
Plans layouts, sections and elevations
Contact: Richard Jones richimjones@hotmail.com; www.apexarchitecture.com

**Neil J. Dransfield** PPCIAT, MCIAT, MCIArb, FCIOB (Chartered Architectural Technologist)
Guidance on The Party Wall Act and Rights of Light
Contact: Neil J. Dransfield architectural@technologist.com; www.dransfield.org.uk

**Walter Leach**
Electrical consultant
Tel: 07976 608108

**Alan Williams Drainage**
Drainage consultant
Contact: Alan Glass Tel: 01792 390309

**Simon Moore** MRICS
Chartered Surveyor
Tel: 01594 840521

**Mark Saunders** MRICS, **Simon Drake** BSc, MRICS, **Aldo Giovanelli** MBEng, **Rob Dickinson** MBEng

**Ian Childs** BSc, MRICS, MIFireE, Chartered Building Control Surveyor (Corporate Approved Inspector)

**Adrian Birch** Senior Lecturer at the University of the West of England

# Notes to the Reader

## Crown Copyright

Crown copyright material (the Building Regulations and Approved Documents) re-used in this guidance has been adapted and/or reproduced under the terms required by Directgov at: www.direct.gov.uk/en/SiteInformation/DG_020460. Information was sourced from http://www.planningportal.gov.uk/buildingregulations/ (last accessed July 2012).

## Approved Documents

The author has reproduced/modified the details contained in the Approved Documents into his own interpretation as contained in this Guide. Where necessary, he has provided additional information that is not available in the Approved Documents. None of the values that are contained within the Approved Documents have been changed. For each table and diagram used or modified, the author has reproduced only the values and information that in his opinion are more commonly used, but he has made it clear that the reader should fully refer to the particular table and diagram in the relevant Approved Document.

The current Approved Documents are available to view on the Department for Communities and Local Government website: www.communities.gov.uk, or to purchase from The Stationery Office (TSO) online at www.tsoshop.co.uk or by telephone: 0870 600 5522.

## Span tables

This Guide uses span tables drawn up by Paul Smith of Geomex (www.geomex.co.uk). However, readers please note that TRADA Technology span tables are available from: www.trada.co.uk/bookshop.

## Disclaimer

The publisher and the author make no representations or warranties with respect to the accuracy or completeness of the contents of this work and specifically disclaim all warranties, including without limitation warranties of fitness for a particular purpose. No warranty may be created or extended by sales or promotional materials. The advice and strategies contained herein may not be suitable for every situation. This work is sold with the understanding that the publisher is not engaged in rendering legal, accounting or other professional services. If professional assistance is required, the services of a competent professional person should be sought. Neither the publisher nor the author shall be liable for damages arising herefrom. The fact that an organisation or website is referred to in this work as a citation and/or a potential source of further information does not mean that the author or the publisher endorses the information the organisation or website may provide or recommendations it may make. Further, readers should be aware that internet websites listed in this work may have changed or disappeared between when this work was written and when it is read.

# Expected Changes to the Building Regulations to Come into Force in 2013

**Part B:** Guidance updated in relation to lighting diffusers in line with the consultation together with changes to classification of wall coverings to align with European classifications (April 2013).

**Parts K, M & N:** Changes to address areas of conflict and overlap to be reflected in a new Approved Document to part K with amendment slips for M & N. Improved guidance on the use of access statements to promote a more proportionate risk-based approach (April 2013).

**Part P:** Reduction in notifiable work but retaining a duty for non-notifiable work to comply with safety provisions of the regulations which have been updated. Regulations to allow third party certification of electrical work will not be introduced until later in 2013.

**Regulation 7:** A new Approved Document will be published to update information on materials and workmanship in line with the European Construction Products Regulations to be implemented on 1st July 2013.

**Fire Safety:** Local Acts: regulations to repeal unnecessary fire provisions in local acts.

**Warranty Link Rule:** Applicable to Approved Inspectors for construction of new dwellings to be removed.

Approved Documents with the full revisions to parts B, K, M, N, P and Regulation 7 above will be available to purchase from The Stationery Office (TSO) online at http://www.tsoshop.co.uk or telephone: 0870 600 5522.

**Matters to be announced in 2013:**
- referencing of British Standards for structural design based on Eurocodes (ADA)
- additional radon protection measures (ADC)
- energy efficiency of buildings (ADL).

**Further information can be found at:**
https://www.gov.uk/government/speeches/changes-to-the-building-regulations
https://www.gov.uk/government/consultations/building-regulations-access-statements-security-changing-places-toilets-and-regulation-7
http://www.planningportal.gov.uk/buildingregulations/approveddocuments/
https://www.gov.uk/government/speeches/minor-consequential-improvements

# Geomex Ltd

**Structural Engineers and Architectural Design Consultants**

Tel:   01886 832810       Email: geomex@fsmail.net
       01594 860109
                          Website: www.geomex.co.uk

We undertake work on residential and commercial properties, from small internal alterations to new build.

*'A prompt and reliable service, keeping costs to a minimum through the innovation and application of sensible, cost-effective solutions'*

- Structural design
- Structural calculations
- Problem solving
- Building Regulation submissions
- Building specifications
- Planning applications
- Architectural plans and drawings
- Structural and building surveys
- Topographical surveys
- Earth retaining structures and slope stability
- Project management and contract management
- Design and Build.

# Section 1  General information

| | |
|---|---|
| Introduction | 1.3 |
| The Building Act 1984 and the Building Regulations 2010 | 1.3 |
| Approved Documents | 1.4 |
|     Approved Documents and sections they cover | 1.4 |
|         Additional requirements for the conservation of fuel and power | 1.4 |
|     Materials and workmanship | 1.5 |
| Other ways of satisfying the Building Regulations requirements | 1.5 |
| Technical and condensation risks | 1.5 |
| Timber-sizing tables independently calculated by GEOMEX for solid timber members | 1.6 |
| Engaging a property professional | 1.6 |
| Obtaining Building Regulations approval | 1.7 |
|     Option 1: Local Authority route | 1.7 |
|         Regularisation certificates | 1.7 |
|         Relaxation of Building Regulations requirements | 1.8 |
|         Contraventions | 1.8 |
|     Option 2: Approved Inspector route | 1.8 |
|         Contraventions | 1.8 |
| Notices of stages of works | 1.9 |
| Exempt buildings and work | 1.9 |
|         Greenhouses and agricultural buildings | 1.9 |
|         Temporary buildings | 1.10 |
|         Ancillary buildings | 1.10 |
|         Small detached buildings (garages, workshops or sheds) | 1.10 |
|         Conservatory, porch, covered yard/way and carports | 1.10 |
|     Option 3: Competent Person Schemes | 1.10 |
| Preliminary works | 1.11 |
|     Site assessment | 1.11 |
|     Demolitions | 1.12 |
|     Statutory service authorities | 1.12 |
|     Public sewers | 1.12 |
|     Existing services | 1.12 |
|     Structural timber | 1.13 |
|     Opening up of the existing structure | 1.13 |
|     Protection of Bats | 1.13 |
|     Protection of the works | 1.13 |
|     Japanese knotweed | 1.14 |

---

*Guide to Building Control: For Domestic Buildings*, First Edition. Anthony Gwynne.
© 2013 John Wiley & Sons, Ltd. Published 2013 by John Wiley & Sons, Ltd.

## 1.2 General information

| | |
|---|---:|
| Matters related to the Building Regulations | 1.14 |
|     Planning Permission, listed building and conservation area consents | 1.14 |
|     Health and safety at work | 1.15 |
|         The Health and Safety at Work etc Act 1974 | 1.15 |
|         Construction (Design and Management) Regulations 2007 (CDM) | 1.15 |
|         Asbestos, contaminated materials, lead paint, etc. | 1.15 |
| The Party Wall Act 1996 | 1.15 |
|     Introduction | 1.15 |
|     Where the Act applies | 1.16 |
|     Disputes under the Act | 1.16 |
|         An Award | 1.17 |
|     Other items | 1.17 |
|     Reference sources | 1.17 |
| Rights of Light | 1.17 |
|     Introduction | 1.17 |
|     Does a Right of Light exist? | 1.17 |
|     What is a Right of Light? | 1.18 |
|     So what is an injury? | 1.18 |
|         How do you establish if there is an injury? | 1.18 |
|         What happens where there is an Injury? | 1.18 |
|     Reference sources | 1.19 |

## Introduction

This document has been produced for home owners/occupiers, students, builders, designers and other property professionals who have a basic knowledge of building construction and require easy-to-understand guidance on the building regulations for domestic building projects in England and Wales.

The document intends to provide education and guidance on how some of the more common technical design and construction requirements of the building regulations can be achieved and met for single-occupancy domestic extensions, new dwellings, loft conversions and conversions of existing buildings, up to three storeys in height, as well as single-storey garages.

Typical details, tables and illustrations have been provided in this guidance document for the more common construction methods used in dwellings; they have been adapted from the technical details contained within the Approved Documents of the Building Regulations and from experience gained by the author. The diagrams and details produced in the document are for guidance only and are *only* the author's interpretation of how the requirements of the building regulations can be met. The actual diagrams and details *must* be agreed and approved by building control at an early stage and before work commences. You must comply with the requirements of the Building Regulations, so you are advised to fully refer to the Approved Documents and contact a suitably qualified and experienced property professional for details and specifications for the most suitable form and method of construction for your project.

Please note that details, values, standards, documents, products, manufacturers, etc. contained in this guidance may have changed, been superseded, or disappeared altogether between the time when it was written and when it was read; they should be checked by the person using the guidance.

## The Building Act 1984 and the Building Regulations 2010

The power to make building regulations is contained within Section 1 of the Building Act 1984 and deals with the powers of the Secretary of State to make building regulations for the following purposes:

- securing the health, safety, welfare and convenience of people in or about buildings
- conservation of fuel and power
- preventing waste, undue consumption, and misuse or contamination of water.

The current building regulations are the Building Regulations 2010 and The Building (Approved Inspectors etc.) Regulations 2010, which came into force on 1 October 2010 and apply to England and Wales (a separate system of building control will apply in Wales from 2013). A separate system of building control applies in Scotland and Northern Ireland. The 2010 Regulations in both cases consolidate the Building Regulations 2000 and the Building (Approved Inspectors etc.) Regulations 2000, incorporating amendments since 2000.

The Building Regulations are very short, contain no technical details and are expressed as functional requirements, so they are difficult to interpret or understand. For this reason, the Department for Communities and Local Government publishes guidance on meeting the requirements in a series of documents known as 'Approved Documents'.

## 1.4 General information

## Approved Documents

The Approved Documents are intended to provide guidance on how to achieve the requirements of the building regulations, and they make reference to other guidance and standards. In themselves the Approved Documents are not mandatory and there is no obligation to adopt any particular solution contained within them if the required result can be achieved in some other way. In all cases it is the responsibility of the designer, applicant/owner and contractor to ensure the works are carried out in compliance with the building regulations.

The current Approved Documents are listed below and are available to view on the Department for Communities and Local Government website: www.communities.gov.uk or to purchase from The Stationery Office (TSO) on line at www.tsoshop.co.uk or by telephone on 0870 600 5522.

TRADA Technology span tables are available from www.trada.co.uk/bookshop.

### Approved Documents and sections they cover

- A: Structure (2004 edition with 2010 amendments), including span tables for solid timber members in floors, ceilings and roofs for dwellings (2nd edition) and Eurocode 5 span tables for solid timber members in floors, ceilings and roofs for dwellings (3rd edition), published by TRADA Technology
- B1: Fire safety in dwelling houses (2006 edition with 2010 amendments)
- C: Site preparation and resistance to contaminants and moisture (2004 edition with 2010 amendments)
- D: Toxic substances (1992 with 2002 and 2010 amendments)
- E: Resistance to the passage of sound (2003 with 2004 and 2010 amendments)
- F: Ventilation (2010 edition with further amendments)
- G: Sanitation, hot-water safety and water efficiency (2010 edition with further amendments)
- H: Drainage and waste disposal (2002 edition with 2010 amendments)
- J: Combustion appliances and fuel storage systems (2010 edition with further amendments);
- K: Protection from falling, collision and impact (1998 with 2000 and 2010 amendments)
- L1A: Conservation of fuel and power in new dwellings (2010 edition with further amendments)
- L1B: Conservation of fuel and power in existing dwellings (2010 edition with further 2010 and 2011 amendments)
- M: Access to and use of buildings (2004 edition with 2010 amendments)
- N: Glazing – safety in relation to impact, opening and cleaning (1998 with 2000 and 2010 amendments)
- P: Electrical safety (2006 edition with further 2010 amendments)

Regulation 7: Materials and workmanship (1999 edition with 2010 amendments).

**Note:** References made in this guidance to Approved Documents are abbreviated as AD – for example, reference to Approved Document A: Structure (2004 edition with 2010 amendments) will be abbreviated to ADA.

### *Additional requirements for the conservation of fuel and power*

It's important to note that many local authority planning departments are now imposing planning conditions that require energy-efficiency standards in buildings that are above the minimum standards stipulated under the building regulations. Since 31 December 2011 the Welsh Assembly Government requires that all new residential properties in Wales meet an 8 per cent improvement over the 2010 Code level 3 for sustainable homes (ENE.1). Guidance on the code for sustainable homes is contained in Section 3 of this document. You are advised to contact your local planning department at an early stage for their specific requirements.

### Materials and workmanship

All materials used for a specific purpose should be assessed for suitability using the following aids (see Approved Document: Regulation 7 for full details):

- British Standards or European Standards (or other acceptable national and international technical specifications and technical approvals)
- Product Certification schemes (Kite marks)
- Quality Assurance schemes
- British Board of Agreement Certificates (BBA)
- Construction Product Directives (CE Marks)
- Local Authority National Type Approvals (System Approval Certification)
- In certain circumstances, materials (and workmanship) can be assessed by past experience – for example, a building already in use, providing it is capable of performing the function for which it was intended – subject to building control approval.

All materials must be fixed in strict accordance with manufacturer's printed details. Workmanship should be in strict accordance with Regulation 7 and BS 8000: Workmanship on Building Sites – Parts 1 to 16. Where materials, products and workmanship are not fully specified or described, they are to be 'fit for purpose', stated or inferred, and in accordance with recognised best practice.

## Other ways of satisfying the Building Regulations requirements

The Building Regulations requirements may be satisfied in other ways, or in non-standard ways, by calculations or test details from a manufacturer, supplier, specialist, or by an approved third-party method of certification such as a British Board of Agreement (BBA or other third-party-accredited) Certification.

## Technical and condensation risks

The technical details in this guidance document should be read in conjunction with the BRE publication 'Thermal Insulation Avoiding Risks', which explains the technical risks and condensation risks that may be associated with meeting the building regulation requirements for thermal insulation for the major elements of the building. A copy of the publication can be obtained from www.brebookshop.com.

A condensation risk analysis (including interstitial condensation risk) should be carried out for the details and diagrams produced in this guidance for particular situations and construction projects, following the procedures set out in BS 5250:2002 (Code of practice for the control of condensation in buildings). The insulation manufacturer's technical services department will normally carry out this service.

## Timber-sizing tables independently calculated by GEOMEX for solid timber members

The timber-sizing tables in this guidance have been independently calculated by Geomex Ltd (Consulting Structural Engineers) and have been carried out totally independently of TRADA Technology's span tables.

The timber sizes stated in the tables in this guidance are commonly available for solid timber members used in the construction of floors, ceilings, cut roofs (excluding manufactured trusses) and flat roofs for single-occupancy dwellings up to three storeys in height (measured above ground level). Normally, two grades of timber are commercially available: strength grades C16 and C24 (grade C24 being stronger than C16).

Grade C24 timber has been used for the calculation of all values for particular imposed and dead loadings as contained in timber-sizing tables in this guidance. Each case should be separately analysed and assessed, since site parameters may change, including wind and snow loadings for particular geographical areas.

Where possible the calculations have been performed using current timber Eurocodes based on the latest release of TEDDS design software. The TEDDS design software is the design package employed to undertake the calculations. However, where the software does not include the Eurocode standards, British Standards have been used. These are still recognised as design standards and we understand that they will remain acceptable for most building control bodies until 2013. Please note that the TRADA Technology span tables have not been reproduced in this guidance.

## Engaging a property professional

The design and construction of extensions, garages and new dwellings, and the conversions of existing buildings, are normally complex projects, so unless you are experienced in design and construction you are advised to get some professional advice and help as follows:

1. Appoint a suitably qualified and experienced property professional who will prepare drawings and designs for your proposal, obtain the necessary approvals and, if required, will also help you to find a suitable builder and manage the project for you, or,
2. Appoint a specialist company who can offer a complete design-and-build package for your proposal. They can usually prepare drawings and designs for your proposal, obtain the necessary approvals and carry out all the necessary construction works and project management to complete the project.
3. Use an experienced builder.

## Obtaining Building Regulations approval

There are three alternative routes available to the applicant to obtain Building Regulations approval, as detailed below. Option 1 is the local authority route, option 2 is an Approved Inspector route and is a private system of certification and option 3 is a Competent Person Scheme.

### Option 1: Local Authority route

The building owner or agent must make a Building Regulations application and pay a fee for the construction of new works. All works must comply with the 2010 Building Regulations.

The person carrying out the building works must liaise with and meet the requirements of the Local Authority Building Control and give the required notice for certain key stages of works, as detailed in the guidance below

**There are two methods of making a Building Regulations application, as follows.**

*(i) Full Plans application*

This is often thought of as the traditional way of applying for Building Regulations approval. The building designer will draw up detailed plans, specification and supporting information for the proposed scheme and will send them to the local authority, together with a completed application form and the necessary fee. The authority will then check the details and, following any necessary consultations and liaisons with the building designer, a Building Regulations approval or conditional approval will be issued. The approvals can also be dealt with in stages when design information becomes available; this can be on a rolling programme agreed between the parties as the information becomes available. Applications can be rejected in certain instances.

Work can start at any time after the application has been submitted, together with the correct fee, has been accepted as a valid application, although it is wise to wait until the scheme has had its initial check under the Building Regulations, which usually takes between two and three weeks. The building control surveyor will normally liaise with the builder/owner and inspect the work as it progresses on site. When the project is satisfactorily completed a Building Regulations Completion Certificate will normally be Issued.

*(ii) Building Notice application*

This system is best suited to minor domestic work carried out by a competent builder. Under this scheme no formal approval of plans is issued and work is approved on site as it progresses.

To use the Building Notice process, the owner or agent will need to submit a completed Building Notice application form, together with a site location plan and the required fee. Work can commence 48 hours after the notice has been received. When work does commence, the person carrying out the works should contact the council's surveyor to discuss the proposals, to agree how the work should be carried out and when it will need to be inspected, and to establish whether any further information will be required, e.g. drawings, specifications or other information. When the project is satisfactorily completed, a Building Regulations Completion Certificate will normally be Issued.

### *Regularisation certificates*

For unauthorised works, an application can be made to the local authority in certain instances to regularise the works, which is a retrospective form of application for unauthorised works carried out on or after 11 November 1985; please contact your local authority's building control department for more information.

### Relaxation of Building Regulations requirements

In certain circumstances, local authorities have powers to dispense with or relax regulation requirements. However, a majority of the regulation requirements cannot be relaxed because they require something to be adequate or reasonable, and to grant a relaxation could mean acceptance of something that was inadequate or unreasonable. For more advice please contact your local authority building control department.

### Contraventions

Where works are carried out in contravention of the Building Regulations, the local authority may require their alteration or removal within a period of time by serving notice on the building owner. Failure to comply with the notice can result in the work being carried out by the local authority, who can recover their expenses from the defaulter. The person who contravened the building regulations also renders themselves liable to prosecution for the offence in a magistrate's court.

**To find your local authority building control in England and Wales, contact Local Authority Building Control (LABC) at: www.labc.uk.com.**

## Option 2: Approved Inspector route

The applicant can employ an approved inspector, who must be approved by the Construction Industry Council (CIC), either corporately or individually to carry out the functions of an approved inspector. The inspector must give to the local authority an initial notice in a prescribed form before the work commences on site.

The approved inspector should ensure that all the relevant information is provided in the prescribed form, because if the local authority is not satisfied that the notice contains sufficient information, or if the works start before they receive it, they can reject it within five working days and it is of no effect.

Once the notice has been accepted, or is deemed to have been accepted by the passing of five days, the approved inspector is responsible for inspecting the works and issuing the appropriate certificates to the Client and local authority as required under the Building (Approved Inspectors etc.) Regulations 2010.

The building designer will draw up detailed plans, a specification and supporting information for the proposed scheme and will send them to the approved inspector; that can be done on a rolling programme agreed between the parties as the information becomes available. When the project is satisfactorily completed a Building Regulations Completion Certificate has to be issued to the applicant and local authority.

### Contraventions

Unlike the local authority, the Approved Inspector has no direct power to enforce the Building Regulations if the works are in contravention of those regulations. If the Approved Inspector is not satisfied with the works and cannot resolve the matter, the inspector will not issue the 'final certificate' and will cancel the initial notice, thereby terminating the inspector's involvement in the project. Cancelling the initial notice results in the building control function being taken on by the local authority, which has enforcement powers to ensure the works comply.

**A list of approved inspectors is available from the Construction Industry Council's website at: www.cic.org.uk.**

## Notices of stages of works

Site inspections are normally carried out by building control at key stages to ensure the works are being carried out in compliance with the building regulations. The period of notice required for site inspections is to be agreed with building control and the number of site inspections will depend on the type and complexity of the works being carried out.

The key stages of work typically include:

- Commencement of works
- Foundation excavations before any concrete is laid
- Over-site covering to ground floors before any concrete is laid
- Below-ground foul and surface water drainage before any pipes are covered over
- Structural elements and components (i.e. upper-storey floor joists, structural beams/columns/connections and roof structure, etc.) before any coverings are fixed
- Any other area of work as required by building control, including unusual design or methods of construction
- Completion of building prior to occupation.

More than one inspection may be carried out for each key stage and where possible additional items for inspection are normally carried out at the same time as the key stages – for example:

- Fire safety and means of escape
- Hidden areas of works
- Any other area of work as required by building control.

Check with building control how they accept notices of stages of work (typically by telephone or e-mail). Building control do not supervise the works, or provide a quality check, and specialists should be independently employed by the building owner if this is required.

## Exempt buildings and work

The following list is a brief extract of the more common buildings and works that are exempt from the Building Regulations; for full details see Regulation 9 and Schedule 2 of the Building Regulations 2010. Note: although these works may be exempt, ADP may apply to any electrical installations – see Part P – Electrical Installations in section 2 of this guidance.

### *Greenhouses and agricultural buildings*

Buildings used for agriculture, including horticulture (i.e. growing of fruit, vegetables, plants, seeds and fish farming) or principally for the keeping of animals; providing in each case that:

- no part of the building is used as a dwelling;
- the building is at least one and a half times its height from a building that contains sleeping accommodation;
- the maximum distance to a fire exit or point of escape from the building is 30m;
- the building is not used for retailing, packing or exhibiting.

### Temporary buildings

These are buildings that are not intended to remain where they are erected for more than 28 days.

### Ancillary buildings

This covers buildings used only in connection with the sale of buildings or plots on that site; or on a site of construction or civil-engineering works that is intended to be used only during the course of those works and contains no sleeping accommodation; or a building, other than a building containing a dwelling or used as an office or showroom, erected for use on the site of and in connection with a mine or quarry.

### Small detached buildings (garages, workshops or sheds)

These would be detached, single-storey buildings with less than $30\,m^2$ internal floor area, with no sleeping accommodation. If constructed substantially of combustible materials, such a building must be positioned at least one metre from the boundary of its curtilage. A detached building with less than $15\,m^2$ internal floor area, with no sleeping accommodation, does not have any boundary restrictions.

### Conservatory, porch, covered yard/way and carports

This would be the extension of a building by the addition of a single-storey building at ground level of:

(a) a conservatory, porch, covered yard or covered way; or
(b) a car port open on at least two sides;

where the floor area of that extension is less than $30\,m^2$ internal floor area, and providing the glazed area satisfies the requirements of ADN for safety glazing. (Please note that as there is no definition of 'conservatory' in the Building Regulations 2010, and owing to the variation in interpretation of the building regulations, building control may require a percentage of the walls and roof in a conservatory formed from translucent materials to be exempt – typically 75 per cent of the roof and 50 per cent of the walls. You are advised to contact your building control provider for their specific requirements.) Additional requirements: existing walls/doors/windows of the building separating the conservatory or porch are to be retained or, if removed, are to be replaced with elements that meet the energy-efficiency requirements of ADL1B; and the heating system of the dwelling must not be extended into the conservatory or porch.

## Option 3: Competent Person Schemes

Certain works can be carried out by an installer who is registered with a Competent Persons Scheme and will not require Building Regulations approval.

Competent Person Schemes (CPS) were introduced by the UK Government to allow individuals and enterprises to self-certify that their work complies with the Building Regulations, as an alternative to submitting a building notice or using an approved inspector. A Competent Person must be registered with a scheme that has been approved by the Department for Communities and

Local Government (DCLG). Schemes authorised by the DCLG are listed on its website at http://www.communities.gov.uk.

An installer registered with a Competent Person Scheme will notify the local authority on your behalf and will issue a certificate on completion, which can be used as proof of compliance. It will also show up on a solicitor's local authority search.

To understand why you should use a Competent Person, a consumer booklet can be downloaded from the DCLG website above, which has been developed by a collaboration of all the approved scheme providers so as to provide the consumer with the ability to search for a Competent Person registered with one of the schemes.

Schemes authorised include:

- Installation of cavity wall insulation
- Installation of gas appliances
- Installation or replacement of hot-water and heating systems connected to gas appliances
- Installation or replacement of oil-fired boilers, tanks and associated hot-water and heating systems
- Installation or replacement of solid-fuel burners and associated hot-water and heating systems
- Installation of fixed air-conditioning or mechanical ventilation systems
- Electrical work in dwellings
- Electrical work only in association with other work (e.g. kitchen installations, boiler installations)
- Replacement windows, doors, roof windows or roof lights in dwellings
- Installation of plumbing and water-supply systems, bathrooms and sanitary ware
- Replacement of roof coverings on pitched and flat roofs (not including solar panels)
- Installation of micro-generation or renewable technologies.

This list can be altered at any time; for a current list of all registered scheme members go to the DCLG website above.

## Preliminary works

Certain works should be considered or undertaken before submitting a building regulation application as follows:

### Site assessment

A desk study and initial walk-over of the site and surrounding area should be carried out by a suitable person to identify any potential hazards and problems at an early stage. Items to be taken into account should include:

- Geology of the area, including any protection measures required for radon ground gas
- Landfill and tipping, including any protection measures required for methane and carbon monoxide ground gases and foundation design requirements
- Surface and ground water, including flooding
- Soils and previous industrial, commercial or agricultural uses, including any protection measures required for ground contaminates
- Mining and quarrying, including any special foundation design requirements.

Further guidance on site preparation and the resistance to contaminants and moisture is provided in ADC and Part C of this guidance. Typical construction details in Part A of this guidance contain information on how to achieve basic and full radon protection in sub-structures.

Sources of information include: local authority (building control, planning departments, environmental health departments), the Environment Agency, the Coal Authority, utility companies, the Health Protection Agency, the British Geological Survey, Ordinance Survey Maps, etc.

Where hazards are suspected, a detailed site investigation should be carried out by a specialist.

## Demolitions

Where the demolition of a structure or part of a structure exceeds $50\,m^3$, a notice of the proposed demolition must be sent to the local authority's building control (and planning) department before works commence even if using an approved inspector. For further information please contact your local authority's planning and building control departments. The Construction (Design and Management) Regulations 2007 will apply to demolition works (see guidance below for details).

## Statutory service authorities

Prior to and during works, the person carrying out the works is to liaise with and meet the requirements of the relevant statutory service authorities/utility companies, including the provision and protection of new services and sewers, and the location and protection of all existing services/sewers as necessary.

## Public sewers

The owner/developer of a building being constructed, extended or underpinned within 3 m of a public sewer, as indicated on the relevant water authority's sewer maps, is required to consult with the water authority and, where necessary, obtain consent and enter into an agreement to build close to or over the public sewer before works commence on site. Further information is provided in Part H of this Guidance for domestic extensions

The owner/developer of a building with new drainage connections or indirect drainage connections being made to a public sewer, as indicated on the relevant water authority's sewer maps, is required to consult with the water authority and, where necessary, obtain consent before works commence on site. Protection of the sewer pipe and systems is to be carried out in compliance with the relevant water authority's requirements.

Since the implementation of the Private Sewer Transfer Regulations on 1 October 2011, all lateral drains and sewers, i.e. those serving two or more properties that connect to the public sewer network, will be adopted by the relevant water authority, and the above requirements for building over/close to and/or making new connections to public sewers will apply. Further information is available from www.defra.gov.uk/environment/quality/water/sewage/sewers and www.water.org.uk/home/policy/private-sewer-transfer.

## Existing services

Plumbing, drainage, heating appliances, electrical services, etc. that need to be altered, modified, adjusted or re-sited to facilitate the new building works should be carried out by suitably qualified and experienced specialists or registered competent persons, with tested and appropriate certification issued where necessary. Existing services should be located, altered, modified or relocated as necessary, including sealing up, capping off, disconnecting and removing redundant services where that is required.

### Structural timber

All structural timber should be stress graded as either C16 or C24 to BS 4978, and sawn to BS 4471. C16-graded timber has a lesser strength than C24-graded timber, and *C24 timber has been used for the calculation of all values contained in Geomex timber-sizing tables in this guidance*. All timber is to be protected on site to minimise moisture content, which must not exceed 22 per cent. Preservative treatment of timber should be in accordance with the requirements of BS 8417, and treatment against house longhorn beetles should be carried out in certain geographical areas in accordance with Table 1 of ADA.

### Opening up of the existing structure

The builder should open up the existing structure where required for inspection purposes in areas or locations as requested by building control or a structural engineer, and should allow for making good all disturbed structures and finishes to match existing ones on completion. For example, the exposure and inspection of the existing foundations and/or lintels of a building may be required to determine whether they are adequate to support the increased loadings of a new storey.

### Protection of Bats

The protection of bats is required when undertaking all works, including demolition, conversions, extensions and/or alterations that would involve changes to the roof space. Please note that all bat species are protected under Schedule 5 of the Wildlife and Countryside Act 1981, and also under Schedule 2 of the Conservation of Habitats and Species Regulations 2010. It is an offence to: intentionally or recklessly kill, injure or capture (take) bats; intentionally or to recklessly disturb bats (whether in a roost or not); or to damage, destroy or obstruct access to bat roosts. If you think that bats may be using the property, or you discover a bat while development work is being undertaken, stop the work immediately and contact the National Bat Helpline on 0845 1300 228.

### Protection of the works

Adequate precautions should be taken on site to protect the work, particularly the laying of concrete and other wet trades or processes in accordance with product manufacturers' details or specialists' requirements, in the following circumstances:

- When the air temperature is below or likely to fall below 2°C (additional consideration should be made for wind chill and freezing conditions).
- No concrete should be placed into or onto frozen surfaces or excavations.
- Ready-mixed concrete should be delivered to site at a minimum temperature of 5°C, in accordance with BS 5328.
- No frozen materials should be used in the works.
- Works should not continue until the site is free of frost and frozen materials.

- When there is a possibility that new work will be affected by frost or freezing before it has set. Curing periods may need to be extended in accordance with product manufacturers' details.

Short- and long-term protection and storage of material on site should be in accordance with product manufacturers' details. The use of admixtures must also be carried out in accordance with product manufacturers' details.

Adequate precautions should be taken to protect the works, in accordance with product manufacturers' details or specialists' requirements. Typically, polythene sheeting or hessian should be used to protect works in progress from becoming saturated, and to prevent drying out from direct winds and sun. Wetting may also be required to ensure that mortars, rendering, plastering, screeds, slabs, etc. do not dry out too quickly and cause failures.

### Japanese knotweed

Japanese knotweed is an invasive weed and it spreads through its crown, rhizome (underground stem) and stem segments, rather than its seeds. The weed can grow rapidly causing heave below concrete and tarmac, coming up through the resulting cracks and damaging buildings and roads. A small section of rhizome (stem) can produce a new plant in 10 days. Rhizome segments can remain dormant in soil for 20 years before producing new plants.

Under the Wildlife and Country Act 1981, it is an offence to plant or cause Japanese knotweed to spread and all waste containing Japanese knotweed comes under the control of Part II of the Environmental Protection Act 1990.

Details on Japanese knotweed and how to control and dispose of it can be found via the following link: http://www.environment-agency.gov.uk/homeandleisure/wildlife/130079.aspx. The knotweed code of practice can be found on the following link: http://www.environment-agency.gov.uk/static/documents/Leisure/Knotweed_CoP.pdf. For more information telephone: 03708 506 506 or email enquiries@environment-agency.gov.uk

## Matters related to the Building Regulations

The following are related to, but are not enforced under, the requirements of the Building Regulations.

### Planning Permission, listed building and conservation area consents

Planning permission or listed building/conservation area consents may be required for the proposed development and no works should be commenced until approval has been given by the relevant local authority planning department.

If the requirements of the Building Regulations will unacceptably alter the character or appearance of a historic or listed building, ancient monument or building within a conservation area, then the requirements may be exempt or relaxed to what is reasonably practical or acceptable, while ensuring – in consultation with the local planning authorities conservation officer – that any exemption or relaxation would not increase the risk of deterioration of the building fabric or fittings. Any such exemption or relaxation must be approved before works commence. For further information, please contact your local authority planning department and building control body.

## Health and safety at work

All necessary health and safety requirements must be provided, including all necessary personal protective equipment; site security; scaffolding; access ladders; material hoists; temporary protection and working platforms, etc., which are to be erected, maintained, certificated, dismantled and removed by suitably qualified and insured specialists.

### *The Health and Safety at Work etc Act 1974*

The Health and Safety at Work etc. Act 1974 (HSWA) is enforced by the Health and Safety Executive. The HSWA requires persons in control of premises to make broad provisions for the health, safety and welfare of people, including visitors and other users of the premises.

The HSWA also requires all persons at work (i.e. contractors) to ensure, so far as is reasonably practical, the health and safety of themselves and any other people who may be affected by their work.

### *Construction (Design and Management) Regulations 2007 (CDM)*

The Construction (Design and Management) Regulations 2007 apply to most construction projects. If you are about to undertake construction work – which could include alterations, extensions, routine maintenance, new build or demolitions – then you need to know to what extent these Regulations will apply to you and whether you are a duty holder under the Regulations.

With non-domestic* projects expected to last longer than 30 days, or more than 500 man-hours, you will require the assistance of an advisor called a CDM coordinator, who should be appointed at the earliest opportunity, before detailed design work is complete. If you are a client thinking of commissioning work, a designer appointed to work on a project, or a builder/developer about to undertake work, you should be aware of your responsibilities or duties under CDM 2007.

(*Non-domestic clients are people who commission building works related to a trade or business, whether for profit or not. This work can be carried out on a domestic property; it is not the type of property that matters, but the type of client – for example, a private landlord.)

### *Asbestos, contaminated materials, lead paint, etc.*

Any suspected asbestos, contaminated material or soil, or lead paint is to be inspected by a specialist and any necessary remedial works are to be carried out in accordance with their requirements by a specialist. Asbestos is to be removed and disposed of off-site by a specialist licensed contractor, as required under the Control of Asbestos Regulations 2012.

**Further information regarding the above can be obtained from the Health and Safety Executive at: www.hse.gov.uk.**

## The Party Wall Act 1996

### Introduction

The Party Wall etc. Act 1996 (the Act) is a law that must be followed in certain circumstances. The Act does not apply to all building work, but its requirements are quite separate from those of Planning and Building Regulations. Professional advice about the Act should be considered by both building owners and neighbours (neighbours affected are called 'adjoining owners' under the Act).

## Where the Act applies

The following are examples of where a building owner is required by law to serve a formal notice on adjoining owners. A notice must show the details of the relevant proposals and other necessary information, and is valid only for one year.

***When building work is planned on a boundary with a neighbouring property***
Examples are building a garden wall, or the outside wall of a new building or extension, at the boundary. Section 1 of the Act applies and a 'Line of Junction Notice' must be served at least one month in advance of the work.

***When work is planned directly to an existing wall or other structure which is shared with another property***
This includes party walls, and the outside wall of a neighbour's building, but also separating floors between flats and garden walls built astride the boundary. Examples are underpinning or thickening of foundations, repair, inserting a damp-proof course or flashing, cutting off projections, strengthening, opening up and exposing the structure. Section 2 of the Act applies and a 'Party Structure Notice' must be served under Section 3 of the Act at least two months in advance. These notices frequently occur in roof-space conversions, building in (or removing) beams, removing chimney breasts, altering chimneys, roofs or floors, demolitions, and sometimes in extensions.

***When an excavation is planned within three metres of a neighbour's building or other structure, where it will be to a lower level than the underside of the neighbour's foundation***
Examples are foundations to a building or extension, but include excavations for drain or services trenches within three metres. Section 6(1) of the Act applies: An 'Adjacent Excavation and Construction Notice' must be served at least one month in advance. These types of notice frequently occur in new building work and in extensions, but can apply to structural alterations.

***When an excavation is planned within six metres of a neighbour's building or other structure, where that excavation would cut a line drawn downwards at 45° from the underside of the neighbour's foundation***
Examples are especially deep foundations or drains within six metres. Section 6(2) of the Act applies: An 'Adjacent Excavation and Construction Notice' must be served, again at least one month in advance. Again, these types of notice frequently occur in new building work, extensions and structural alterations.

## Disputes under the Act

Agreeing in writing to a notice allows the work to proceed in due course. However, if a neighbour does not agree (including not replying in writing within 14 days) a dispute arises. For a dispute, Section 10 of the Act (Resolution of disputes) applies, necessitating the appointment of surveyors. The building owner and adjoining owner must either:

(a) agree to appoint one surveyor (an 'agreed surveyor'), or
(b) each appoint their own surveyor. (Those two surveyors then select a third surveyor, but only in case of a dispute between themselves.)

The dispute procedure may well be longer than the period required for the notice, and in complex cases can be several months.

### An Award

By appointing surveyors, the dispute is resolved by them on behalf of the owners, and the result is the service of an 'Award' for each dispute. An Award is a legal document describing when, where and how the work subject to the Act is to be carried out. An Award cannot deal with matters outside the Act, and cannot deal with other work on site. Once served, both the building owner and the adjoining owner each have a right to appeal the Award in the county court, but only for a period of 14 days. After that the Award is *totally* binding *and shall not be questioned in any court.* This is a very powerful provision and must be most carefully considered by all involved.

### Other Items

The Act cannot be used to resolve a boundary dispute, and neighbours cannot use it to prevent approved work from being carried out. The Act deals with many matters not covered above and only the Act should be relied on for the scope and meaning of any item. There are many guides available relating to the Act, but even they should not be relied on in preference to the Act.

### Reference sources

The Party Wall etc Act 1996 (published by HMSO, ISBN 0-10-544096-5) http://www.legislation.gov.uk/ukpga/1996/40/contents

The Party Wall etc. Act 1996 Explanatory Booklet (published by the Department for Communities and Local Government) http://www.communities.gov.uk/documents/planningandbuilding/pdf/133214.pdf

(Wording provided by and reproduced by permission of Neil J. Dransfield.)

## Rights of Light

### Introduction

Rights of Light are rights enjoyed by a building. Put simply, they are the rights of a building to have light cross someone else's land. They are protected by law, and court action can be brought against someone who has 'injured' them or, quite importantly, someone who intends to injure them. The general principles summarised below are not an entire list of all the principles involved, and are not legal definitions. Expert advice is usually needed for any particular situation, and legal advice is usually necessary relating to Deeds. None of the following involves a much quoted, and quite incorrect, '45-degree rule'.

### Does a Right of Light exist?

A room within a building may enjoy a right of light, but it does not necessarily do so. Usually a right of light has to be gained over time, unless it has been granted, for example, by a Deed. Normally, the law sets a period of 20 years for a room to gain a right of light through its window(s). However, the ability to gain a right can be prevented (for all time) by restrictive wording in the Deeds, or temporarily by obstruction during the 20-year period, or by other legal means. A room cannot gain a right to a view or a right to sunlight. Also, open spaces such as garden areas cannot gain Rights of Light.

## What is a Right of Light?

Where a right of light exists the building has an Easement over the land the light crosses. In confined spaces this may not be just an immediate neighbour, but in most instances it is the adjoining property. Even with a right of light in place there is nothing to prohibit a new building or extension reducing the amount of light entering a room, so long as a certain minimal standard remains. However, if that minimal standard is transgressed, an 'injury' results, which is actionable at law. Further, the right to light is a property right and so action can be taken in advance of a transgression, so as to prevent the injury.

## So what is an Injury?

It is accepted by the courts that the amount of natural light on a working plane (table-top level) in a room should not be less than one lumen. That is a fairly low level of natural light, and is equivalent to one candle held one foot away from a surface, with no other light input. Even then there is a 'working rule' that an 'injury' would only be proved if the room in question were to be left with less than about half its area (50/50) receiving one lumen.

Where domestic properties are concerned, the 'working rule' of 50/50 has been slightly modified, and it has been shown that up to 55 per cent of the working plane should remain 'well lit' by natural light to the one-lumen standard. If less than 50 per cent remains, an injury would be proved, but if 50–55 per cent remains well lit, the injury is not so clear-cut. Over 55 per cent remaining well lit (again to the one-lumen standard) shows no injury. Artificial lighting cannot be argued as a substitute for natural light.

### How do you establish if there is an Injury?

The method is to have a recognised expert carry out calculations, either by detailed diagrams or computer programs, to create a line, drawn on the plan of the room, where just one lumen of natural light is received on the working plane.

To do this the expert knows that the whole hemisphere of the sky provides 500 lumens (accepted by the courts as being equivalent to an overcast winter sky). Therefore, for the one-lumen level, a point on the working plane must receive just one five-hundredth of that amount. That is, a point on the working plane must be able to 'see' just 0.2 per cent of the sky – quite a small amount.

These techniques map the window(s) to a room, and the obstructions outside, onto a grid of the sky so as to correct for curvature. Examining the results allows a line to be drawn on a plan of the room, at working-plane level, showing the limits of where the one-lumen requirement is achieved. The expert then calculates what percentage of the room receives one lumen or more, and what does not.

If a new obstruction, such as a building or extension, is proposed near a room's windows and can be shown by these calculations to reduce the amount of light to below the working rule level (the 50/50 or 55/45 rule), only then an 'injury' can be proved in court.

### What happens where there is an Injury?

A right of light is protected by law and if any owner injures a right in the way described, an Injunction may be obtained through the courts. This means the building work can be prevented or, if it is under way, stopped. Even more dramatic is the likelihood of the offending part of a new building having to be knocked down.

The courts may hold, but only in a minority of circumstances, that damages are more appropriate than an Injunction. An award of damages does allow the 'offending' development to remain (or to proceed), but awards compensation.

## Reference sources

For further information and interpretation of all the above expert, and probably legal, advice is required. See 'Anstey's Rights of Light', published by the Royal Institution of Chartered Surveyors (RICS Books, ISBN 9781842192221).

(Wording provided by and reproduced by permission of Neil J. Dransfield.)

# Section 2  Domestic extensions

| | |
|---|---:|
| Part A: Structure | 2.9 |
| A1: Sub-structure | 2.9 |
| Foundations | 2.9 |
|    Concrete mixes for foundations | 2.9 |
|    Foundation types | 2.10 |
|       Strip foundations | 2.10 |
|       Trench-fill foundations | 2.11 |
|    Building near trees, hedges, shrubs or in clay subsoils | 2.12 |
|    Alternative foundation designs | 2.13 |
|       Raft foundations | 2.16 |
|       Piled foundations | 2.16 |
|    Retaining walls and basements | 2.16 |
|    Basements and tanking systems | 2.18 |
|       Site investigation and risk assessment | 2.18 |
|       Basement sub-structures | 2.18 |
|       Tanking systems | 2.18 |
| Ground floors and sub-structure walls | 2.18 |
|    Sub-structure walls | 2.18 |
|    Ground floors | 2.19 |
|       Ground-bearing solid concrete floors | 2.20 |
|       Suspended reinforced in-situ concrete ground floor slab supported on internal walls | 2.21 |

*Guide to Building Control: For Domestic Buildings*, First Edition. Anthony Gwynne.
© 2013 John Wiley & Sons, Ltd. Published 2013 by John Wiley & Sons, Ltd.

| | |
|---|---:|
| Suspended beam-and-block ground floors | 2.22 |
| Proprietary under-floor heating system | 2.24 |
| Floating floor | 2.24 |
| Suspended timber ground floor | 2.25 |
| Garage ground-bearing concrete floor | 2.27 |
| A2: Superstructure | 2.28 |
| Minimum headroom heights | 2.28 |
| External walls | 2.28 |
|     Sizes and proportions of residential buildings up to three storeys | 2.28 |
|     Measuring storey heights | 2.28 |
|     Measuring wall heights | 2.28 |
|     Measuring external/compartment/separating wall lengths | 2.28 |
|         Vertical lateral restraint to walls | 2.28 |
|     Thickness of walls | 2.29 |
|         Minimum thickness of certain external walls, compartment walls and separating walls constructed of coursed brick or block work | 2.29 |
|         Minimum thickness of internal load-bearing walls | 2.30 |
|         Buttressing wall design | 2.30 |
|         Pier and chimney design providing restraint | 2.32 |
|     Compressive strength of masonry units | 2.32 |
| External cavity wall construction | 2.35 |
|     Cavity wall construction | 2.35 |
|         Natural stone-faced cavity walls | 2.35 |
|     Buttressing, sizes of openings and recesses in cavity walls | 2.37 |
|     Walls between heated and unheated areas | 2.37 |
|     External timber-framed walls with separate brick or block finish | 2.41 |
|     External timber-framed stud walls | 2.41 |
|         Notches/holes/cuts in structural timbers | 2.41 |
|         External boarding | 2.41 |
|         Breather membrane | 2.43 |
|         Thermal insulation and fire resistance | 2.43 |
|     External walls | 2.43 |
|         Proprietary steel lintels | 2.43 |
|         Separation of combustible materials from solid-fuel fireplaces and flues | 2.43 |
|     External timber-framed walls with render finish | 2.44 |
|     External timber-framed walls with cladding finish | 2.45 |
|         Detached garage (or similar single-storey building) with SINGLE-SKIN external walls | 2.47 |
|         Wall abutments | 2.47 |
|         Lintels and weep holes | 2.48 |
|         Structural columns/beams etc. | 2.48 |
|         Movement joints | 2.49 |
|         Cavity closers | 2.49 |
|         Proprietary dpc trays (new and existing walls) | 2.49 |
| Lateral restraint strapping of upper floors to walls | 2.49 |
|     (i)  Strapping of floor joists parallel to walls | 2.49 |
|     (ii)  Strapping of floor joists at right angles to walls | 2.50 |

| | |
|---|---|
| Lateral restraint strapping of roofs to walls | 2.51 |
|     Strapping of roofs to gable-end walls | 2.51 |
|     Strapping of wall plates and roofs at eaves level | 2.52 |
| Lateral restraint strapping of walls at ceiling level | 2.52 |
| A3: Separating walls and floors | 2.53 |
| Masonry party walls separating dwellings | 2.53 |
| Double-leaf timber-frame party walls separating dwellings | 2.54 |
|     Upgrading sound insulation of existing party walls separating dwellings | 2.55 |
|         Party floors separating buildings | 2.55 |
|         Sound-testing requirements | 2.55 |
| A4: Internal partitions | 2.56 |
|   Internal load-bearing masonry partitions | 2.56 |
|   Internal load-bearing timber stud partitions | 2.56 |
|   Internal masonry non-load-bearing partitions | 2.56 |
|   Internal timber studwork non-load-bearing partitions | 2.56 |
| A5: Intermediate upper floor(s) | 2.56 |
| Floor joists | 2.56 |
| Trimming and trimmer joists | 2.58 |
|     Notching and drilling of structural timbers | 2.59 |
|     Sound insulation to floors within the dwelling | 2.60 |
|         Soil-and-vent pipe (SVP) boxing internally | 2.60 |
|         Exposed intermediate upper floors | 2.60 |
| A6: Pitched roofs | 2.60 |
| Pitched roof coverings | 2.61 |
| Pitched roof structure | 2.62 |
|     (i) Roof trusses (including attic and girder trusses) | 2.62 |
|     (ii) Cut roof construction | 2.62 |
|         Notching and drilling of structural timbers | 2.63 |
|         Roof restraint | 2.63 |
|         Roof insulation and ventilation gaps | 2.65 |
|     Pitched roof ventilation requirements when using a non-breathable roof membrane | 2.66 |
|         Proprietary vapour-permeable roof membrane | 2.68 |
|         Valleys and lead work | 2.68 |
|         Roof abutment with cavity wall | 2.68 |
|         Lofts hatches, doors and light wells to roof spaces | 2.69 |
| A7: Flat-roof construction | 2.70 |
| Option 1: Flat roof with 'cold deck' | 2.70 |
|   Flat roofs with unlimited access/habitable use | 2.72 |
| Option 2: Flat roof with 'warm deck' | 2.72 |
|   Flat roofs with unlimited access/habitable use | 2.74 |
| Option 3: Flat roof with inverted 'warm deck' (insulation on top of waterproof coverings) | 2.74 |
| Option 4: Flat roof with green roof on 'warm deck' (either intensive or extensive) | 2.74 |
|     Intensive green roof | 2.74 |
|     Extensive green roof | 2.75 |
| The design, workmanship and selection of materials for flat roofs | 2.75 |
|   Cavity closers | 2.75 |

| | |
|---|---:|
| A8: Mortars, renders and gypsum plasters | 2.75 |
| Cement mortars and renders | 2.75 |
| Gypsum plasters | 2.77 |
| Part B: Fire safety and means of escape | 2.78 |
| Fire detection and fire alarm systems | 2.78 |
|     New houses and extensions | 2.78 |
|     Large houses | 2.78 |
|         Large two-storey house (excluding basement) | 2.78 |
|         Large three-storey house (excluding basements) | 2.78 |
| Means of escape | 2.79 |
|     1. Means of escape from the ground storey in dwellings | 2.79 |
|     2. Means of escape from a two-storey dwelling with a floor not more than 4.5 m above ground level | 2.79 |
|     3. Means of escape from three-storey dwellings with one upper floor more than 4.5 m above ground level | 2.81 |
|         Option 1: Protected stairway | 2.81 |
|         Option 2: Fire-separated third storey with alternative external/internal fire exit | 2.83 |
|     General provisions for means of escape | 2.83 |
|         Fire doors to protected stairway enclosures | 2.83 |
|         Protected stairway enclosures | 2.84 |
|         Limitations on the use of uninsulated glazed elements | 2.84 |
|         Fire resistance to upper floors and elements of structure | 2.84 |
|         Means-of-escape windows and external doors | 2.85 |
|         Galleries | 2.85 |
|         Basements | 2.85 |
|         Passenger lifts | 2.86 |
|         Replacement windows (excludes repairs) | 2.86 |
|         Fire separation between an integral garage and dwelling | 2.86 |
|         Protection of openings, fire stopping and cavity barriers | 2.86 |
|         Fire resistance of areas adjacent to external fire exit stairs | 2.86 |
|         Circulation systems in houses with a floor more than 4.5 m above ground level | 2.87 |
|         Residential sprinkler systems for means of escape | 2.87 |
| Surface spread of flame: internal wall and ceiling linings including roof lights | 2.89 |
|     Fire resistance to elements of structure | 2.89 |
|     External wall construction in relation to a boundary | 2.89 |
|     Compartment walls and floors separating buildings | 2.91 |
|     Permitted building openings in relation to a boundary | 2.91 |
|         Openings within 1.0 m of a boundary | 2.91 |
|         Openings more than 1.0 m from a boundary | 2.91 |
|     Designation of roof covering and minimum distance to boundary | 2.91 |
|     Typical fire and rescue service vehicle access route specification for dwellings | 2.92 |
| Part C: Site preparation and resistance to contaminants and moisture | 2.92 |
| C1: Resistance to contaminants | 2.92 |
| Radon gas | 2.93 |
|     What is radon gas? | 2.93 |
|     Protective measures | 2.93 |

| | |
|---|---|
| Basic radon protection | 2.93 |
| Full radon protection | 2.93 |
|     Number and position of sumps | 2.94 |
|     Sump construction | 2.94 |
|     Radon fan locations | 2.95 |
|     Stepped foundations and retaining walls | 2.95 |
| Further details | 2.95 |
| Methane and other ground gas protection | 2.96 |
|     Landfill gas and radon | 2.96 |
| C2: Resistance to moisture | 2.96 |
|     Horizontal damp-proof courses (dpcs) | 2.96 |
|     Vertical damp-proof courses (dpcs) and damp-proof course trays, etc. | 2.96 |
|     External cavity walls | 2.97 |
|     Tanking systems | 2.97 |
|     Flood risk | 2.97 |
|     Condensation risks | 2.97 |
| Part D: Cavity wall filling with insulation | 2.98 |
| Part E: Resistance to the passage of sound | 2.98 |
|     New party walls and floors in new extension | 2.98 |
|     New internal walls and floors in new extension | 2.98 |
|     Pre-completion sound testing | 2.98 |
|         Remedial works and retesting | 2.98 |
|         Exemptions and relaxations | 2.99 |
| Part F: Ventilation | 2.99 |
| Purge (natural) ventilation | 2.99 |
| Mechanical extract ventilation and fresh air inlets for rooms without purge ventilation | 2.99 |
| Background ventilation | 2.100 |
| Intermittent mechanical extract ventilation | 2.100 |
|     Ventilation systems for basements | 2.100 |
|     Ventilation of a habitable room through another room or conservatory | 2.100 |
|     General requirements for mechanical extract ventilation | 2.100 |
| Part G: Sanitation, hot-water safety and water efficiency | 2.100 |
| Wholesome hot and cold water supply | 2.100 |
|     Scale of provisions | 2.101 |
|     Wash basins and separation of w/c from any food-preparation areas | 2.101 |
|     Water tanks/cisterns base | 2.101 |
|     Pumped small-bore foul-water drainage | 2.101 |
|     Vented and unvented hot-water storage systems | 2.101 |
|     Safety valves, prevention of scalding and energy cut-outs | 2.101 |
|     Discharge pipes from safety devices | 2.102 |
| Solar water heating | 2.102 |
| Electrical water heating | 2.102 |
| Insulation of pipework to prevent freezing | 2.102 |
|     Commissioning certificates | 2.103 |
| Supply (Water Fittings) Regulations 1999 | 2.103 |
| Part H: Drainage and waste disposal | 2.103 |

## 2.6 Domestic extensions

| | |
|---|---|
| H1: Foul- and storm-water drainage | 2.103 |
|     Foul-, rain- and storm-water drainage systems (single dwellings) | 2.103 |
|         Bedding and backfilling requirements for rigid and flexible pipes | 2.104 |
|     Pipes penetrating through walls | 2.105 |
|         Drain trenches near buildings | 2.106 |
|         Inspection chambers, gullies and access fittings etc. | 2.106 |
|         Manholes | 2.108 |
|         Foul-water disposal | 2.108 |
|         Waste pipes | 2.108 |
|         Soil-and-vent pipes (discharge stack) | 2.110 |
|         Waste-pipe connections to soil-and-vent pipes (discharge stack) – to prevent cross-flow | 2.112 |
|         Stub stacks | 2.112 |
|         Air-admittance valves | 2.112 |
|         Airtightness and testing | 2.112 |
|         Pumping installations | 2.113 |
| H2: Septic tanks, sewage treatment systems and cesspools | 2.113 |
|     Existing septic tank and effluent drainage | 2.113 |
|     Non-mains foul-drainage waste water treatment systems | 2.113 |
| Septic tanks | 2.113 |
| Sewage treatment systems | 2.114 |
|     Disposal of sewage from septic tanks and sewage treatment systems | 2.115 |
|         Drainage fields | 2.115 |
|         Drainage mounds | 2.115 |
|         Wetlands/reed beds | 2.115 |
| Percolation tests | 2.116 |
|     Ground conditions | 2.116 |
|     Percolation test method | 2.116 |
| Cesspools | 2.118 |
| H3: Rainwater drainage and harvesting | 2.119 |
|     Rainwater gutters and down pipes | 2.119 |
|     Rainwater/grey water harvesting storage tanks and systems | 2.120 |
|     Surface water drainage around the building | 2.120 |
|     Rain-/surface-water disposal | 2.120 |
|         Existing soak-aways | 2.121 |
|         New soak-aways | 2.121 |
|     Oil/fuel separators | 2.121 |
| H4: Building over or close to, and connections to, public sewers | 2.122 |
| Building over or close to a public sewer | 2.122 |
|     Locating a public sewer | 2.122 |
|         Options | 2.122 |
|         The build-over process | 2.123 |
|         Typical procedure | 2.123 |
|     Private Sewer Transfer Regulations | 2.123 |
|     Protection | 2.123 |
|     Further information | 2.123 |
| Connections to public sewers | 2.123 |

| | |
|---|---|
| H5: Separate systems of drainage | 2.124 |
| H6: Solid waste storage | 2.124 |
| Part J: Combustion appliances and fuel storage systems | 2.124 |
|     Space and hot-water/heat-producing appliances in general | 2.124 |
| Solid fuel appliances up to 50 kW rated output | 2.125 |
|     Construction of open fire recessed and hearth | 2.125 |
|     Construction of solid-fuel masonry chimneys | 2.125 |
|     Free-standing solid-fuel stove and hearth | 2.126 |
|         Flue pipe connections to free-standing stove and chimneys | 2.129 |
|     Construction of factory-made, insulated, twin-walled metal chimneys | 2.129 |
|     Carbon monoxide alarms | 2.129 |
|     Air supply (ventilation) to solid-fuel appliances | 2.129 |
|         Construction of factory-made flue block chimneys | 2.130 |
|         Configuration of flues serving open-flue appliances | 2.130 |
|         Inspection and cleaning openings in chimneys and flues | 2.130 |
|         Interaction of mechanical extract vents and open-flue combustion appliances | 2.131 |
|     Chimney/flue heights | 2.132 |
|         Repair/relining of existing flues | 2.132 |
|         Notice plates for hearths and flues | 2.133 |
| Appliances other than solid fuel | 2.133 |
|     Gas heating appliances up to 70 kW | 2.133 |
|         Interaction of mechanical extract vents and open-flue gas combustion appliances | 2.133 |
|     Oil heating appliances up to 45 kW | 2.133 |
|         Interaction of mechanical extract vents and open-flue oil combustion appliances | 2.133 |
| Fuel storage tanks | 2.134 |
|     LPG tanks and cylinders up to 1.1 tonnes | 2.134 |
|     Oil tanks up to 3500 litres | 2.134 |
| Renewable energy/micro regeneration installations | 2.134 |
|     Further information | 2.135 |
|         Provision of information – commissioning certificates (testing) | 2.135 |
| Part K: Protection from falling, collision and impact | 2.135 |
| Internal stairs, guarding and landings for changes in level of 600 mm or more | 2.135 |
|     Stair pitch | 2.135 |
|     Headroom | 2.135 |
|     Rise and going | 2.136 |
|     Landings | 2.136 |
|         Stair width | 2.136 |
|         Handrails | 2.136 |
|     Guarding | 2.136 |
|         Length of flights | 2.137 |
|         Typical internal staircase construction details | 2.138 |
| External stairs, guarding and landings for changes in level of 600 mm or more | 2.138 |
|     External stairs and landings | 2.138 |
|     External guarding | 2.138 |

## 2.8 Domestic extensions

| | |
|---|---|
| Guarding to upper-storey window openings/other openings within 800 mm of floor level | 2.139 |
| Loft conversion stairs | 2.139 |
|     Reduced headroom to stairs in loft conversions | 2.139 |
|     Alternating tread stairs for loft conversions | 2.139 |
|         Fixed ladders for loft conversions | 2.140 |
| Ramps | 2.140 |
| Part L: Conservation of fuel and power in existing dwellings | 2.141 |
|     Listed buildings, conservation areas and ancient monuments | 2.141 |
| Areas of external windows, roof windows and doors | 2.141 |
| New thermal elements | 2.141 |
|     External glazing | 2.141 |
|         Closing around window and door openings | 2.141 |
|         Sealing and draught-proofing measures | 2.142 |
| Energy-efficient lighting | 2.143 |
|     Fixed internal lighting | 2.143 |
|     Fixed external lighting | 2.143 |
| Insulation of pipework to prevent freezing | 2.143 |
| External walls, roofs, floors and swimming-pool basin | 2.143 |
| Renovation/upgrading of existing thermal elements | 2.144 |
|     Consequential improvements (applies to existing buildings with a total useful floor area exceeding 1000 m$^2$) | 2.144 |
|         Commissioning of fixed building services | 2.144 |
|         Providing information – building log book | 2.145 |
| Part M: Access to and use of buildings for disabled | 2.145 |
| Part N: Safety glazing, opening and cleaning | 2.146 |
| Safety glass and glazing | 2.146 |
|     Glazing in small panes | 2.147 |
|     New system of marking | 2.147 |
| Part P: Electrical safety | 2.147 |
| Electrical installations | 2.147 |
|     Additional notes (as ADP) | 2.148 |
| External works – paths, private drives, patios and gardens | 2.149 |
|     Concrete areas and paths etc. | 2.149 |
|         Tarmac areas | 2.150 |
|         Block pavers | 2.150 |
|         Precast concrete or natural stone slabs | 2.150 |
|         Gravel | 2.150 |
|     Drainage of paved areas | 2.150 |

# PART A: STRUCTURE

Please refer fully to the relevant Approved Document.

## A1: Sub-structure

**Note:** Selected diagrams in this guidance contain details on how to achieve basic and full protection against radon gas in sub-structures. Further guidance on radon and other ground gases, contaminates and hazards can be found in ADC and in Part C of this guidance below.

## Foundations

Foundations should be designed to safely sustain and transfer the combined total dead and imposed loads of a building to a suitable subsoil to prevent settlement or other movement in any part of the building, or any adjoining building or works, in accordance with ADA A1/2. Type and design of foundation depend on total building loads and the nature and bearing capacity of the subsoil. The foundation design should support the total load (dead and imposed loads) of the load-bearing walls of the building. Typically, strip and trench-fill foundations are used for most domestic buildings up to three storeys in height, as detailed in the guidance below. Raft and pile foundations are designed for use on soft or loose soils, or filled areas with low bearing capacity, and should be designed by a suitably qualified person.

### Concrete mixes for foundations

Foundation work should comply with BS 8000:1, 2 and 5 and BS 8004. General-purpose concrete mixes for non-hazardous conditions need to comply with BS 8500 and BS EN 206-1.

(i) ***Site-mixed concrete (Standardised Prescribed mix ST)***
This Is a site-mixed concrete suitable for most domestic construction activities in accordance with the guidance table below, designed using materials and mix proportions given in BS 5328:1 Section 4. Standard mixes should not be used in soils, ground waters or adjoining materials containing sulfates or other aggressive chemicals.

(ii) ***Ready-mixed concrete (Designated mix GEN, RC and FND)***
This Is a ready-mixed concrete designed and specified in accordance with BS 5328:1 Section 5, produced and mixed under quality-controlled conditions in accordance with BS EN ISO 9001. GEN is a concrete used for general purposes, RC is used for reinforced concrete and FND is used in soils containing sulfates, in accordance with the guidance table below.

(iii) ***Hand-mixed concrete***
Hand-mixed concrete proportions *must* be agreed with building control before works commence on site.

## 2.10 Domestic extensions

**Table 2.1:** Concrete mixes for foundations

| Application | Site-mixed (standardised prescribed mix) | Ready-mixed (designated mix) | Compaction method |
|---|---|---|---|
| Blinding, Strip foundations, trench-fill foundations and mass concrete fill | ST2 | GEN1 | Mechanical vibration/poker or tamping by hand |
| Reinforced concrete foundations (i.e. raft foundation) | Designed by a suitably qualified specialist | RC35 – designed by a suitably qualified specialist | Designed and specified by a suitably qualified specialist |
| Foundations in sulfate conditions* | n/a | FND – designed by a suitably qualified specialist | Designed and specified by a suitably qualified specialist |

**Note:** *Foundations in sulfate conditions are to be in accordance with BS 5328:1 Table 7a.

## Foundation types

### *Strip foundations* (see Figure 2.1)

Strip foundations should have a minimum width in accordance with Table 2.2 below (typically 600 mm wide for 300 mm-thick cavity walls and 450 mm wide for 100 mm-thick walls); a minimum thickness of 150 mm (typically 225 mm in practice); and a minimum depth below ground level in accordance with Table 2.2 below and as required by building control. Maximum depth is normally restricted to 1.0 m deep for health and safety reasons for persons working in trenches. Foundations for three-storey buildings should be designed by a suitably qualified person, i.e. a structural engineer. Any services should pass through the sub-structure walls protected by precast concrete lintels and not through the foundation – for more details, see pipes penetrating though walls in Part H of this guidance. Foundations should be located centrally under all external and internal walls and taken to a depth below the influence of drains and/or surrounding trees, and taken to natural undisturbed ground of adequate ground-bearing capacity.

Cavity wall W as guidance details should be central on foundation

Ground level

600 wide x 150mm minimum thick (225mm in practice) concrete strip foundation, 750-1000mm deep depending on sub soil conditions. Actual sizes and depth to be agreed on site with building control surveyor.

Foundation depths in relation to trees to be in accordance with NHBC tables- contact building control for guidance.

The minimum thickness of the foundation T should be either P or 150mm which ever is the greater.

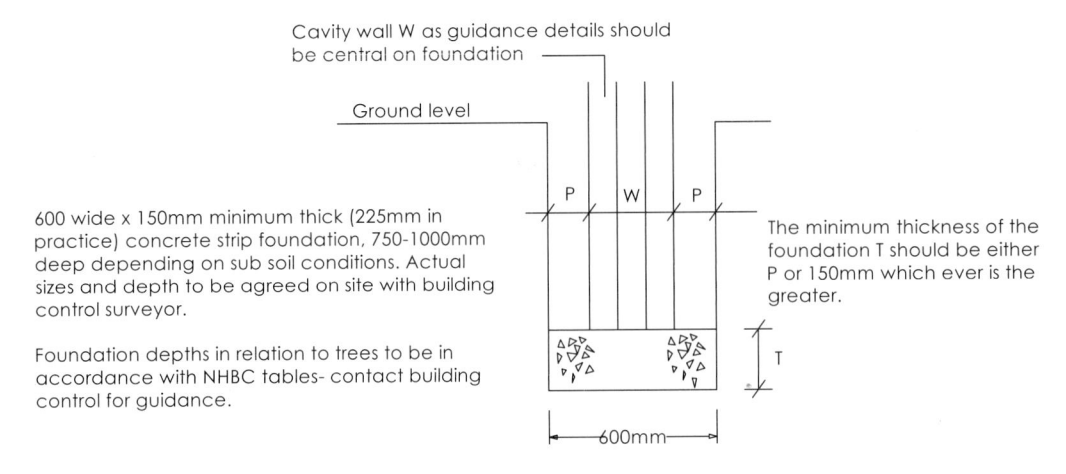

**Figure 2.1:** Strip foundation section detail *(not to scale)*
See Diagram 23 of ADA for full details

**Table 2.2:** Minimum width of strip/trench-fill foundations
(See Table 10 of ADA for full details.)

| Type of ground (including engineered fill) | Condition of ground | Field test applicable | Total load of load-bearing walling not more than (kN/linear metre) | | | | | |
|---|---|---|---|---|---|---|---|---|
| | | | 20 | 30 | 40 | 50 | 60 | 70 |
| | | | Minimum width of strip/trench-fill foundation (mm) | | | | | |
| I Rock | Not inferior to sandstone, limestone Or firm chalk | Requires at least a pneumatic or other mechanically operated pick for excavation | In each case equal to the width of the wall | | | | | |
| II Gravel or sand | Medium dense | Requires pick for excavation. Wooden peg 50 mm square cross section hard to drive beyond 150 mm | 250 | 300 | 400 | 500 | 600 | 650 |
| III Clay Sandy clay | Stiff Stiff | Can be indented slightly by thumb | 250 | 300 | 400 | 500 | 600 | 650 |
| IV Clay Sandy clay | Firm Firm | Thumb makes impression easily | 300 | 350 | 450 | 600 | 750 | 850 |
| V Sand Silty sand Clayey sand | Loose Loose Loose | Can be excavated with a spade. Wooden peg 50 mm square cross section can be easily driven | 400 | 600 | Note: Foundations on soil type V do not fall within the provisions of this section if the total load exceeds 30 kN/m | | | |
| VI Silt Clay Sandy clay Clay or silt | Soft Soft Soft Soft | Finger pushed in up to 10 mm | 450 | 650 | Note: Foundations on soil type V I do not fall within the provisions of this section if the total load exceeds 30 kN/m | | | |
| VII Silt Clay Sandy clay Clay or silt | Very soft Very soft Very soft Very soft | Finger pushed in up to 25 mm | Refer to specialists advice | | | | | |

The table is applicable only within the strict terms of the criteria described within it.

### Trench-fill foundations

Trench-fill foundations typically should have a minimum width of 450/500 mm, a minimum thickness and a minimum depth below external ground level in accordance with Figure 2.2 and the Tables below and as required by building control. Foundation for three-storey buildings or depths in excess of 2.5 m should be designed by a suitably qualified person, i.e. a structural engineer. Any services passing through trench-fill concrete should be ducted, sleeved or wrapped in flexible material (e.g.

fibreglass and polythene sheet fixed around drainage pipes or services to allow space for movement and to prevent differential movement damaging the services). Pipes through foundations should have flexible joints either side of the foundation – see pipes penetrating though walls in Part H of this guidance for more details. Foundations should be located centrally under all external and internal walls and taken to a depth below the influence of drains and/or surrounding trees, and taken to natural undisturbed ground of adequate ground-bearing capacity.

Steps in strip foundations should not exceed the foundation thickness and should have a minimum overlap equal to twice the height of the step, or thickness of the foundation, or 300 mm, whichever is greater, as detailed in the simplified diagram below (see Diagram 21 of ADA for full details).

### Building near trees, hedges, shrubs or in clay subsoils

Foundations in shrinkable, cohesive clay soils should be taken to a depth below the influence of any existing or proposed trees, hedges or shrubs near the building that could take moisture from the ground causing significant volume changes, resulting in possible ground settlement and damage to the foundations and building.

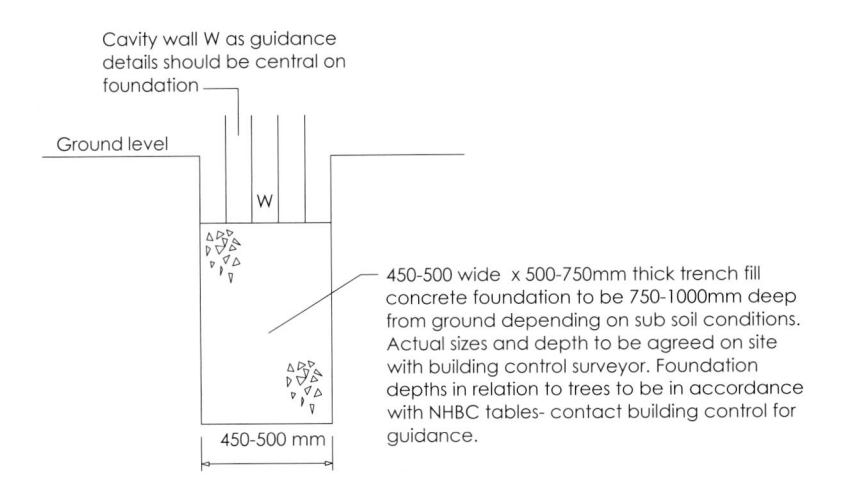

**Figure 2.2:** Trench-fill foundation section detail *(not to scale)*

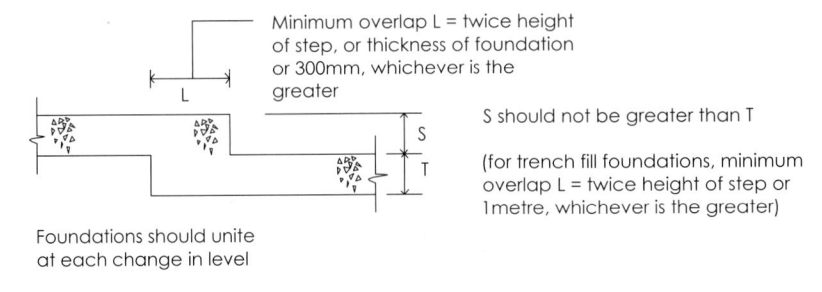

**Figure 2.3:** Stepped foundation section detail *(not to scale)*

Projection X should not be less than P

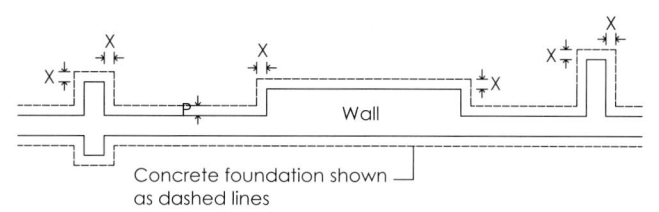

**Figure 2.4:** Foundation projections to piers, buttresses and chimneys (plan detail – *not to scale*) (See Diagram 22 of ADA for full details.)

**Table 2.3:** Minimum depth of strip/trench-fill foundations
(to be in compliance with paragraph 2E4 of ADA)

| Ground condition | Minimum foundation depth[1 and 4] |
|---|---|
| Rock or low-shrinkage firm natural gravel, sand or chalk subsoils (not clays or silts) | 450 mm minimum in frost susceptible soils[4] |
| Low-shrinkage clay subsoils[2] | 750 mm[4] |
| Medium- to high-shrinkage clay subsoils[3] | 900–1000 mm[4] |

**Note 1:** Minimum foundation depth is taken from external ground level to formation level (trench bottom). If finished ground level is above existing ground level and freezing is likely to occur, the foundation depth should be taken from the existing ground level and not the finished levels.
**Note 2:** Clay with a Modified Plasticity Index of less than 20% has a **low** volume-change potential, in accordance with BRE Digest 240.
**Note 3:** Clay with a Modified Plasticity Index of 20 to 40% has a **medium** volume-change potential, and with 40 to 60% has a **high** volume-change potential, in accordance with BRE Digest 240.
**Note 4:** Foundations should be taken to a depth below the influence of drains and or surrounding trees and taken to natural undisturbed ground of adequate ground-bearing capacity to support the total loads of the building to the approval of the building control surveyor.

Foundation depths should be in accordance with the NHBC foundation depth calculator (or other foundation depth calculator acceptable by building control), which calculates the foundation depth from the type of subsoil and tree type, including the mature height and water demand. Calculators can be obtained at a charge from www.nhbc.co.uk.

Foundations, substructure and services should incorporate adequate precautions to prevent excessive movement due to ground heave in shrinkable clay subsoils, in accordance with design details from a suitably qualified specialist. Typical heave precautions for trench-fill foundations with suspended floors in shrinkable subsoils should be carried out in accordance with requirements of building control and the Figures and Tables below.

## Alternative foundation designs

Alternative foundation designs, i.e. raft foundations (as detailed below) and pile foundations are designed for use on soft/ loose soils or filled areas which have low bearing capacity and should be designed for the particular project by a suitably qualified person; design should be approved by building control before works commence on site. Note: insulation details in this guidance are to be read in conjunction with Part L of this guidance.

## 2.14 Domestic extensions

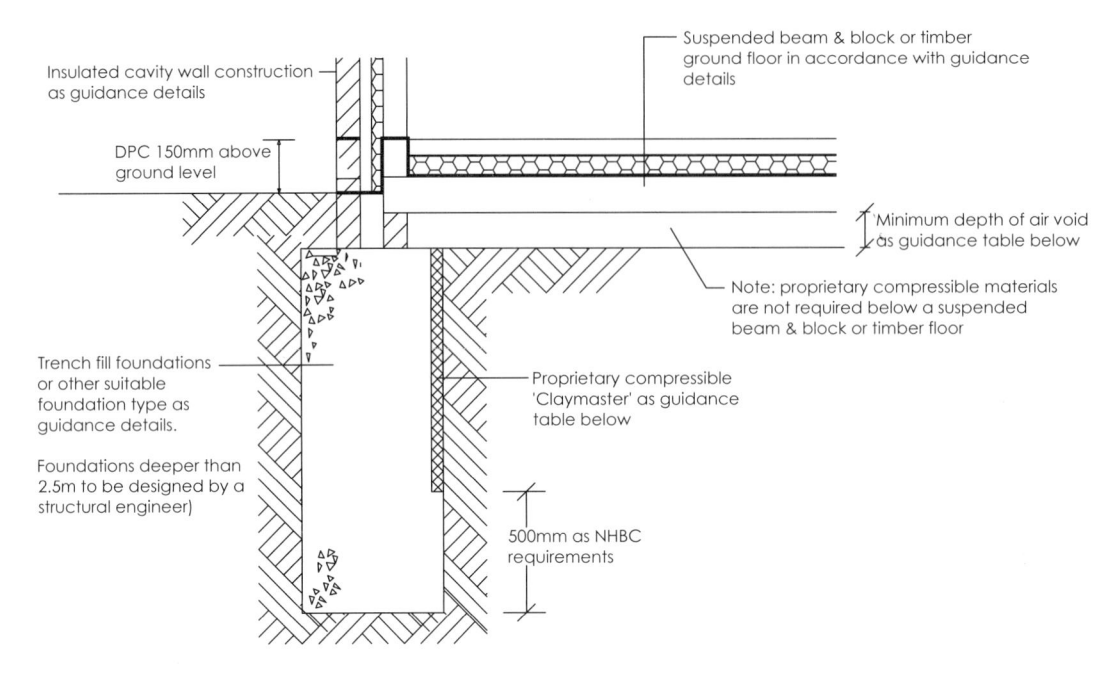

**Figure 2.5.1:** Heave precautions for trench-fill foundations with suspended beam and block floors in shrinkable clay subsoils (section detail – *not to scale*)

**Table 2.4.1:** Minimum void dimensions and clay heave protection for foundations and suspended beam and block floors

| Volume change potential in soil | Minimum NHBC void dimension against side of foundation (mm)[1] | Thickness of 'Claymaster' against side of foundation (mm)[2] | Minimum NHBC void dimension under suspended beam and block floors (mm)[1] |
|---|---|---|---|
| Low-shrinkage clay subsoils (10–20%) | 0 | Not required | 200 |
| Medium-shrinkage clay subsoils (20–40%) | 25 | 50 | 250 |
| High-shrinkage clay subsoils (40–60%) | 35 | 75 | 300 |

**Notes:**

[1] The void dimension is measured from the underside of the beam/joist to the top of the ground level under the floor (includes 150 mm ventilated void). Where the void beneath suspended floors is liable to flooding, drainage is to be provided.

[2] Compressible 'Claymaster' products are to be installed in accordance with the manufacturer's details. Information and products can be obtained from: www.cordek.com or other approved compressible products with BBA or other approved technical accreditation.

Source: includes information from The National House Building Council (NHBC – for more information see www.nhbc.co.uk) and based on information from Cordek Ltd. Reproduced by kind permission of Cordek Ltd.

Domestic extensions 2.15

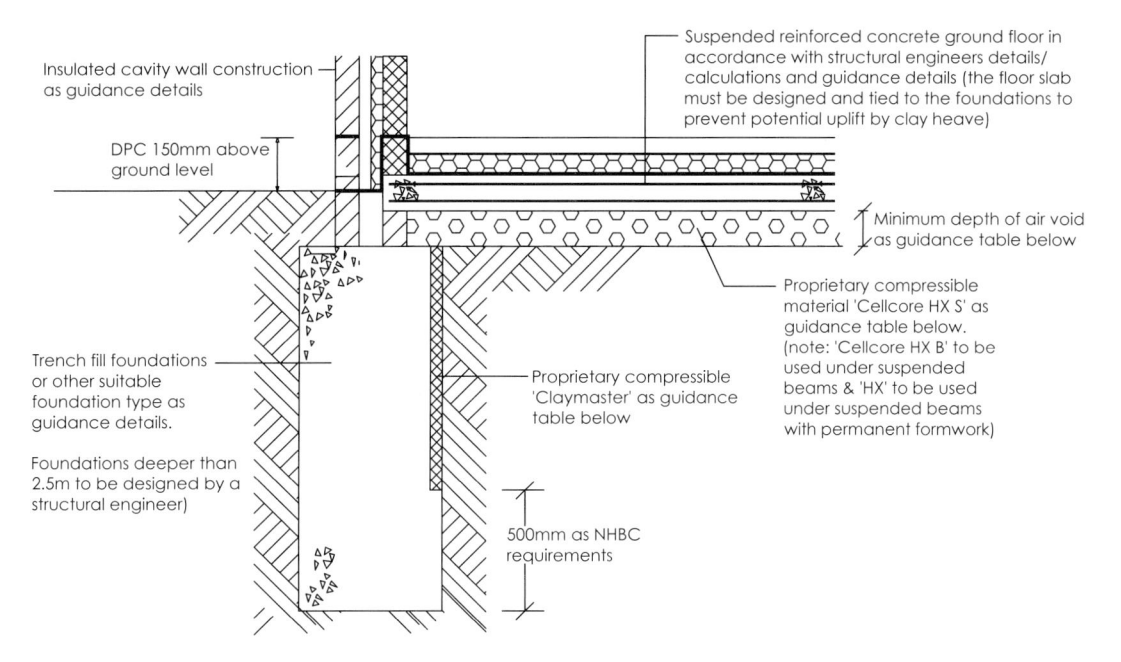

**Figure 2.5.2:** Heave precautions for trench-fill foundations with suspended cast in-situ reinforced concrete floor in shrinkable clay subsoils (section detail – *not to scale*)

**Table 2.4.2:** Minimum void dimensions and clay heave protection for foundations and suspended in-situ reinforced concrete floors and beams

| Volume change potential in soil | Minimum NHBC void dimension against side of foundation (mm)[1] | Thickness of 'Claymaster' against side of foundation (mm)[2] | Minimum NHBC void dimension under suspended in-situ reinforced concrete floors and beams (mm)[1] | Thickness of 'Cellcore' under suspended in-situ reinforced concrete floors and beams (mm)[2] |
|---|---|---|---|---|
| Low shrinkage clay sub soils (10-20%) | 0 | Not required | 50 | 90 |
| Medium shrinkage clay sub soils (20-40%) | 25 | 50 | 100 | 160 |
| High shrinkage clay sub soils (40-60%) | 35 | 75 | 150 | 225 |

**Note 1:** The void dimension should be able to accommodate the clay heave and compressible product. The void dimension shown is the minimum void dimension after collapse of the compressible product. Compressible products are to be installed in accordance with the manufacturer's details. Where the void beneath suspended floors is liable to flooding, drainage is to be provided.

**Note 2:** Compressible 'Claymaster' and 'Cellcore' products are to be installed in accordance with the manufacturer's details. Information and products can be obtained from www.cordek.com, or other approved compressible products with BBA or other approved technical accreditation may be used.

Source: includes information from The National House Building Council (NHBC – for more information see www.nhbc.co.uk) and based on information from Cordek Ltd. Reproduced by kind permission of Cordek Ltd.

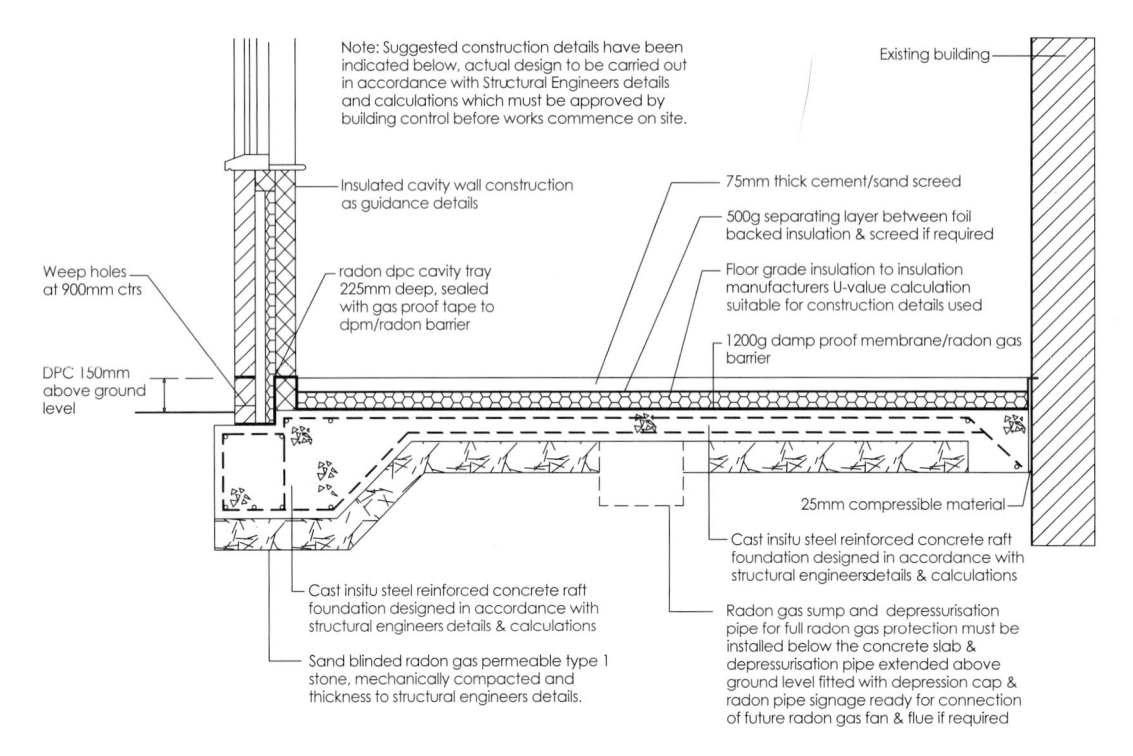

**Figure 2.6:** Raft foundation section detail *(not to scale)*
*Suitable for basic and full radon protection* (U-value 0.22 W/m$^2$.k)

### Raft foundations

Raft foundations can be used instead of strip or trench-fill foundations on soft/loose soils or filled areas that have low bearing capacity; they are designed to spread the building loads over the entire area of the structure and have the advantage of reducing differential settlements, as the reinforced concrete slab resists differential movements between loading positions. Raft foundations should be designed for the particular project by a suitably qualified specialist (i.e. a structural engineer) and constructed normally as a cast in-situ reinforced concrete slab with thickened edges, normally 450 mm below ground level – typically as detailed in Figure 2.6.

### Piled foundations

Piled foundations should be designed by a suitably qualified specialist (i.e. a structural engineer) for a particular project and are outside the scope of this guidance

## Retaining walls and basements

Retaining walls and basements (typically as detailed below) should be designed for the particular project by a suitably qualified person, and the design should be approved by building control before works commence on site.

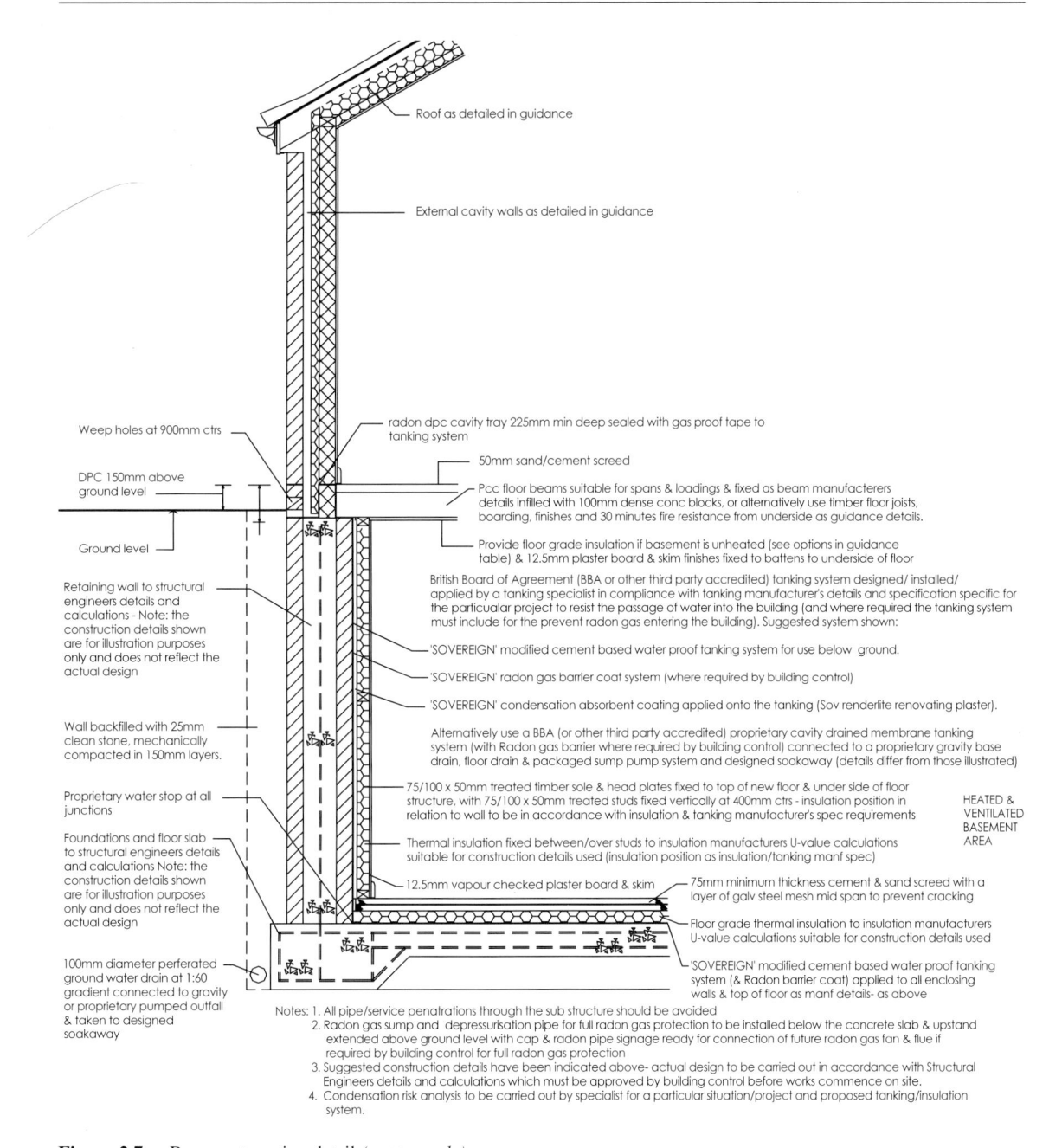

**Figure 2.7:** Basement section detail *(not to scale)*
*Suitable for basic and full radon protection* (U-value $0.22\,\text{W/m}^2.\text{k}$)

Figure 2.7 below contains suggested construction details only and the actual details must be in accordance with a structural engineer's details and calculations. The tanking and insulation details in the diagram are suggested details only; the actual details must be in accordance with a tanking specialist's and an insulation specialist's details and specifications, which have been produced for the particular project, as detailed in the guidance commentary below the diagram.

## Basements and tanking systems

### Site investigation and risk assessment

These are to be carried out before works commence to establish ground conditions, water-table levels, the presence of any contaminates and radon gas, and the location of drains and services, etc.

### Basement sub-structures

Are to be constructed in compliance with a structural engineer's details and calculations suitable for the ground conditions, loadings and proposed tanking system.

Provide all necessary temporary protection, support, shoring and working platforms, etc. in compliance with current health and safety requirements and structural engineer's details. All are to be erected, maintained, certificated, dismantled and removed by suitably qualified and insured specialists.

### Tanking systems

Tanking systems providing either barrier, structural or drained protection to the building must be assessed, designed and installed for the particular project in compliance with BS 8102: 2009 – Code of Practice for Protection of Below Ground Structures Against Water from The Ground. The systems can be installed either internally or externally, in accordance with a tanking specialist's details.

The illustrated tanking section details above are suggested ones only, and actual details must be approved by building control before works commence on site. Forms of tanking include bonded sheet materials; liquid applied membranes; mastic asphalt; drained cavity membranes; and cementitious crystallisation and cementitious multi-coat renders.

Suitable tanking systems need to have British Board of Agreement (BBA or other approved third-party) accreditation and be individually assessed by a tanking specialist as suitable for the proposed situation.

Tanking systems must be designed, installed and applied by a tanking specialist for the particular project, in compliance with the tanking manufacturer's details, to resist the passage of water into the building and prevent condensation and mould growth within the building – also, where required, to prevent radon gas and other ground gases from entering the building.

Tanking systems must be properly connected to and made continuous with wall damp-proof courses (dpc) and radon dpc trays. Perforation of the tanking system by service entry pipes etc. should be avoided or carried out strictly in accordance with the tanking manufacturer's details.

**Important note:** The risk of condensation with any tanking system should be assessed by a specialist. A condensation risk analysis should be carried out for the particular project and the tanking and thermal insulation system should be designed and installed to prevent any potential condensation or interstitial condensation problems.

## Ground floors and sub-structure walls

### Sub-structure walls

Foundations are to be provided centrally positioned under all exterior, party and interior load-bearing walls, as detailed in Figure 2.8 below. Steel mesh reinforcement is to be provided in foundations where required by a structural engineer or building control.

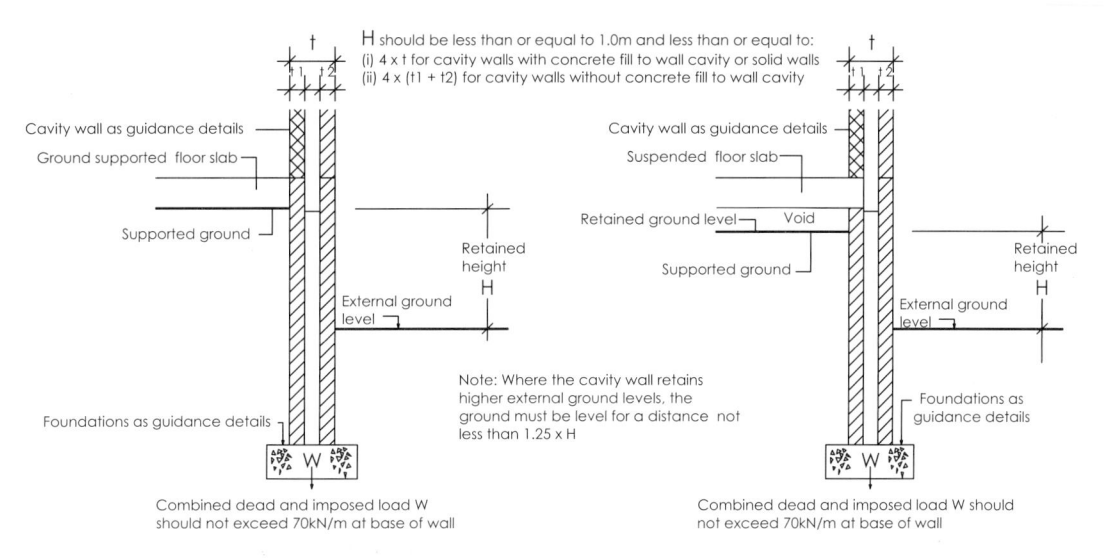

**Figure 2.8:** Walls supporting differences in ground levels *(not to scale)*
(See Diagram 11 of ADA for full details.)

Walls below dpc level up to 1 m deep are to be constructed with two skins of 7 N/mm² 100 mm (or 140 mm if over 1 m deep) concrete blocks of 1:3–4 cement mortar with plasticiser and in-filled with concrete to a maximum of 225 mm below dpc level. Block and cavity width and wall tie spacing, etc., are to be the same as the wall above, but with a row of wall ties to support the cavity wall insulation below dpc level. All materials must be frost-resistant. If sulfates are present in the ground, use sulfate-resisting cement.

## Ground floors

Ground floors can be either be ground-bearing or suspended, as detailed in the guidance details and Figures below, or they can be an integral part of alternative foundation design (i.e. raft or pile-and-beam foundation, as detailed in the guidance above). Radon gas and Insulation details illustrated are to be read in conjunction with Parts C and L of this guidance.

Ground-bearing floors rely on the direct support of the ground over the footprint of the building and are suitable for areas requiring basic or no radon protection, but they are *not* suitable for areas requiring full radon protection in accordance with Part C of this guidance. Ground-bearing slabs are not suitable for use over clay subsoils, which can shrink or heave and damage the floor.

Suspended ground floors, consisting of either cast in-situ reinforced concrete, beam-and-block or timber joist, are supported on the cavity walls and designed to span over the building. They are suitable for full radon protection, as they will prevent settlement of the ground floor at the external wall junction and prevent rupture of the radon-proof barrier, in accordance with Part C of this guidance. Suspended cast in-situ concrete and beam-and-block ground floors are suitable for use over clay subsoils that can shrink or heave, in accordance with the details and diagrams in this guidance. Suspended timber floors with sleeper walls and sub-floor slabs with radon barriers bearing on the ground are *not* suitable for clay subsoils that can shrink or heave, rupture the radon membrane and damage the floor.

### Ground-bearing solid concrete floors

(U-value 0.22 W/m$^2$.k)

*These are suitable for areas where basic radon protection is required, but are not suitable where full radon protection is required or for use over clay subsoils that can shrink or heave and damage the floor.*

Topsoil and vegetable matter are to be cleared from the site and floor area is to be in-filled between walls with minimum 150 mm (maximum 600 mm) of clean, well-graded inert hardcore, sand-blinded, and mechanically compacted in 150 mm maximum deep layers.

1200 g (300 micrometre) continuous polythene damp-proof membrane (dpm) and radon gas-proof barrier is to be laid over the sand-blinded hardcore, lapped and sealed at all joints and linked to damp-proof courses in walls. To provide basic protection from radon gas, the dpc within the cavity wall should be in the form of a cavity tray and sealed with dpm to prevent radon from entering the building through the cavity. Sealing of joints in the barrier and sealing around service penetrations with radon gas-proof tape are also required, in compliance with part C of this guidance.

Lay floor-grade insulation over dpm, minimum thickness and type in accordance with Table 2.5 below, including 25 mm-thick insulated up-stands between slab and external walls.

Lay minimum 100 mm-thick ST2, or Gen1 concrete floor slab with a trowel-smooth surface ready for finishes over insulation (note that 500 g polythene separating layer is to be installed between the concrete slab and insulation if using a foil-faced polyurethane/PIR-type insulation board). Insulation

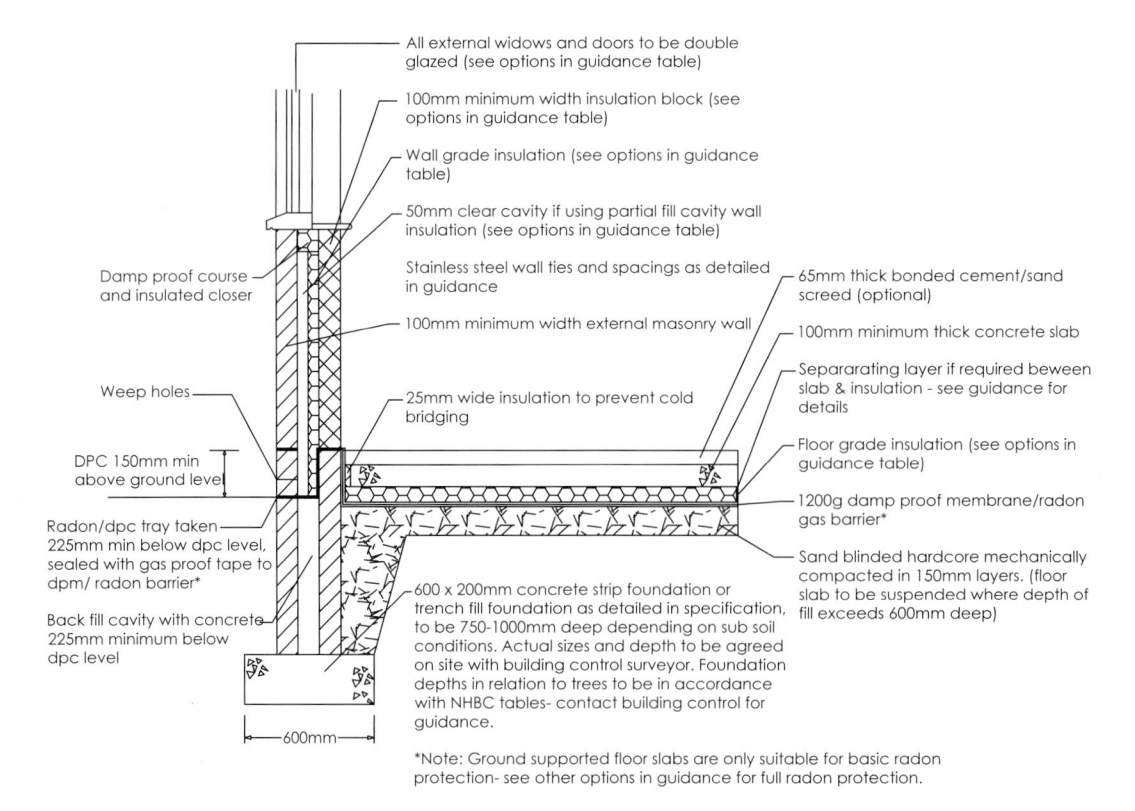

**Figure 2.9:** Typical section through a ground-bearing solid concrete floor and foundation *(not to scale)*

**Table 2.5:** Examples of insulation for ground-bearing floor slabs
(U-value no worse than 0.22 W/m²k)
**NOTE: Where P/A ratio has not been calculated, use insulation thickness stated in 1.0.***

| Insulation product | K value | Required thickness of insulation (mm) | | | | | | | | | |
|---|---|---|---|---|---|---|---|---|---|---|---|
| | | Calculated Perimeter/Area ratio (P/A) | | | | | | | | | |
| | | 1.0* | 0.9 | 0.8 | 0.7 | 0.6 | 0.5 | 0.4 | 0.3 | 0.2 | 0.1 |
| Kingspan Kooltherm K3 Floorboard | 0.020–0.023 | 75 | 70 | 70 | 65 | 65 | 60 | 55 | 45 | 30 | 20 |
| Celotex GA4000 | 0.022 | 80 | 75 | 75 | 70 | 70 | 65 | 60 | 50 | 30 | 12 |
| Jablite Jabfloor Premium | 0.030 | 105 | 105 | 100 | 95 | 95 | 95 | 80 | 65 | 50 | 20 |
| Styrofoam Floormate 300A | 0.035 | 110 | 100 | 100 | 100 | 90 | 90 | 80 | 70 | 50 | 50 |
| Rockwool Rockfloor | 0.038 (50–100 mm) 0.040 (25–40 mm) | 130 | 130 | 125 | 120 | 120 | 110 | 100 | 80 | 50 | 50 |

**Note 1:** Figures indicated above should be rounded up to the insulation manufacturer's nearest thickness.
**Note 2;** *Where P/A ratio has not been calculated, use insulation thickness stated in 1.0* above.
**Note 3:** Insulation to be installed in accordance with manufacturer's details.
Source: a representative selection of values taken from *Technical Note 10, U-Values of Elements (Approved Document L1B 2010)* produced by Hertfordshire Technical Forum for Building Control. Reproduced by permission of Hertfordshire Technical Forum for Building Control.

is to be omitted and concrete thickness increased in areas where non-load-bearing partitions are to be built off the floor slab (load-bearing partitions should be built off a foundation). Where area of fill exceeds 600 mm the floor is to be suspended, as detailed in this guidance.

### *Suspended reinforced in-situ concrete ground floor slab supported on internal walls*

(U-value 0.22 W/m².k) *This is suitable for areas where basic or full radon protection is required, and is also suitable for use over clay subsoils that can shrink or heave.*

Topsoil and vegetable matter are to be cleared from the site and floor area is to be in-filled between walls with minimum 150 mm deep (maximum 600 mm) of clean, well-graded inert hardcore, sand-blinded, and mechanically compacted in 150 mm maximum deep layers. Minimum void dimensions and compressible materials below slab in clay subsoils are to be in accordance with Figure 2.10 and Table 2.4.2 (in clay subsoils, the floor slab should be designed and restrained to prevent uplift from the compressible materials). Where full radon protection is required, a sub-floor sump and depressurisation pipe (with an external up-stand for future connection to an in-line fan and dispersal pipe) are to be positioned below the concrete floor slab in radon gas-permeable hardcore – in accordance with the sump manufacturer's details and part C of this guidance.

Shutter and cast reinforced concrete floor slab are to be supported on the inner leaf of the cavity wall, in accordance with the structural engineer's details and calculations, to prevent settlement of the slab and rupture of the radon-proof barrier. 1200 g (300 micrometre) of continuous polythene

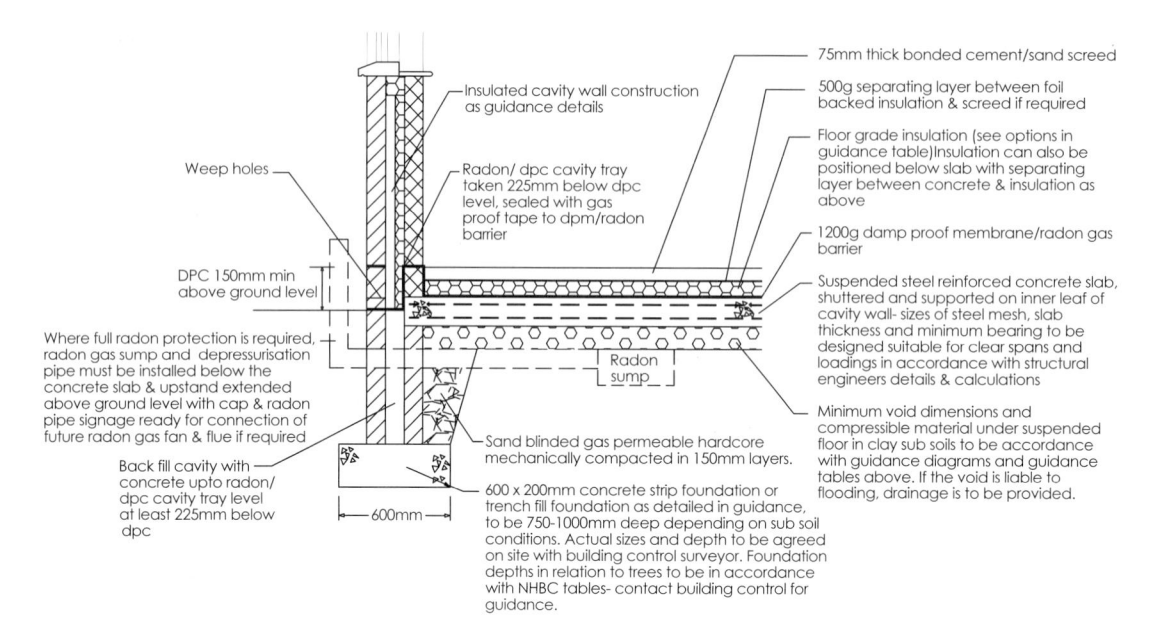

**Figure 2.10:** Typical section through a suspended reinforced in-situ concrete ground floor slab supported on internal walls *(not to scale)*

damp-proof membrane (dpm)/radon gas-proof barrier is to be laid over the concrete slab surface, lapped and sealed at all joints and linked to damp-proof courses in the walls. To provide basic protection from radon gas, the dpc within the cavity wall should be in the form of a cavity tray and sealed by dpm to prevent radon from entering the building through the cavity. Sealing of joints in the barrier and sealing around service penetrations with radon gas-proof tape are also required.

Floor-grade insulation is to be laid over DPM, minimum thickness and type in accordance with Table 2.5, including 25 mm-thick insulated up-stands between slab and external walls. 75 mm sand/cement thick structural screed (mix between 1:3 and 1:4½) should be laid over insulation with trowel-smooth finish ready for finishes; screed area should be limited to room sizes; floor areas exceeding 40 m$^2$ should have expansion/contraction joints as detailed in the note below. Screed should be laid over insulation with a trowel-smooth surface, ready for finishes. (A 500 g polythene separating layer is to be installed between the screed and insulation if using a foil-faced polyurethane/PIR-type insulation board.) Insulation is to be omitted in areas where non-load-bearing partitions are built off the floor slab.

See guidance below for installation of proprietary under-floor heating system

### Suspended beam-and-block ground floors

(U-value 0.22 W/m$^2$.k) *This is suitable for areas where basic and full radon protection is required, and is also suitable for use over clay sub soils that can shrink or heave.*

Ground is to be stripped of all top soil and vegetation and laid to 1:80 falls to the outside of the building; minimum void dimension below underside of suspended floor is to be in accordance with the guidance above. PCC beams to be supplied and fixed in accordance with the beam manufacturer's plan layouts and details (copies are to be sent to Building Control and approved before works commence on site).

Typically, for domestic loading, prestressed beams are to have 100 mm minimum bearing onto the dpc and load-bearing walls. All garage floors are to be designed suitable for loadings. Wet and grout all joints with 1:4 cement: sand mix. Provide double beams below non-load-bearing parallel partitions. Sub-structure void is to be vented on opposing sides to provide cross-ventilation, using 225 × 150 mm proprietary ventilators at 2 m centres and 450 mm from wall corners; sub floor level is to be above external ground levels; and if the floor void is liable to flood, drainage is to be provided. 1200 g (300 micrometre) continuous polythene damp-proof membrane (DPM) and radon gas-proof barrier are to be laid over beam-and-block floor, taken across cavity, cut back from face of masonry wall by at least 15 mm to avoid capillary action and ingress of water, lapped and sealed at all joints and linked to damp-proof courses in walls. To provide basic radon gas protection, the dpc within the cavity wall should be in the form of a cavity tray and sealed to the dpm to prevent radon gas from entering the building through the cavity. Sealing of joints in the barrier and sealing around service penetrations with radon gas-proof tape are also required. Where full radon protection is required, provision is to be made for connection of future depressurisation pipe to vented floor and up-stand, in accordance with the manufacturer's details.

Floor-grade insulation is to be laid over dpm, with minimum thickness and type of insulation to be in accordance with Table 2.6 below, including 25 mm-thick insulated up-stands between screed and external walls. 75 mm sand/cement thick structural screed (mix between 1:3 and 1:4½) is to be laid over insulation with trowel-smooth finish, ready for finishes; screed area should be limited to room sizes; floor areas exceeding 40 m$^2$ should have expansion/contraction joints as detailed in the note below (500 g polythene separating layer is to be installed between the concrete slab and insulation if using a foil-faced polyurethane/PIR-type insulation board).

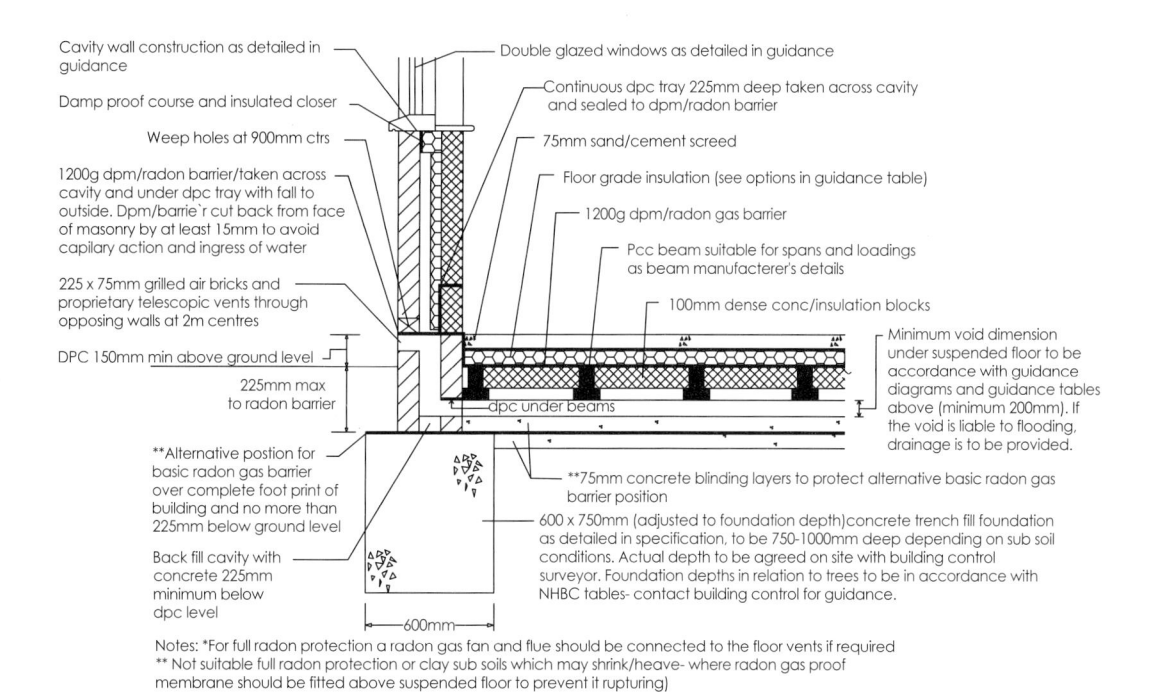

**Figure 2.11:** Typical section through a suspended beam-and-block ground floor *(not to scale)*

**Table 2.6:** Examples of insulation for suspended beam-and-block ground floors
(U-value no worse than 0.22 W/m²k. Block K value = 0.18)
**NOTE: Where P/A ratio has not been calculated, use insulation thickness stated in 1.0.***

| Insulation product | K value | Required thickness of insulation (mm) Calculated Perimeter/Area ratio (P/A) | | | | | | | | | |
|---|---|---|---|---|---|---|---|---|---|---|---|
| | | 1.0* | 0.9 | 0.8 | 0.7 | 0.6 | 0.5 | 0.4 | 0.3 | 0.2 | 0.1 |
| Kingspan Kooltherm K3 Floorboard | 0.020–0.023 | 65 | 65 | 65 | 60 | 60 | 55 | 55 | 50 | 40 | 20 |
| Celotex GA4000 | 0.022 | 75 | 75 | 70 | 70 | 65 | 65 | 65 | 55 | 40 | 12 |
| Jablite Jabfloor Premium | 0.030 | 95 | 95 | 95 | 90 | 90 | 85 | 80 | 70 | 55 | 25 |
| Styrofoam Floormate 300A | 0.035 | 100 | 100 | 100 | 100 | 90 | 90 | 90 | 80 | 60 | 50 |
| Rockwool Rockfloor | 0.038 (50–100 mm) 0.040 (25–40 mm) | 130 | 125 | 125 | 125 | 120 | 110 | 100 | 90 | 70 | 25 |

**Note 1:** Figures indicated above should be rounded up to the insulation manufacturer's nearest thickness.
**Note 2:** *Where P/A ratio has not been calculated, use insulation thickness stated in 1.0* above.
**Note 3:** Insulation to be installed in accordance with manufacturer's details.
Source: a representative selection of values taken from *Technical Note 10, U-Values of Elements (Approved Document L1B 2010)*, produced by Hertfordshire Technical Forum for Building Control. Reproduced by permission of Hertfordshire Technical Forum for Building Control.

Insulation is to be omitted where non-load-bearing partitions are built off the beams to the beam manufacturer's design details. See guidance below for installation of proprietary under-floor heating system.

### *Proprietary under-floor heating system*

Where a proprietary under-floor heating system is installed it should be fixed above insulation and under screed layer, in compliance with the heating pipe manufacturer's/heating specialist's details. Screeds over under-floor heating should be subdivided into bays not exceeding 40 m² in area. Expansion/contraction joints in screeds should be consistent with joints in slabs, and pipes protected in accordance with the heating pipe manufacturer's/heating specialist's details.

### *Floating floor*

Alternatively, instead of cement/sand floor screed, a floating timber-board floor can be laid over the insulation using minimum 22 mm-thick, moisture-resistant, tongue-and-grooved timber floorboard sheets, with all joints glued, pinned and secured at the perimeters by skirting boards, with allowance for expansion joints in compliance with the floorboard manufacturer's details (typically 10–15 mm) and current BS EN standards. Minimum thickness and type of insulation are to be in accordance with Table 2.7 below.

**Table 2.7:** Examples of insulation for floating floors
(U-value no worse than 0.22 W/m²k)
**NOTE: Where P/A ratio has not been calculated, use insulation thickness stated in 1.0.***

| Insulation product | K-value | Required thickness of insulation (mm) |||||||||
|---|---|---|---|---|---|---|---|---|---|
| | | Calculated Perimeter/Area ratio (P/A) |||||||||
| | | 1.0* | 0.9 | 0.8 | 0.7 | 0.6 | 0.5 | 0.4 | 0.3 | 0.2 |
| Kingspan Kooltherm K3 Floorboard | 0.020–0.023 | 95 | 95 | 90 | 85 | 85 | 75 | 70 | 60 | 35 |
| Kingspan Thermafloor TF70 and Celotex GA4000 | 0.022 | 75 | 75 | 70 | 70 | 65 | 60 | 55 | 45 | 30 |

**Note 1:** Figures indicated above should be rounded up to the insulation manufacturer's nearest Thickness.
**Note 2:** *Where P/A ratio has not been calculated, use insulation thickness stated in 1.0* above.
**Note 3:** Insulation to be installed in accordance with the manufacturer's details.
Source: a representative selection of values taken from *Technical Note 10, U-Values of Elements (Approved Document L1B 2010)*, produced by Hertfordshire Technical Forum for Building Control. Reproduced by permission of Hertfordshire Technical Forum for Building Control.

### *Suspended timber ground floor*

(U-value 0.22 W/m².k) *This is suitable for areas where basic radon protection is required, but is* **not** *suitable in clay subsoils, which can heave and rupture the sub-floor radon membrane and damage the floor.*

Topsoil and vegetable matter to be cleared from site and floor area to be in-filled between walls with minimum 150 mm deep (maximum 600 mm) of clean well-graded inert hardcore, sand blinded, and mechanically compacted in 150 mm maximum deep layers; lay minimum 100 mm-thick concrete over site at 1:80 gradient to outside of building (concrete mix should be in accordance with BS 8110, BS 5328, mix type ST2 or GEN1 or RC grade if reinforcement is required), on 1200 g (300 micrometre) damp-proof membrane (dpm)/radon gas-proof membrane, which should extend across footprint of building and cavity wall for basic radon gas protection. The radon barrier should be no more than 225 mm below external ground level and positioned to prevent water collection. A sub-floor sump and depressurisation pipe with up-stand are to be positioned below the over-site concrete floor slab in radon gas-permeable hardcore – in accordance with the sump manufacturer's details and Part C of this guidance below.

Allow a ventilated air space at least 75 mm from the top of the over-site concrete to the underside of any wall plates, and at least 150 mm to the underside of the suspended timber floor or insulation. Provide sub-floor ventilation using 225 × 75 mm grilled air bricks and proprietary telescopic vents through two opposing external walls at 2 m centres and 450 mm from wall corners, to vent all parts of the floor void. If the floor void is liable to flood, a beam-and-block floor should be used instead of timber. Joists are to be built into walls off the dpc and sealed with silicon or supported off proprietary, heavy-duty, galvanised joist hangers, built into new masonry walls or fixed to treated timber wall plate (same size as joists), resin-bolted to existing walls at 600 mm centres using 16 mm-diameter high-tensile bolts. Where necessary, floor joists can be supported in the span on treated wall plates and damp-proof course (dpc) onto masonry honeycombed sleeper walls built off over-site concrete. Floor joist sizes should be in accordance with Table 2.20 (depth to be increased where necessary to match floor levels). Proprietary galvanised-steel strutting is to be fixed at mid-span for

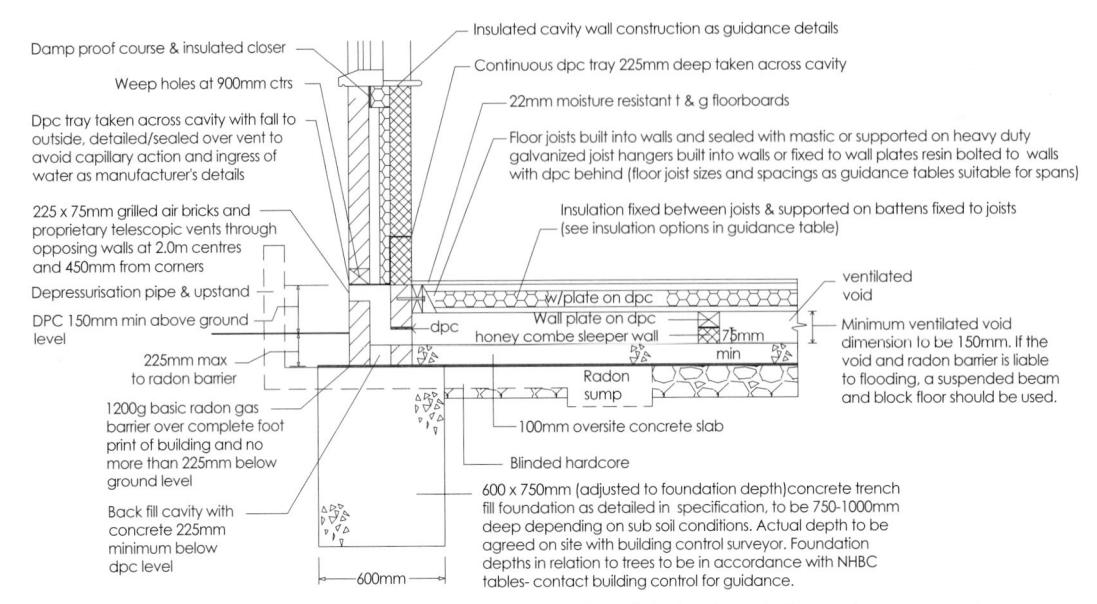

## 2.26 Domestic extensions

**Figure 2.12:** Typical section through a suspended timber ground floor *(not to scale)*

**Table 2.8:** Examples of insulation for suspended timber ground floors
(U-value no worse than 0.22 W/m²k.)
**NOTE:** Where P/A ratio has not been calculated, use insulation thickness stated in 1.0.*

| Insulation product | K value | Required thickness of insulation (mm) — Calculated Perimeter/Area ratio (P/A) | | | | | | | | |
|---|---|---|---|---|---|---|---|---|---|---|
| | | 1.0* | 0.9 | 0.8 | 0.7 | 0.6 | 0.5 | 0.4 | 0.3 | 0.2 |
| Kingspan Thermafloor TF70 | 0.022 | 75 | 75 | 75 | 70 | 70 | 65 | 60 | 50 | 35 |
| Celotex FR4000 | 0.022 | 75 | 75 | 75 | 70 | 70 | 65 | 60 | 50 | 35 |
| Jablite Jabfloor Premium 70 | 0.030 | 140 | 140 | 130 | 130 | 125 | 120 | 115 | 105 | 80 |
| Jablite Jabfloor 70 | 0.038 | 160 | 160 | 160 | 155 | 150 | 145 | 135 | 120 | 100 |
| Rockwool Flexi | 0.038 | 160 | 160 | 160 | 140 | 140 | 140 | 140 | 120 | 90 |
| Knauf Earthwool loft roll 40 and loft roll 44 | 0.040 | 170 | 170 | 170 | 170 | 170 | 150 | 150 | 150 | 100 |
| | 0.044 | 200 | 200 | 200 | 170 | 170 | 170 | 150 | 150 | 100 |

**Note 1:** Figures indicated above should be rounded up to the insulation manufacturer's nearest thickness.
**Note 2:** *Where P/A ratio has not been calculated, use insulation thickness stated in 1.0* above.
**Note 3:** Insulation is to be installed in accordance with the manufacturer's details.
Source: a representative selection of values taken from *Technical Note 10, U-Values of Elements (Approved Document L1B 2010)*, produced by Hertfordshire Technical Forum for Building Control. Reproduced by permission of Hertfordshire Technical Forum for Building Control.

2.5–4.5 m span, and 2 rows at 1/3rd points for spans over 4.5 m. Floor is to be insulated in accordance with Table 2.8 and friction-fixed between joists. Fix 22 mm-thick, moisture-resistant, tongue-and-grooved timber floorboards laid with joints staggered, long edges fixed across the joists and all joints positioned over joists/noggins. All boards are to be glued and screwed to floor joists, with all joints glued (using waterproof glue) and pinned, in accordance with the floorboard manufacturer's details and current BS EN standards. Allow an expansion gap around wall perimeters as in the manufacturer's details (typically 10–15 mm).

### *Garage ground-bearing concrete floor*

Use power-floated, 150 mm-thick concrete slab (concrete mix should be in accordance with BS 8500 and BS EN 206-1, mix type ST4 or GEN3 for non-hazardous conditions, with 1 layer anti-crack steel mesh positioned mid-depth of the slab where required – typically A193 or A252) on 1200 g polythene damp-proof course/radon barrier on minimum 150 mm deep (maximum 600 mm) of clean well-graded inert hardcore, sand-blinded, and mechanically compacted in 150 mm maximum deep layers in between walls (ensuring all topsoil and vegetable matter is removed and taken from site).

1:80 fall is required on the floor from the back of the garage to the front garage-door opening; the floor is to be thickened to 300 mm at the the garage entrance.

Provide 25 mm polystyrene compressible clay board to perimeter of walls. Where area of fill exceeds 600 mm, the floor is to be suspended in compliance with the structural engineer's details and calculations, which must be approved by building control.

A 100 mm-high, non-combustible step or ramp down into the garage (including FD30 fire door as guidance details) is to be provided at doorways from attached domestic accommodation.

Radon gas protection is to be provided in garages integral with the dwelling, in accordance with the above ground-floor details and depending on the level of radon protection required.

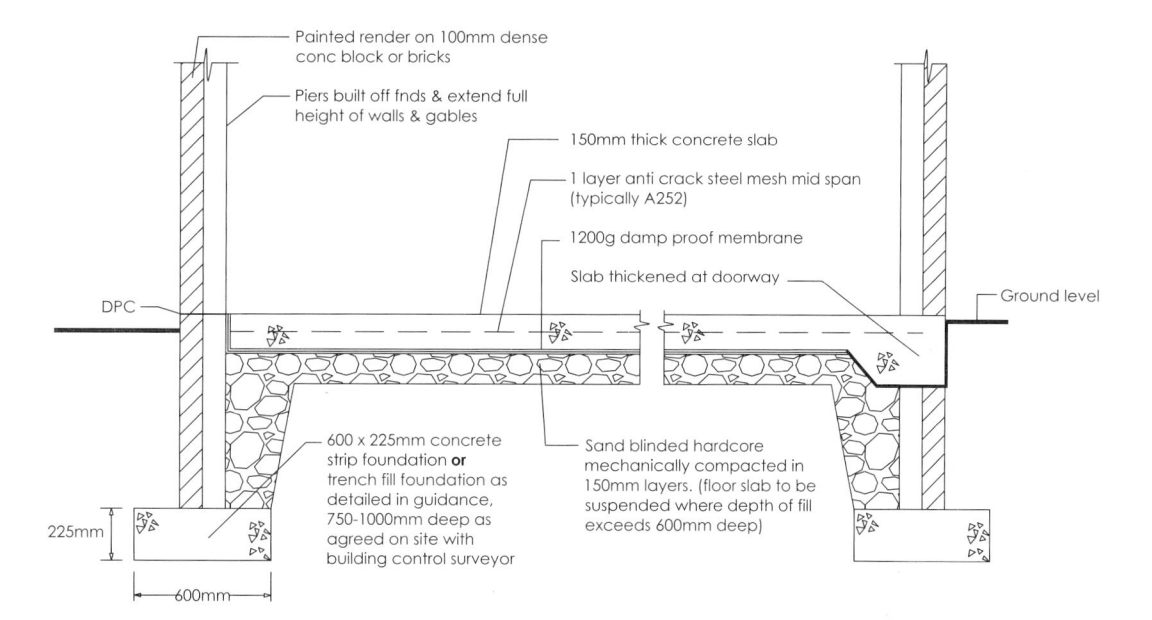

**Figure 2.13:** Typical section through a ground-bearing garage floor and foundation *(not to scale)*

## A2: Superstructure

## Minimum headroom heights

There are no minimum headroom height requirements in the building regulations for habitable rooms in single-occupancy dwellings, except for stairs and ramps (see Part K of this guidance); however, a minimum ceiling height of 2.3 m is recommended.

## External walls

### Sizes and proportions of residential buildings up to three storeys

(See paragraph 2C4 and Diagram 1 of ADA.)

(i) The maximum height of the building constructed of coursed brick or block work measured from the lowest finished ground level adjoining the building to the highest point of any wall or roof should not be greater than 15 m, and may need to be less than that subject to paragraph 2C16 of ADA, which provides the maximum heights of buildings ranging from 3 m to 15 m as in Tables a, b and c of Diagram 7 of ADA, correlating to various site exposure conditions and wind speeds. A map showing wind speeds and topographic zones is given in Diagram 6 of ADA.
(ii) The height of the building H should not exceed twice the least width of the building W1.
(iii) The height of wing H2 should not exceed twice the least width of the wing W2 where the projection P exceeds twice the width W2.
(iv) Floor area limit should not exceed the following: 70 m$^2$ where floor is bounded by walls on all four sides, and 36 m$^2$ where floor is bounded by walls on three sides.

### Measuring storey heights

Storey heights should not exceed those stated in Diagram 8 of ADA, and illustrated in the simplified Figure 2.15 below, for buildings constructed of coursed brick or block work.

### Measuring wall heights

The height of a wall should be measured in accordance with Diagram 8 of ADA for buildings constructed of coursed brick or block work, as illustrated in the simplified Figure 2.16 below.

### Measuring external/compartment/separating wall lengths

Depending on the height and thickness of certain external walls, compartment walls and separating walls, wall lengths should not exceed 12 m for buildings constructed of coursed brick or block work, in accordance with Paragraph 2C17 of ADA (also see Table 3 of ADA) and as illustrated in the simplified diagram below (Figure 2.17).

#### Vertical lateral restraint to walls

The ends of every wall should be bonded or otherwise securely tied throughout their full height to a buttressing wall, pier or chimney. Long walls may be provided with intermediate buttressing walls,

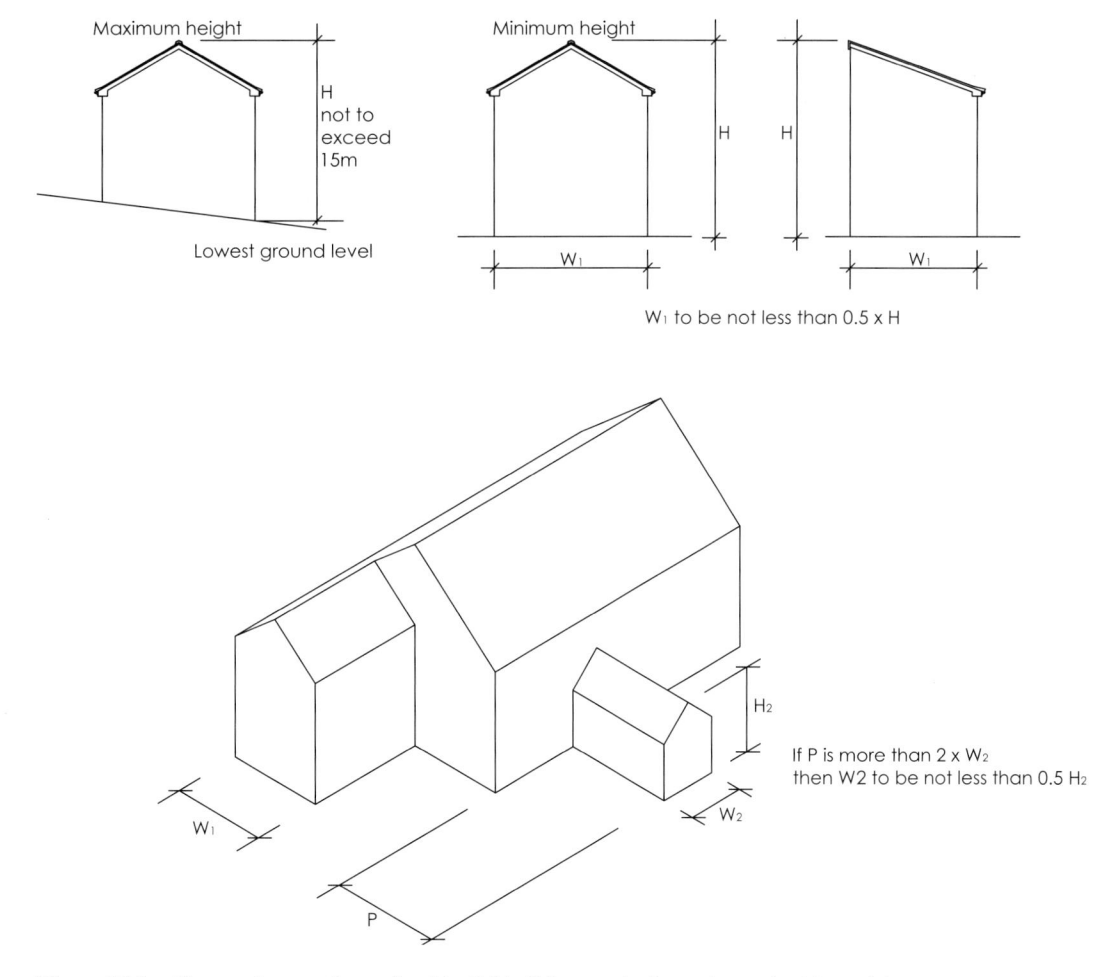

**Figure 2.14:** Sizes and proportions of residential buildings up to three storeys (*not to scale*) (See diagram 1 of ADA for full details.)

Diagram annotations:
- Maximum height: H not to exceed 15m, measured from Lowest ground level
- Minimum height: H
- $W_1$ to be not less than $0.5 \times H$
- If P is more than $2 \times W_2$ then W2 to be not less than $0.5\, H_2$

piers or chimneys dividing the wall into distinct lengths within each storey; each distinct length is a supported wall for the purposes of this section. The intermediate buttressing walls, piers or chimneys should provide lateral restraint to the full height of the supported wall, but they may be staggered at each storey.

## Thickness of walls

### *Minimum thickness of certain external walls, compartment walls and separating walls constructed of coursed brick or block work*

This is to be carried out in accordance with the Table 2.9 below and paragraphs 2C5 to 2C18 and Table 3 of ADA.

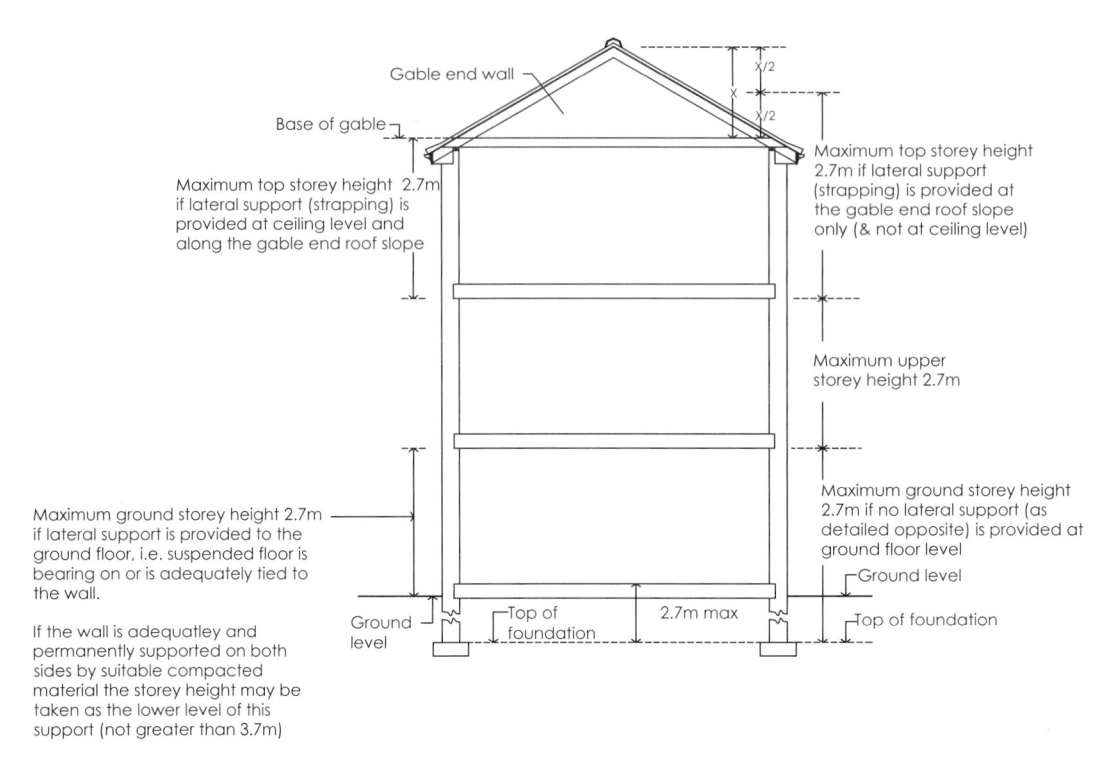

**Figure 2.15:** Measuring storey heights *(section detail not to scale)*

### Minimum thickness of internal load-bearing walls

All internal load-bearing walls (except compartment walls and separating walls) should have a thickness (in accordance with paragraph 2C10 of ADA) of not less than:

$$\frac{\textit{Minimum thickness of wall from guidance table 2.9 (as Table 3 of ADA)}}{2} - 5 \text{ mm}$$

### Buttressing wall design

If the buttressing wall is not itself a supported wall, its thickness should not be less than:

(a) half the thickness required by this section for an external or separating wall of similar height and length less 5 mm; or
(b) 75 mm if the wall forms part of a dwelling house and does not exceed 6 m in total height and 10 m in length; and
(c) 90 mm in other cases.

The length of the buttressing wall should be at least one-sixth of the overall height of the supported wall, be bonded or securely tied to the supporting wall, and at the other end to a buttressing wall, pier or chimney. The size of any opening in the buttressing wall should be restricted, as shown in Diagram 12 of ADA.

Domestic extensions **2.31**

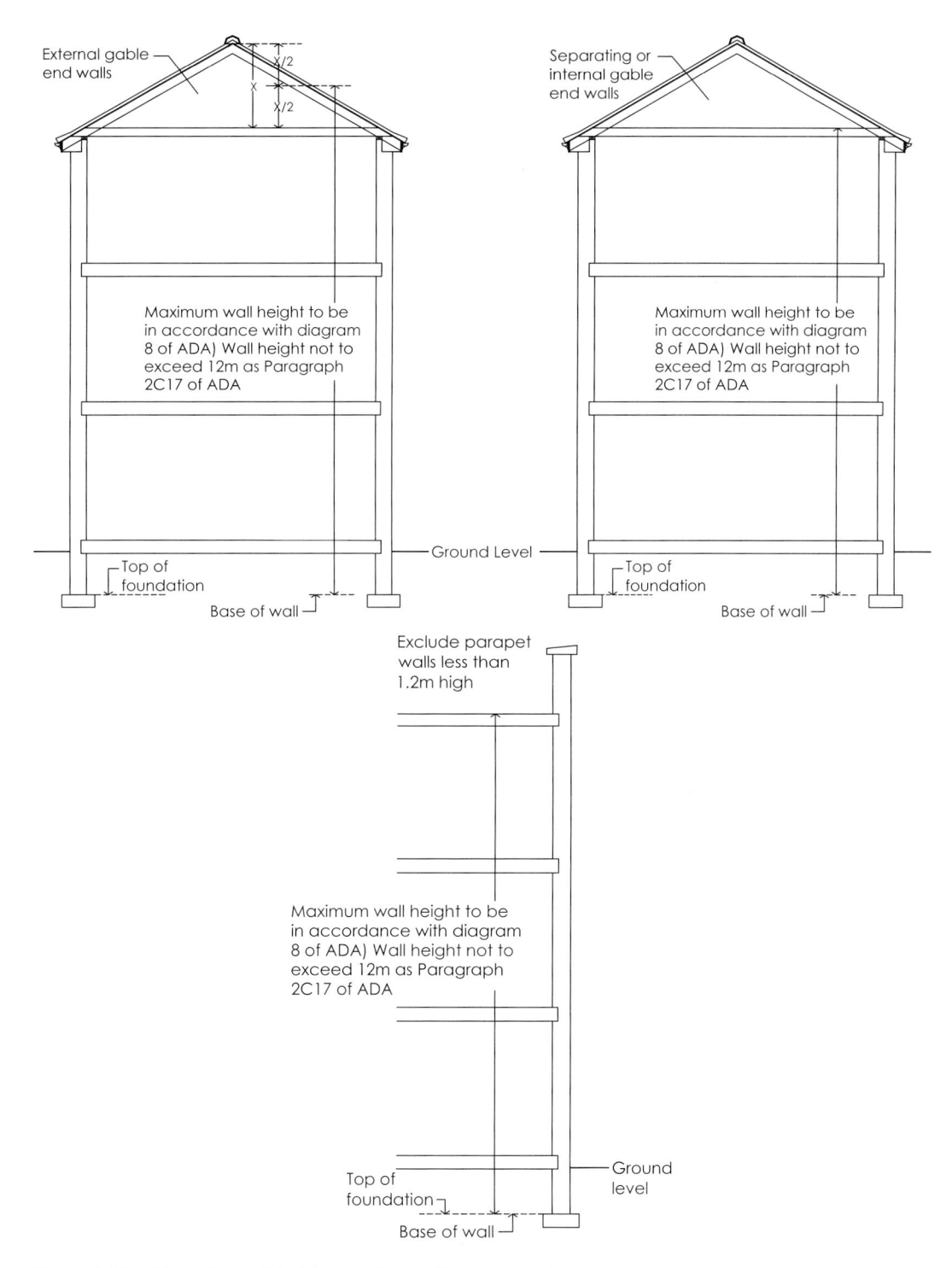

**Figure 2.16:** Measuring wall heights (section detail – *not to scale*)

## 2.32 Domestic extensions

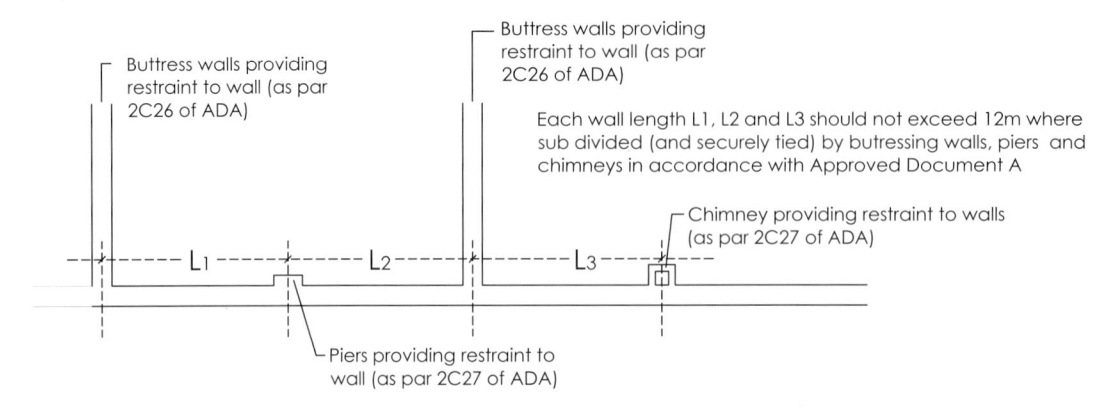

**Figure 2.17:** Measuring wall lengths (plan detail – *not to scale*)

Annotations in figure:
- Buttress walls providing restraint to wall (as par 2C26 of ADA)
- Buttress walls providing restraint to wall (as par 2C26 of ADA)
- Each wall length L1, L2 and L3 should not exceed 12m where sub divided (and securely tied) by buttressing walls, piers and chimneys in accordance with Approved Document A
- Chimney providing restraint to walls (as par 2C27 of ADA)
- Piers providing restraint to wall (as par 2C27 of ADA)

**Table 2.9:** Minimum thickness of certain external walls, compartment walls and separating walls constructed of coursed brick or block work

| Height of wall | Length of wall | Minimum thickness of wall |
|---|---|---|
| Up to 3.5 m | Up to 12 m | 190 mm for whole of its height |
| 3.5 m–9 m | Up to 9 m | 190 mm for whole of its height |
| | 9–12 m | 290 mm from the base for the height of one storey and 190 mm for the rest of its height |
| 9 m–12 m | Up to 9 m | 290 mm from the base for the height of one storey and 190 mm for the rest of its height |
| | 9–12 m | 290 mm from the base for the height of two storeys and 190 mm for the rest of its height |

### *Pier and chimney design providing restraint*

(a) Piers should measure at least three times the thickness of the supported wall, and chimneys twice the thickness, measured at right angles to the wall. Piers should have a minimum width of 190 mm (see Diagram 13 of ADA.);
(b) The sectional area on plan of chimneys (excluding openings for fireplaces and flues) should be not less than the area required for a pier in the same wall, and the overall thickness should not be less than twice the required thickness of the supported wall (see Diagram 13 of ADA).

### Compressive strength of masonry units

The declared compressive strength of masonry units is to be in compliance with Figure 2.18 and Table 2.10 below and Paragraph 2C21, Diagram 9 and Tables 6 and 7 of ADA.

Domestic extensions **2.33**

**Key**

- Condition A (see guidance table for details)
- Condition B (see guidance table for details)
- Condition C (see guidance table for details)
- $H_f$ less than or equal to 1.0m Condition A
- $H_f$ more than 1.0m Condition B

Notes:
1. If $H_s$ is not greater than 2.7m the compressive strength of bricks or blocks should be used in walls as indicated by the key above
2. If $H_s$ is greater than 2.7m, the compressive strength of bricks or blocks shall either be at least Condition B or as indicated by the key, whichever is the greater
3. If the external wall is solid construction, the masonry units should have a compressive strength of at least that shown for internal leaf of a cavity wall in the same position
4. The guidance given in the diagram for walls of two and three storey buildings should only be used to determine the compressive strength of the masonry units where the roof construction is of timber.

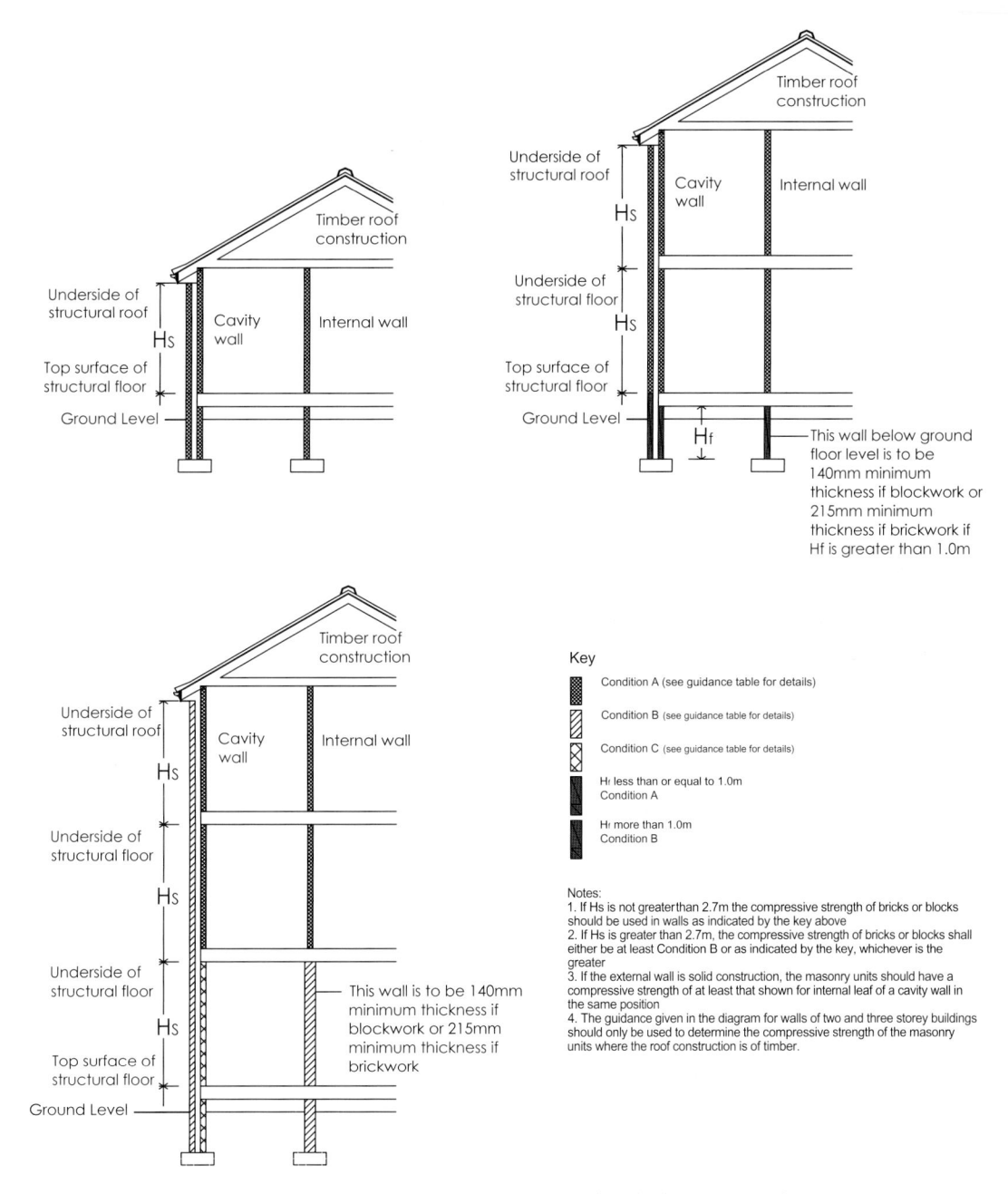

**Figure 2.18:** Declared compressive strength of masonry *units* (section detail – *not to scale*)

**Table 2.10:** Declared compressive strength of masonry units in walls complying with BS EN 771-1 to 5. (N/mm$^2$) (See Table 6 of ADA for full details Also see Table 7 of ADA for normalised compressive strength of masonry units of clay and calcium silicate blocks complying with BS EN 771-1 and 2 (N/mm$^2$).)

| Masonry unit | Clay Masonry to BS EN 771-1 | | Calcium Silicate masonry to BS EN 771-2 | | Dense Concrete masonry to BS EN 771-3 | Aerated masonry to BS EN 771-4 | Manufactured stone to BS EN 771-5 |
|---|---|---|---|---|---|---|---|
| **Condition A (see diagram above)** | | | | | | | |
| Brick | Group 1 6 N/mm$^2$ | Group 2 9 N/mm$^2$ | Group 1 6 N/mm$^2$ | Group 2 9 N/mm$^2$ | 6 N/mm$^2$ | n/a | Any unit complying to BS EN 771-5 will be acceptable for conditions A, B and C |
| Block | See Table 7 of ADA | See Table 7 of ADA | See Table 7 of ADA | See Table 7 of ADA | 2.9 N/mm$^2$* | 2.9 N/mm$^2$ | |
| **Condition B (see diagram above)** | | | | | | | |
| Brick | Group 1 9 N/mm$^2$ | Group 2 13 N/mm$^2$ | Group 1 9 N/mm$^2$ | Group 2 13 N/mm$^2$ | 9 N/mm$^2$ | n/a | Any unit complying to BS EN 771-5 will be acceptable for conditions A,B and C |
| Block | See Table 7 of ADA | See Table 7 of ADA | See Table 7 of ADA | See Table 7 of ADA | 7.3 N/mm$^2$* | 7.3 N/mm$^2$ | |
| **Condition C (see diagram above)** | | | | | | | |
| Brick | Group 1 18 N/mm$^2$ | Group 2 25 N/mm$^2$ | Group 1 18 N/mm$^2$ | Group 2 25 N/mm$^2$ | 18 N/mm$^2$ | n/a | Any unit complying to BS EN 771-5 will be acceptable for conditions A,B and C |
| Block | See Table 7 of ADA | See Table 7 of ADA | See Table 7 of ADA | See Table 7 of ADA | 7.3 N/mm$^2$* | 7.3 N/mm$^2$ | |

\* These values are dry strengths to BS EN 772-1

**Notes:**
1. This table applies to Group 1 and Group 2 units.
2. For the EN 771 series of standards for masonry units, the values of declared compressive strengths (N/mm$^2$) given above (and in Table 6 of ADA) are mean values.
3. Brick: a masonry unit having work sizes not exceeding 337.5 mm in length or 112.5 mm in height.
4. Block: a masonry unit exceeding either of the limiting work sizes of a brick and within a minimum height of 190 mm.
   For blocks with smaller heights, excluding cuts or make-up units, the strength requirements are as for bricks except for solid external walls, where the blocks should have a compressive strength at least equal to that shown for block for an inner leaf of a cavity wall in the same position.
5. Group 1 masonry units have no more than 25% formed voids (20% for frogged bricks). Group 2 masonry units have formed voids greater than 25%, but not more than 55%.

# External cavity wall construction

## Cavity wall construction

(U-value not worse than $0.28\,W/m^2.k$)

External walls are constructed in either 100 mm minimum thickness reconstituted stone facings; facing brickwork; or two-coat render on 100 mm-thick dense concrete block skin with a 100 mm minimum thickness insulation/dense block inner leaf – with either a 15 mm lightweight plaster finish, or 12.5 mm-thick dry lining plasterboard fixed to block work on adhesive dabs and finished with a skim coat of plaster, as in insulation tables below.

Proprietary purpose-made lintels are to be constructed over all external openings in accordance with the lintel manufacturer's details, which should be approved by building control before works commence on site. Walls should be built in 1:5–1:6 cement/sand mortar mix with plasticiser and tied with British Board of Agreement (BBA or other third-party accredited) stainless-steel wall ties suitable for cavity width, as in Table 2.11 below.

Full-fill or partial-fill insulating material can be placed in the cavity between the outer leaf and an inner leaf of masonry walls, subject to the suitability of the cavity wall construction and UK zones for exposure to wind-driven rain, in accordance with Diagram 12 and Table 4 of ADC.

Subject to the suitability of the cavity wall construction, insulation is to be positioned in the wall in compliance with the insulation guidance tables below (and installed to prevent cold bridging and also any possible capillary attraction of water between the insulation and cavity surfaces past the damp-proof courses into the building) and in accordance with the insulation manufacturer's details.

Proprietary British Board of Agreement (BBA or other third-party accredited) acoustic/thermally insulated/fire-resistant cavity closers/cavity barriers are to be provided to all cavity openings/closings, tops of walls and junctions with other properties, in accordance with the manufacturer's details.

Tops of cavity walls can be closed using a proprietary British Board of Agreement (BBA or other third-party accredited) 30 minutes' fire-resistant rigid board or proprietary fire-resistant closers to prevent the passage of fire, fixed in accordance with the manufacturer's details.

Typical cavity wall construction details are indicated in the sub-structure guidance diagrams above.

### *Natural stone-faced cavity walls*

(U-value not worse than $0.28\,W/m^2.k$)

100–150 mm thick natural stone facings are to be fixed against one of the following backing options to form a uniform cavity within the cavity wall:

(i) 100 mm-thick dense concrete block backing course connected together with stainless-steel wall ties, as detailed below, and foundation widths increased to 750 mm as detailed in guidance Figures 2.19 and 2.20 below; or
(ii) British Board of Agreement (BBA or other third-party accredited) proprietary cavity spacer system, installed strictly in accordance with the manufacturer's details and as detailed in guidance Figures 2.21 and 2.22 below; or
(iii) 100 mm-wide clear continuous cavity, shuttered and formed with temporary shuttering in 450 mm vertical stages between wall ties, moved as work proceeds the following day (subject to proposed height of wall and building control approval).

Proprietary manufactured lintels (British Board of Agreement BBA or other third-party accredited) are to be constructed over all external openings using either a combination of proprietary or purpose-made lintels with extended flange to suit thickness of stone – normally 100–150 mm thick. Proprietary lintels suitable for stone faced cavity walls (including proprietary dpc trays and stop ends etc.)

## 2.36 Domestic extensions

**Table 2.11:** Cavity wall tie spacing

| Wall tie position | Maximum spacing of wall tie | |
|---|---|---|
| | Horizontally (mm) | Vertically (mm) |
| Cavity up to 75 mm wide | 900 | 450 |
| Cavities 75–100 mm wide | 750 (may need to be decreased to 600 if retaining partial fill insulation) | 450 |
| Cavities over 100 mm wide | To wall-tie manufacturer's details | To wall-tie manufacturer's details |
| Jamb openings (windows and doors, etc.) and movement joints | Within 225 of opening | Not more than 300 |

**Notes:**
1. Wall ties to be staggered.
2. Wall ties to be built at least 50 mm in to each wall leaf.
3. Wall ties to be built above and below the damp-proof course.
4. All wall ties to be stainless steel, in accordance with British/European Standards and have British Board of Agreement (BBA or other third-party accredited) certification. Wall ties to be installed in accordance with manufacturer's details.

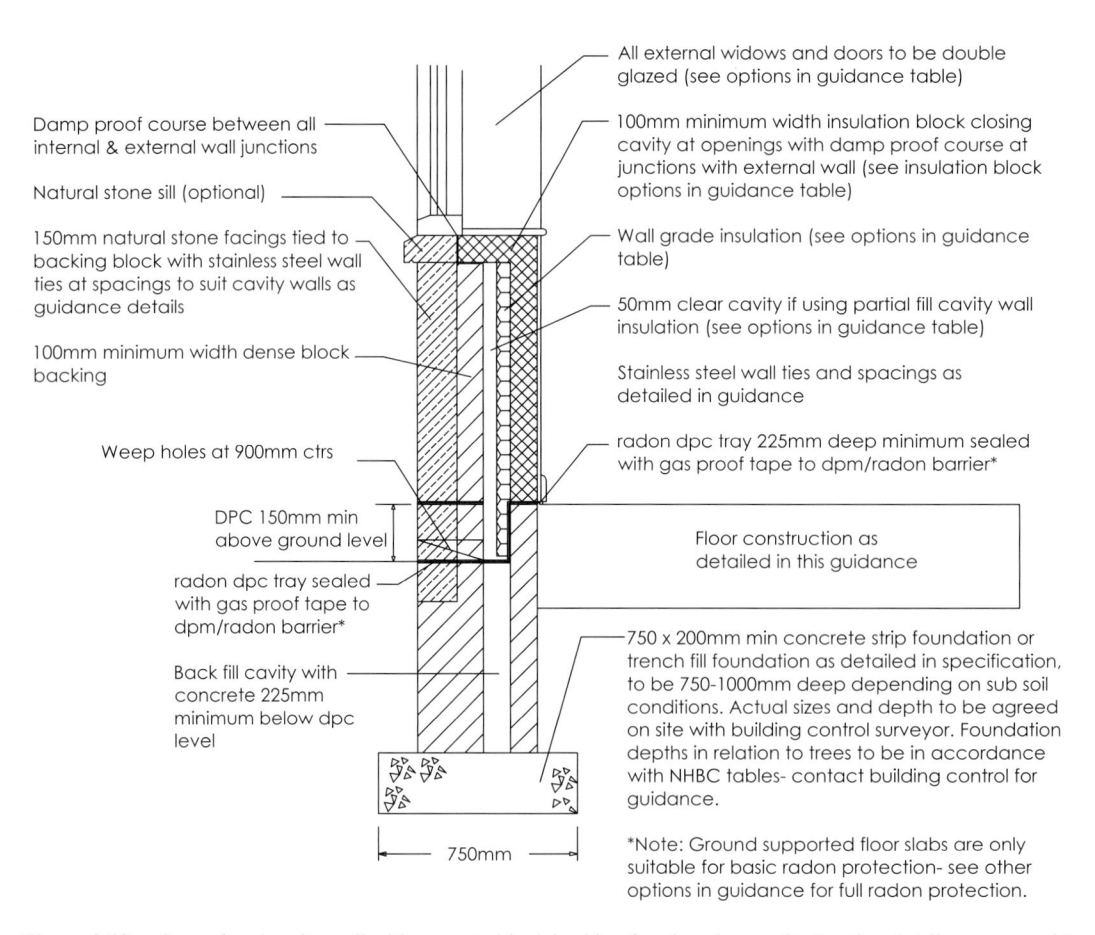

**Figure 2.19:** Stone-faced cavity wall with concrete block backing forming clear cavity (section detail – *not to scale*)

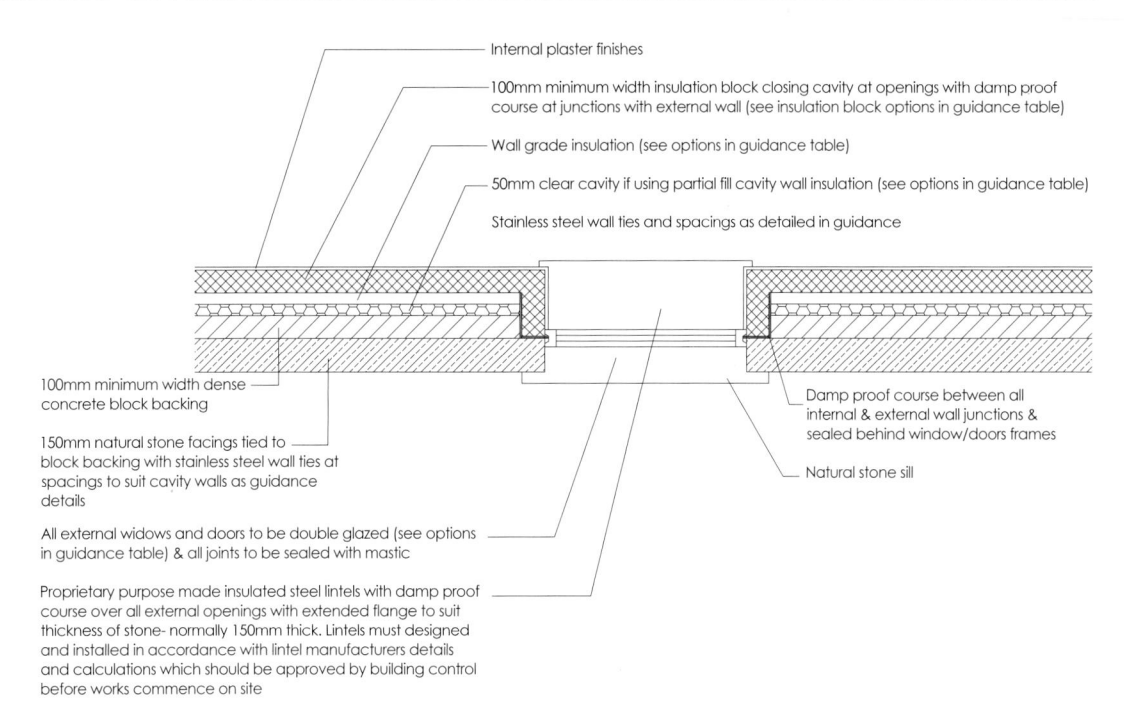

**Figure 2.20:** Stone-faced cavity wall with concrete block backing forming clear cavity (plan detail – *not to scale*)

must be designed and installed in accordance with the lintel manufacturer's details and calculations, which should be approved by building control before works commence on site.

Walls should be built in 1:5–1:6 cement/sand mortar mix with plasticiser and tied with British Board of Agreement (BBA or other third-party accredited) approved stainless-steel wall ties suitable for cavity width, as in Table 2.11 below.

Cavity wall insulation is to be positioned in the wall in compliance with Tables 2.12 to 2.15 below. The wall insulation should be continuous with roof insulation level, and taken below floor insulation levels in compliance with the manufacturer's details.

### Buttressing, sizes of openings and recesses in cavity walls

Openings, buttressing and sizes of openings and recesses should be in accordance with Diagrams 12, 13 and 14 of ADA. Openings exceeding 2.1 m in height, or openings less than 665 mm measured horizontally to an external corner wall, should be in accordance with details and calculations from a suitably qualified person (i.e. a structural engineer).

### Walls between heated and unheated areas

(U-value $0.28 \, W/m^2.k$)

Walls between heated and unheated areas such as garages etc. are to be constructed and insulated as external walls or constructed with $2.8/mm^2$ 215 mm thick insulation blocks (k-value 0.32) with lightweight plaster/plasterboard on dabs finish to one side, $25 \times 50$ mm timber battens to the opposite side, with insulation fixed across face of battens (as detailed in Table 2.16 below), with integral 12.5 mm vapour-checked plasterboard (or 500 g polythene vapour check) and 5 mm skim-coat plaster finish.

## 2.38 Domestic extensions

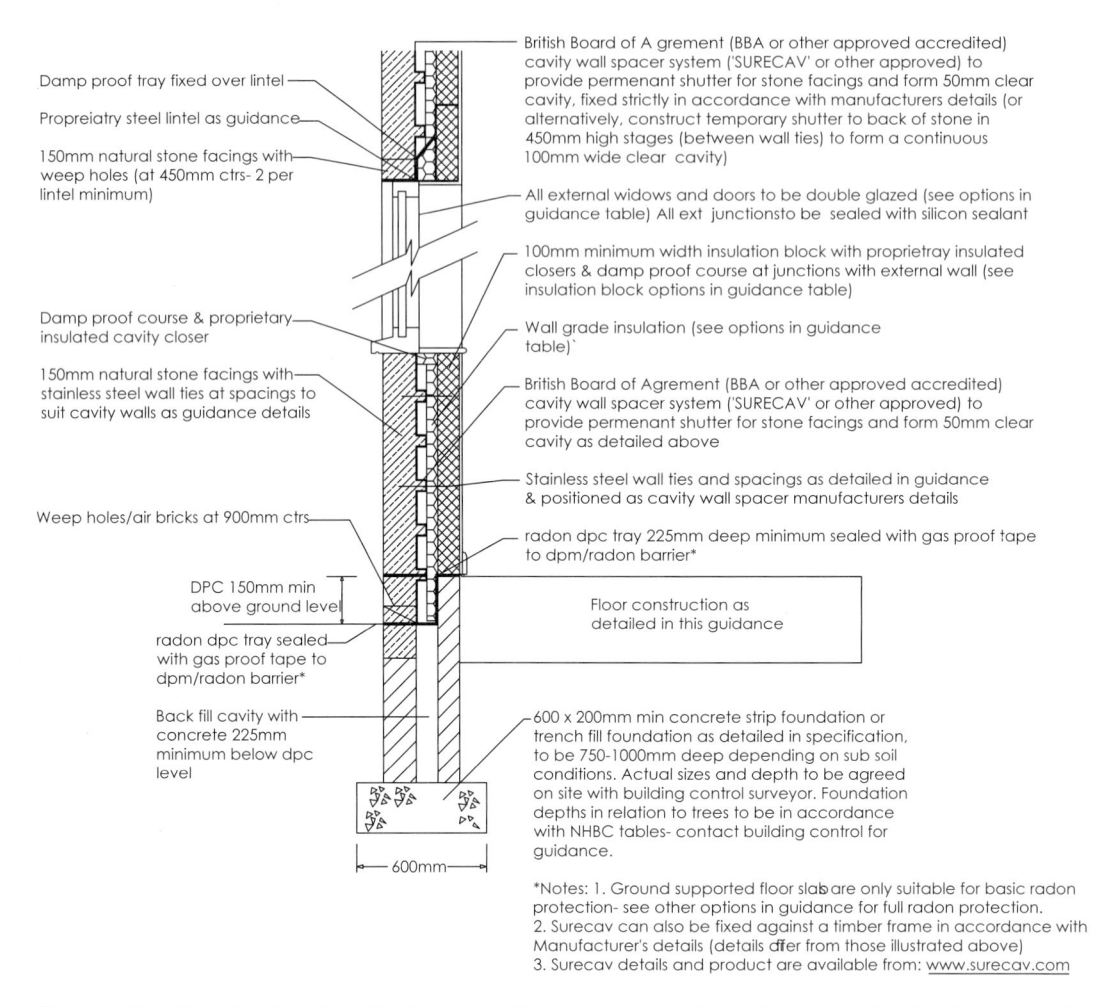

**Figure 2.21:** Stone-faced cavity wall with cavity wall spacer system or shuttered cavity (section detail – *not to scale*)

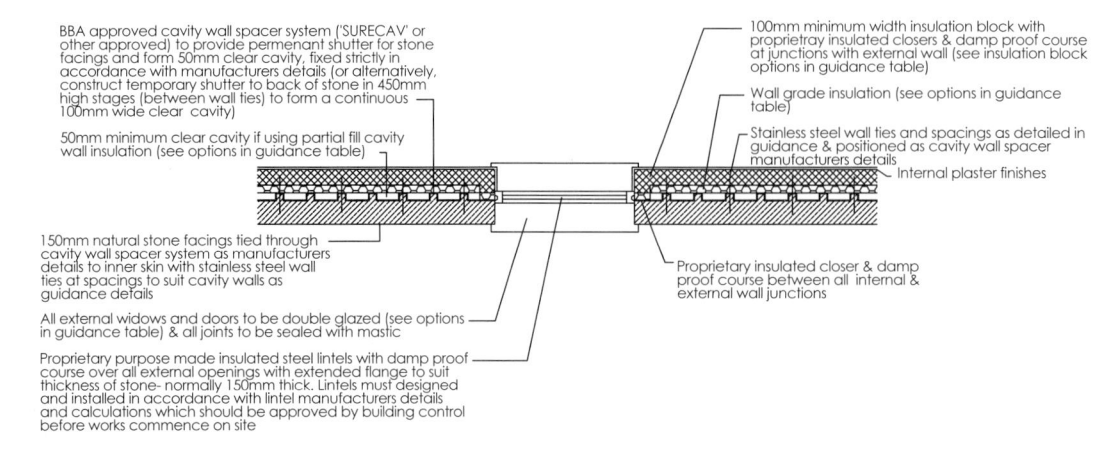

**Figure 2.22:** Stone-faced cavity wall with cavity wall spacer system or shuttered cavity (plan detail – *not to scale*)

**Table 2.12:** Examples of partial cavity-fill insulation for external cavity walls
100 mm dense brick outer leaf, cavity, partial fill insulation, block inner leaf and internal finishes
(U-value no worse than 0.28 W/m$^2$k)

| Clear cavity width required | Insulation type and minimum thickness | Overall cavity width required | Internal block type and thickness |
|---|---|---|---|
| 50 mm[1] | 40 mm Kingspan Kooltherm K8 Cavity Board – K value 0.021 | 90 mm[1] | 100 mm insulation block – K value 0.14 or lower with 13 mm lightweight plaster |
| 50 mm[1] | 40 mm Celotex CW4000 K value 0.022 | 90 mm[1] | 100 mm insulation block – K value 0.12 or lower with 13 mm lightweight plaster |
| 50 mm[1] | 50 mm Kingspan Kooltherm K8 Cavity Board – K value 0.020 **or** 50 mm Celotex CW4000 K value 0.022 | 100 mm[1] | 100 mm dense concrete block (K value 1.13) with 13 mm lightweight plaster |

**Notes:**
1. Clear cavities can be reduced to 25 mm in compliance with certain insulation manufacturer's details – subject to building control approval and any building warranty provider's approval where applicable.
   Insulation is to be installed in accordance with manufacturer's details, subject to the suitability of the cavity wall construction and UK zones for exposure to wind-driven rain and in accordance with Diagram 12 and Table 4 of ADC.
   Source: a representative selection of values taken from *Technical Note 10, U-Values of Elements (Approved Document L1B 2010)*, produced by Hertfordshire Technical Forum for Building Control. Reproduced by permission of Hertfordshire Technical Forum for Building Control.

**Table 2.13:** Examples of partial cavity-fill insulation for external cavity walls
100 mm dense block with render-finished external leaf, cavity, partial-fill insulation, block inner leaf and internal finishes
(U-value no worse than 0.28 W/m$^2$k)

| Clear cavity width required | Insulation type and minimum thickness | Overall cavity width required | Internal block type and thickness |
|---|---|---|---|
| 50 mm[1] | 40 mm Kingspan Kooltherm K8 Cavity Board K value 0.021 **or** 40 mm Celotex CW4000 K value 0.022 | 90 mm[1] | 100 mm insulation block – K value 0.11 or lower with 13 mm dense or lightweight plaster |
| 50 mm[1] | 50 mm Kingspan Kooltherm K8 Cavity Board K value 0.020 **or** 55 mm Celotex CW4000 K value 0.022 | 100 mm[1] 105 mm[1] | 100 mm dense concrete block (K value 1.13) with 12.5 mm plasterboard on dabs and skim |

**Notes:**
1. Clear cavities can be reduced to 25 mm in compliance with certain insulation manufacturer's details – subject to building control approval and any building warranty provider's approval where applicable.
   Insulation is to be installed in accordance with manufacturer's details, subject to the suitability of the cavity wall construction and UK zones for exposure to wind-driven rain and in accordance with Diagram 12 and Table 4 of ADC.
   Source: a representative selection of values taken from *Technical Note 10, U-Values of Elements (Approved Document L1B 2010)*, produced by Hertfordshire Technical Forum for Building Control. Reproduced by permission of Hertfordshire Technical Forum for Building Control.

## 2.40 Domestic extensions

**Table 2.14:** Examples of full cavity-fill insulation for external cavity walls
100 mm dense brick outer leaf, full-fill insulation, block inner leaf and internal finishes
(U-value no worse than 0.28 W/m$^2$k)

| Clear cavity width required | Insulation type and minimum thickness | Overall cavity width required | Internal Block Type and Thickness |
|---|---|---|---|
| n/a | 85 mm Earthwool DriTherm 32 K value 0.032 | 85 mm | 100 mm insulation block – K value 0.15 or lower with 12.5 mm plasterboard on dabs and skim |
| n/a | 100 mm Earthwool DriTherm 37 K value 0.037 | 100 mm | 100 mm insulation block – K value 0.11 or lower with 12.5 mm plasterboard on dabs and skim |
| n/a | 100 mm Earthwool Dritherm 32 K value 0.032 | 100 mm | 100 mm dense concrete block (K value 1.13) with 12.5 mm plasterboard on dabs and skim |

**Note:**
Insulation is to be installed in accordance with manufacturer's details, subject to the suitability of the cavity wall construction and UK zones for exposure to wind-driven rain and in accordance with Diagram 12 and Table 4 of ADC.
Source: a representative selection of values taken from *Technical Note 10, U-Values of Elements (Approved Document L1B 2010)*, produced by Hertfordshire Technical Forum for Building Control. Reproduced by permission of Hertfordshire Technical Forum for Building Control.

**Table 2.15:** Examples of full cavity-fill insulation for external cavity walls
100 mm dense block with render-finished external leaf, full-fill insulation, block inner leaf and internal finishes
(U-value no worse than 0.28 W/m$^2$k)

| Clear cavity width required | Insulation type and Minimum thickness | Overall cavity width required | Internal Block Type and Thickness |
|---|---|---|---|
| n/a | 100 mm Earthwool DriTherm 37 K value 0.037 | 100 mm | 100 mm insulation block – K value 0.15 or lower with 13 mm lightweight plaster |

**Note:**
Insulation is to be installed in accordance with manufacturer's details, subject to the suitability of the cavity wall construction and UK zones for exposure to wind-driven rain and in accordance with Diagram 12 and Table 4 of ADC.
Source: a representative selection of values taken from *Technical Note 10, U-Values of Elements (Approved Document L1B 2010)*, produced by Hertfordshire Technical Forum for Building Control. Reproduced by permission of Hertfordshire Technical Forum for Building Control.

**Table 2.16:** Examples of insulation for solid walls between heated and un-heated areas
215 mm thick insulation block – K value 0.32 with 13 mm lightweight plaster/12.5 mm plaster board on dabs to block wall
(U-value no worse than 0.28 W/m$^2$k)

| Insulation product | Minimum thickness (mm) |
|---|---|
| Kingspan Kooltherm K18 Insulated Plasterboard K value 0.020 | 62.5 fixed over battens |
| Celotex PL 4000 Insulated Plasterboard K value 0.022 | 72.5 fixed over battens |

**Note:** Insulation is to be installed in accordance with manufacturer's details.

### External timber-framed walls with separate brick or block finish (see Figure 2.23)

**General:** Design, manufacture, supply, erection and certification of the complete timber frame including roof, walls, lintels and floors, etc. are to be carried out by a specialist timber-frame manufacturer in compliance with the structural engineer's details and calculations. The shell of the building is to be air-sealed and fitted with protective coverings and measures to prevent condensation within the building. All details must be approved by building control before works commence on site. Moisture content of the timber should not exceed 20 per cent and it is to be kiln-dried and of grade C24; workmanship is to comply to BS 8000:5. All timber is to be treated using an approved system and all fixings are to be stainless steel or other approved material.

**Sole plates:** $38 \times 140$ mm CCA preservative-treated C16 CLS kiln-dried timber, set level and securely fixed to the sub-structure, which must not puncture the DPC/DPM/radon gas barrier and must not overhang or set back from the wall edge by more than 12 mm and must be protected from damp.

### External timber-framed stud walls

These are prefabricated panels – factory-fabricated, timber framing with $38 \times 140$ mm C16/24 CLS kiln-dried, preservative-treated timber studs, secured at 600 mm maximum centres, including sole and head plates and bracing to the structural engineer's details. Panels are to be accurately aligned, plumb and level, and fixed together with suitable rust-resistant fixings. Holes and notches are to be in accordance with the frame manufacturer's/structural engineer's details.

Structural beams, lintels and columns, etc. are to be factory fixed for the timber superstructure only, as dictated by the structural engineer's recommendations. Window/door closers to be $38 \times 89$ mm timber closers/cavity barriers, with dpc fixed around all external openings.

#### *Notches/holes/cuts in structural timbers*

Notches, holes and cuts in structural timbers should be carried out in accordance with BS 5268: 2002 and should not be deeper than 0.125 times the depth of the joists. They should be not closer to the support than 0.07 times the span, and not further away than 0.25 times the span. Holes should have a diameter not greater than 0.25 times the depth of the joist and should be drilled at the joist centre line. They should be not less than 3 diameters (centre to centre) apart and should be located between 0.25 and 0.4 times the span from the support. Notches or holes exceeding the above requirements or cut into other structural members should be checked by a structural engineer.

#### *External boarding*

This should be 12 mm preservative-treated OSB (Orientated Strand Board) or other approved structural sheathing boards to BS EN 622; 634:2; 314; 636 and BS 1982:1, nail-fixed using galvanised or stainless-steel fixings to the timber studwork or in accordance with the board manufacturer's details.

## 2.42 Domestic extensions

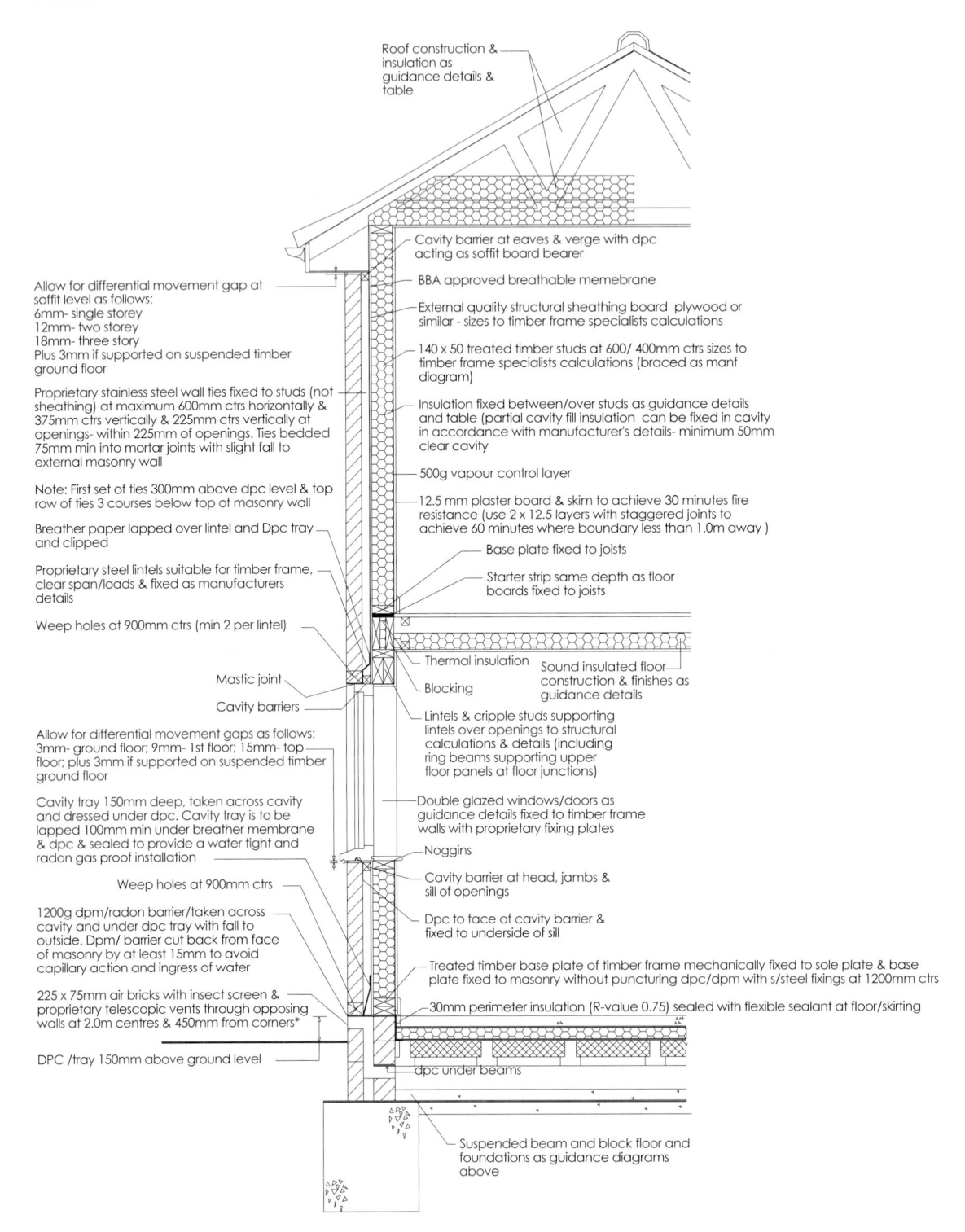

**Labels (left side, top to bottom):**

- Allow for differential movement gap at soffit level as follows:
  6mm- single storey
  12mm- two storey
  18mm- three story
  Plus 3mm if supported on suspended timber ground floor
- Proprietary stainless steel wall ties fixed to studs (not sheathing) at maximum 600mm ctrs horizontally & 375mm ctrs vertically & 225mm ctrs vertically at openings- within 225mm of openings. Ties bedded 75mm min into mortar joints with slight fall to external masonry wall
- Note: First set of ties 300mm above dpc level & top row of ties 3 courses below top of masonry wall
- Breather paper lapped over lintel and Dpc tray and clipped
- Proprietary steel lintels suitable for timber frame, clear span/loads & fixed as manufacturers details
- Weep holes at 900mm ctrs (min 2 per lintel)
- Mastic joint
- Cavity barriers
- Allow for differential movement gaps as follows: 3mm- ground floor; 9mm- 1st floor; 15mm- top floor; plus 3mm if supported on suspended timber ground floor
- Cavity tray 150mm deep, taken across cavity and dressed under dpc. Cavity tray is to be lapped 100mm min under breather membrane & dpc & sealed to provide a water tight and radon gas proof installation
- Weep holes at 900mm ctrs
- 1200g dpm/radon barrier/taken across cavity and under dpc tray with fall to outside. Dpm/ barrier cut back from face of masonry by at least 15mm to avoid capillary action and ingress of water
- 225 x 75mm air bricks with insect screen & proprietary telescopic vents through opposing walls at 2.0m centres & 450mm from corners*
- DPC /tray 150mm above ground level

**Labels (right side, top to bottom):**

- Roof construction & insulation as guidance details & table
- Cavity barrier at eaves & verge with dpc acting as soffit board bearer
- BBA approved breathable memebrane
- External quality structural sheathing board plywood or similar - sizes to timber frame specialists calculations
- 140 x 50 treated timber studs at 600/ 400mm ctrs sizes to timber frame specialists calculations (braced as manf diagram)
- Insulation fixed between/over studs as guidance details and table (partial cavity fill insulation can be fixed in cavity in accordance with manufacturer's details- minimum 50mm clear cavity
- 500g vapour control layer
- 12.5 mm plaster board & skim to achieve 30 minutes fire resistance (use 2 x 12.5 layers with staggered joints to achieve 60 minutes where boundary less than 1.0m away )
- Base plate fixed to joists
- Starter strip same depth as floor boards fixed to joists
- Thermal insulation
- Sound insulated floor construction & finishes as guidance details
- Blocking
- Lintels & cripple studs supporting lintels over openings to structural calculations & details (including ring beams supporting upper floor panels at floor junctions)
- Double glazed windows/doors as guidance details fixed to timber frame walls with proprietary fixing plates
- Noggins
- Cavity barrier at head, jambs & sill of openings
- Dpc to face of cavity barrier & fixed to underside of sill
- Treated timber base plate of timber frame mechanically fixed to sole plate & base plate fixed to masonry without puncturing dpc/dpm with s/steel fixings at 1200mm ctrs
- 30mm perimeter insulation (R-value 0.75) sealed with flexible sealant at floor/skirting
- dpc under beams
- Suspended beam and block floor and foundations as guidance diagrams above

**Figure 2.23:** Typical section through external timber-framed walls with separate external brick or block wall and cavity *(not to scale)*

### Breather membrane

Proprietary British Board of Agreement (BBA or other third-party accredited) breather membrane is to be factory-fixed, as manufacturer's details, to external sheathing by stainless-steel staples fixed through white proprietary tape to distinguish wall-tie positions.

### Thermal insulation and fire resistance

Thermal insulation is to be fitted between studs in accordance with guidance tables and the manufacturer's details, and stud walls are to be finished internally with 500 g sheet polythene vapour check and 12.5 mm thick plasterboard fixed to studs and 3 mm skim coat of finishing plaster (to achieve 30 minutes' fire resistance – within 1.0 m of a boundary, in accordance with part B of this guidance). All junctions are to have water- and air-tight construction; all perimeter joints are to be sealed with tape internally and with silicon sealant externally.

## External walls

These are to be 100 mm minimum thickness brick/reconstituted stone/painted sand and cement render (render to BS 5262), on 100 mm medium-dense external concrete as required. Masonry walls/mortar/render details are contained elsewhere in this guidance.

External masonry skin is to be tied to timber-frame studs (not the sheathing) using British Board of Agreement (BBA or other third-party accredited) proprietary flexible stainless-steel wall ties in compliance with the manufacturer's details, BS 5628 and BS EN 845-1, typically at spacings not exceeding 600 mm horizontally and 375 mm vertically, and 225 mm max at reveals. Wall ties should be embedded in mortar to a minimum depth of 75 mm, with a slight fall towards the external brickwork. Proprietary flexible water- and fire-resistant cavity barriers should be provided at eaves level and gable end walls, vertically at junctions with separating walls and horizontally at separating walls with continuous dpc tray over, installed in compliance with the manufacturer's details.

### Proprietary steel lintels

Lintels to BS EN 845 are to be provided, with 150 mm bearing over all external openings to support external masonry skin, fitted with continuous dpc tray and retaining clips. Lintel type and sizes are to be in accordance with the manufacturer's details and suitable for proposed clear spans and loadings. Weep holes using proprietary insect-proof vents are to be provided at 900 mm spacing at base of wall above the dpc tray and above all lintels (2 weep holes minimum per lintel).

### Separation of combustible materials from solid-fuel fireplaces and flues

The minimum separating distance from combustible materials from a chimney or fireplace should be in compliance with Part J of guidance details and Diagram 21 of ADJ, as follows:

(i) a solid non combustible masonry wall at least 200 mm thick should separate combustible materials from a flue liner;

**Table 2.17:** Examples of insulation for cavity walls with internal timber frame 103 mm dense brick/100 mm dense block, with render-finished external leaf, 50 mm clear cavity with breather membrane, structural board, timber studs at 600 and 400 mm centres, 12.5 mm vapour-checked plasterboard and 3 mm skim internal finish.
(U-value no worse than 0.28 W/m²k)

| External wall | Clear cavity width required | Timber stud (mm) | Insulation type and Minimum thickness |
|---|---|---|---|
| Brick or rendered dense block | 50 mm | 100 | 80 mm Kingspan Kooltherm K12 Framing Board between studs K value 0.020 |
| Brick or rendered dense block | 50 mm | 150 | 70 mm Kingspan Kooltherm K12 Framing Board between studs K value 0.020 |
| Rendered dense block | 50 mm | 100 | 90 mm Celotex FR4000 between studs K value 0.022 |
| Rendered dense block | 50 mm | 150 | 75 mm Celotex FR4000 between studs K value 0.022 |
| Brick | 50 mm | 150 | 140 mm Knauf Earthwool Frame Therm 38 slab between studs K value 0.038 |
| Brick | 50 mm | 150 | 140 mm Rockwool Flexi slab between studs K value 0.035 using 140 mm insulation thickness |

**Note:** Insulation is to be installed in accordance with the manufacturer's details, subject to the suitability of the cavity wall construction and UK zones for exposure to wind-driven rain and in accordance with Diagram 12 and Table 4 of ADC.
Source: a representative selection of values taken from *Technical Note 10, U-Values of Elements (Approved Document L1B 2010)*, produced by Hertfordshire Technical Forum for Building Control. Reproduced by permission of Hertfordshire Technical Forum for Building Control.

(ii) an air gap of at least 40 mm is required between combustible materials and a solid, non-combustible masonry wall which is up to 200 mm thick (but must not be less than 100 mm minimum thickness).

Ensure that **all gaps** and **all voids** are **sealed** to prevent any air leakage.

### External timber-framed walls with render finish (see Figure 2.24)

(U-value 0.28 W/m².k)

Render finish (to comply to BS 5262) – to be applied in three coats at least 16–20 mm thick overall to render lath as detailed below. Typical render mixes for first and second coats 1:3 (cement: sand with plasticiser), final coat 1:6 (cement: sand with plasticiser) – proportions by volume. Render should be finished onto an approved durable render stop, angle beads or jointing sections – stainless steel or other approved material using drilled or shot-fired fixings.

Stainless steel render lath fixed (using stainless-steel staples) to vertical studs at 600 mm maximum centres with all laps wired together at 150 mm centres (mesh to be backed by a water-resistant membrane) and fixed to:

– treated battens: 25 × 50 mm preservative- treated battens fixed vertically to studs at maximum 600 mm centres using 75 mm-long, hot-dipped, galvanised or stainless-steel annular ring nails, fixed to:

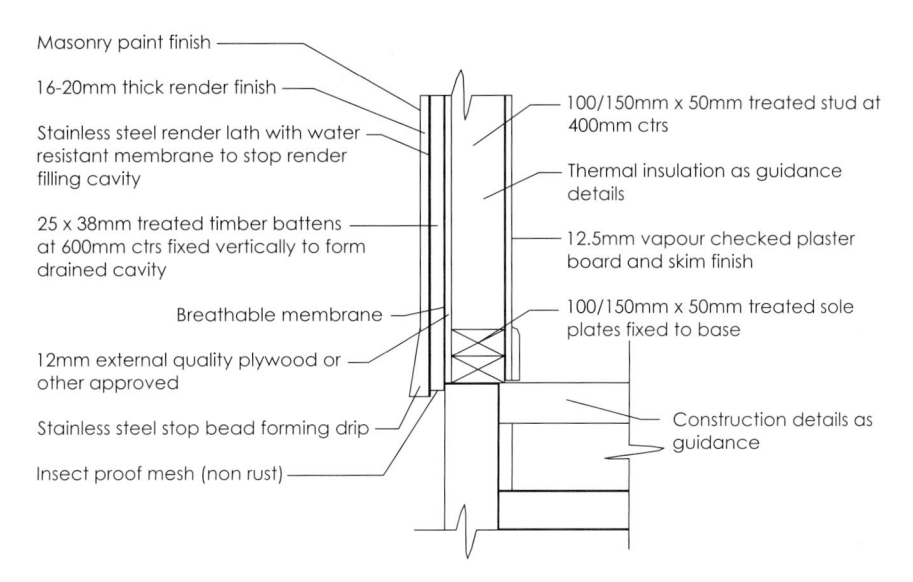

**Figure 2.24:** Typical section through external timber-framed walls with painted render finish *(not to scale)*

– British Board of Agreement (BBA or other third-party accredited) proprietary breathable membrane (suitable for timber-framed walls), fixed according to manufacturer's details to:

– 12 mm external-quality plywood or other approved structural waterproof sheathing (joints covered by dpc and battens), fixed to $100/150 \times 50$ mm timber studs at 400 mm centres with $100/150 \times 50$ mm timber head and sole plates and two rows of noggins and diagonal bracing, as structural engineer's details. Studs exceeding 2.5 m high should be designed by a structural engineer.

Thermal insulation is to be fixed between/over studs in accordance with the insulation guidance table below (Table 2.18), with vapour check and plasterboard fixed to internal face of studs (increase thickness of plasterboard in certain circumstances for increased fire resistance, in accordance with Part B of this guidance), finished with 3 mm skim coat of plaster. All junctions are to have watertight construction; seal all perimeter joints with tape internally and with silicon sealant externally.

### External timber-framed walls with cladding finish (see Figure 2.25)

(U-value $0.28 \text{ W/m}^2.\text{k}$)

Approved timber/uPVC weatherboarding/vertical wall tiling is to be fixed with proprietary, rust-resistant fixings to $50 \times 25$ mm treated battens/counter battens at 400 mm centres, fixed to:

– British Board of Agreement (BBA or other third-party accredited) proprietary breathable membrane (suitable for timber-framed walls), fixed according to manufacturer's details to:

– 12 mm external-quality plywood or other approved structural waterproof sheathing (joints covered by dpc and battens), fixed to $100/150 \times 50$ mm timber studs at 400 mm centres with $100/150 \times 50$ mm timber head and sole plates and two rows of noggins and diagonal bracing, as per structural engineer's details. Studs exceeding 2.5 m high should be designed by the structural engineer.

Thermal insulation is to be fixed between/over studs in accordance with the insulation guidance table below (Table 2.18), with vapour check and plasterboard fixed to internal face of studs (increase thickness of plasterboard in certain circumstances for increased fire resistance, in accordance with

## 2.46 Domestic extensions

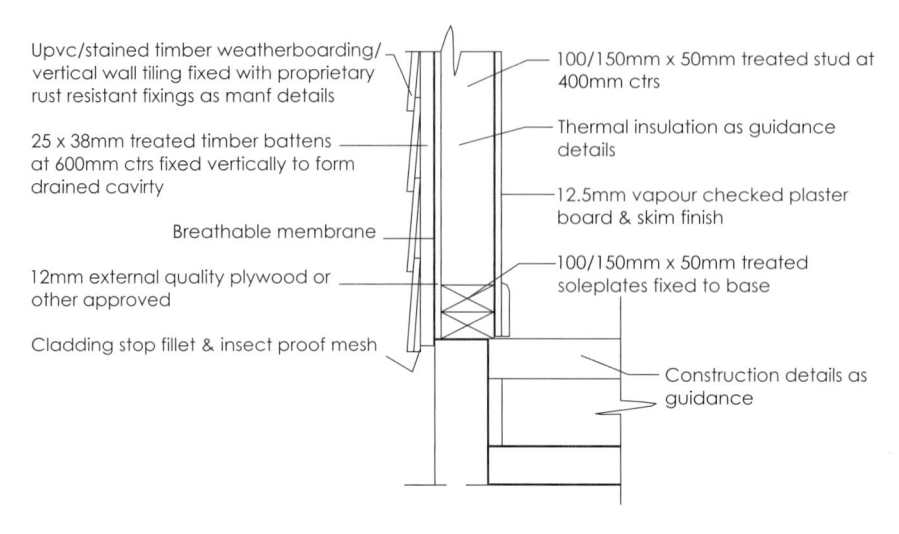

**Figure 2.25:** Typical section through external timber-framed walls with uPVC/timber weatherboard finish *(not to scale)*

Labels on figure:
- Upvc/stained timber weatherboarding/vertical wall tiling fixed with proprietary rust resistant fixings as manf details
- 25 x 38mm treated timber battens at 600mm ctrs fixed vertically to form drained cavirty
- Breathable membrane
- 12mm external quality plywood or other approved
- Cladding stop fillet & insect proof mesh
- 100/150mm x 50mm treated stud at 400mm ctrs
- Thermal insulation as guidance details
- 12.5mm vapour checked plaster board & skim finish
- 100/150mm x 50mm treated soleplates fixed to base
- Construction details as guidance

**Table 2.18:** Examples of insulation for timber-frame walls with external tile/render/cladding finishes
Tiles/render/cladding on battens as guidance; timber studs at 600/400 mm centres with insulation fixed between/over studs with vapour-checked integral/separate plasterboard, as stated below with 3 mm plaster finishes.
(U-value no worse than 0.28 W/m²k)

| Timber stud (mm) | Insulation type and minimum thickness | Internal insulation/finish |
|---|---|---|
| 100 × 50 mm | 50 mm Kingspan Kooltherm K12 Framing Board, K Value 0.20 **or** 50 mm Kingspan Thermawall TW55 K value 0.022, fixed between studs | 32.5 mm Kingspan Kooltherm K18 Insulated plasterboard K value 0.023, fixed over studs |
| 100 × 50 mm | 60 mm Celotex FR4000 K value 0.22, fixed between studs | 37.5 mm Celotex PL4000 K value 0.22, fixed with integral plasterboard with lightweight skim, fixed over studs |
| 125/150 × 50 mm | 85 mm Kingspan Kooltherm K12 Framing Board. K value 0.020, fixed between studs | 12.5 mm plasterboard and 3 mm skim finish fixed over studs |
| 125/150 × 50 mm | 90 mm Celotex FR4000 K value 0.22, fixed between studs | 12.5 mm plasterboard and 3 mm skim finish, fixed over studs |

**Note:** Insulation is to be installed in accordance with manufacturer's details, subject to the suitability of the wall construction and UK zones for exposure to wind-driven rain and in accordance with Diagram 12 and Table 4 of ADC.
Source: a representative selection of values taken from *Technical Note 10, U-Values of Elements (Approved Document L1B 2010)*, produced by Hertfordshire Technical Forum for Building Control. Reproduced by permission of Hertfordshire Technical Forum for Building Control.

Part B of this guidance), finished with 3 mm skim coat of plaster. All junctions are to have watertight construction; seal all perimeter joints with tape internally and with silicon sealant externally.

### *Detached garage (or similar single-storey building) with SINGLE-SKIN external walls* (see Figure 2.26)

External walls are to be constructed using: 100 mm minimum thickness brick/reconstituted stone, or sand and cement render (render to BS 5262), on 100 mm minimum thickness dense concrete blocks with $100 \times 400$ mm minimum sized piers at maximum 3.0 m centres, tied or built into walls, with fair face finish internally. Bricks are to have a minimum compressive strength of $5\,\text{N/mm}^2$ and dense concrete blocks $2.8\,\text{N/mm}^2$ minimum.

Floor area exceeding $36\,\text{m}^2$ will require structural engineer's details and calculations to confirm the stability of the structure. Eaves level should not exceed 3.0 m in height and ridge height should not exceed 3.6 m without structural engineer's details and calculations to confirm the stability of the structure.

Size and proportion of the garage are to comply with paragraph 2C38, and the size and location of openings in the building are to comply with Diagrams 17, 18 and 19 of ADA, briefly as follows:

- Major openings are to be restricted to one wall only (normally at the front entrance).
- Their aggregate width should not exceed 5.0 m and their height should not exceed 2.1 m.
- There should be no other openings within 2.0 m of a wall containing a major opening.
- The aggregate size of openings in a wall **not** containing a major opening should not exceed $2.4\,\text{m}^2$.
- There should not be more than one opening between piers.
- Unless there is a corner pier, the distance from a window or a door to a corner should not be less than 390 mm.
- Isolated central columns between doorways (where applicable) are to be $325 \times 325$ mm minimum.
- Openings other than those stated above will require structural engineer's details and calculations to confirm structural stability.
- Mortar mix to be 1:1:5–6 or as required by the stone/brick/block manufacturer.

### *Wall abutments*

Vertical junctions of new and old walls are to be secured with a proprietary, profiled stainless-steel metal crocodile-type system bolted to the existing wall, with a dpc inserted into a vertical chase cut into the existing wall above the horizontal dpc and pointed with flexible mastic, as per the manufacturer's details. Depth of chase and position of dpc are to be agreed with building control.

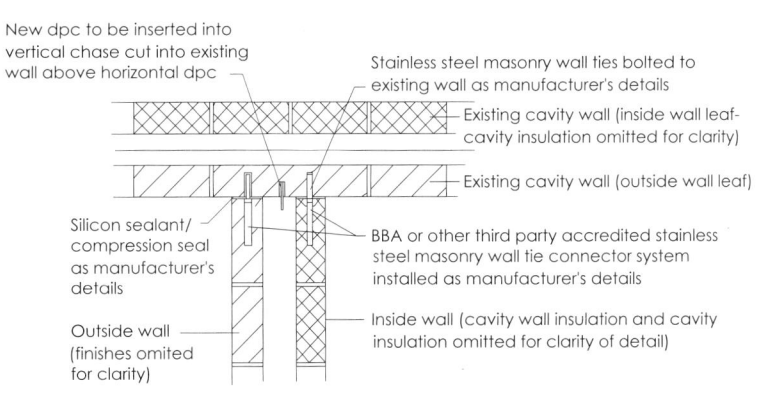

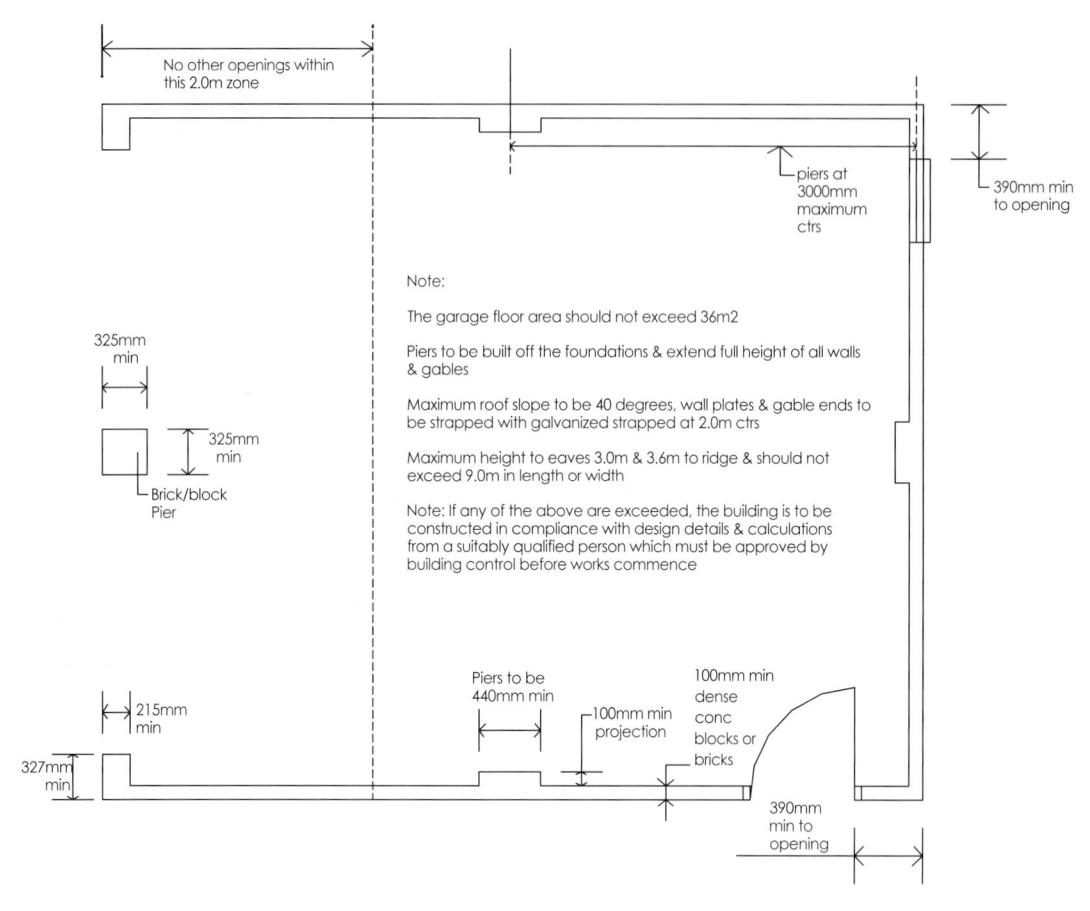

**Figure 2.26:** Design criteria for small, detached, single-storey garages or similar (plan – *not to scale*) (See Para 2C38 and diagrams 17/18/19 of ADA for full details)

### *Lintels and weep holes*

Proprietary manufactured lintels to current British Standards/Euro codes/BBA or other third-party accredited certification (including specialist lintels supporting stone facings) are to be provided over all structural openings. The positions, types, sizes, end bearings, and fixing etc. of lintels must be in compliance with the lintel manufacturer's details and standard tables and suitable for the proposed loadings and clear spans. Stop ends, dpc trays and weep holes etc. are to be provided above all externally located lintels, in compliance with the lintel manufacturer's details. Weep holes are required in porous external walls (i.e. brickwork) at typically 450 mm centres or two minimum per opening.

### *Structural columns/beams etc.*

Non proprietary beams/columns (including pad stones) are to be fabricated and installed, in compliance with details and structural calculations carried out by a suitably qualified and experienced person (i.e. a structural engineer), and must be approved by building control before works commence on site. Dpc trays are to be provided above all externally located beams. Weep holes are required in porous external walls (i.e. brickwork) at 900 mm centres, with at least two per opening.

Table 2.19: Movement joint widths and spacing in walls

| Construction | Movement joint widths | Spacing of movement joints in walls[1] |
|---|---|---|
| Clay bricks | 16 mm | 12 m |
| Calcium silicate bricks | 10 mm | 7.5 m |
| Concrete bricks and blocks | 10 mm | 6 m |

**Key:** [1] The first movement joint should be positioned not more than half the above distance from a wall return and should extend the full storey height of the wall.

### *Movement joints*

The external leaf of a cavity wall should be provided with adequately spaced and sized vertical movement joints in accordance with Table 2.19 below, to minimise the risk of cracking due to the expansion and contraction of the wall and maintain stability, in accordance with the masonry manufacturer's and structural engineer's details. Proprietary wall ties are to be provided on each side of the joint using stainless-steel wall ties, positioned at each block height (225 mm maximum) and the joint sealed externally with a proprietary flexible mastic sealant.

### *Cavity closers*

Proprietary British Board of Agreement (BBA or other third-party accredited) acoustic/thermally insulated/fire-resistant cavity closers, or similar, are to be provided to all cavity openings/closings, tops of walls and junctions with other properties, in accordance with the manufacturer's details.

Tops of cavity walls are to be closed to prevent the passage of fire, using a proprietary British Board of Agreement (BBA or other third-party accredited) 30 minutes' fire-resistant rigid board, fixed in accordance with the manufacturer's details.

### *Proprietary dpc trays (new and existing walls)*

Proprietary dpc trays/stepped dpc trays, stop ends and weep holes etc. for new works or retrospective fitting into existing walls etc. above lintels/beams or at cavity wall/roof abutments etc. should have British Board of Agreement (BBA or other third-party accredited) certification, and are to be fixed in accordance with the manufacturer's details.

## Lateral restraint strapping of upper floors to walls

Upper floors should be connected to walls with lateral restraint straps (in accordance with Paras 2C32–2C37 and Diagram 15 of ADA), fixed horizontally to stiffen and stabilise the walls by restraining their movement in a direction at right angles to the wall length by the provision of pre-galvanised and edge-coated, heavy-duty horizontal lateral restraint straps, with a minimum cross-sectional size of $30 \times 5$ mm and with a tensile strength of 8 kN, in compliance with BS 5268 Part 3 and BS EN 845-1 in the following locations:

### (i) Strapping of floor joists parallel to walls (see Figure 2.27)

Straps should be spaced at maximum 2 m centres with a minimum length of 1200 mm, carried across and fixed to at least three joists by the use of $4 \times 4$ mm $\times 75$ mm round nails or screws into noggins (noggins are to be at least 38 mm wide and 3/4 the depth of the joist or rafter). Any gap between the wall and the timber member is to be packed with timber folding wedges. The bend length should be

## 2.50 Domestic extensions

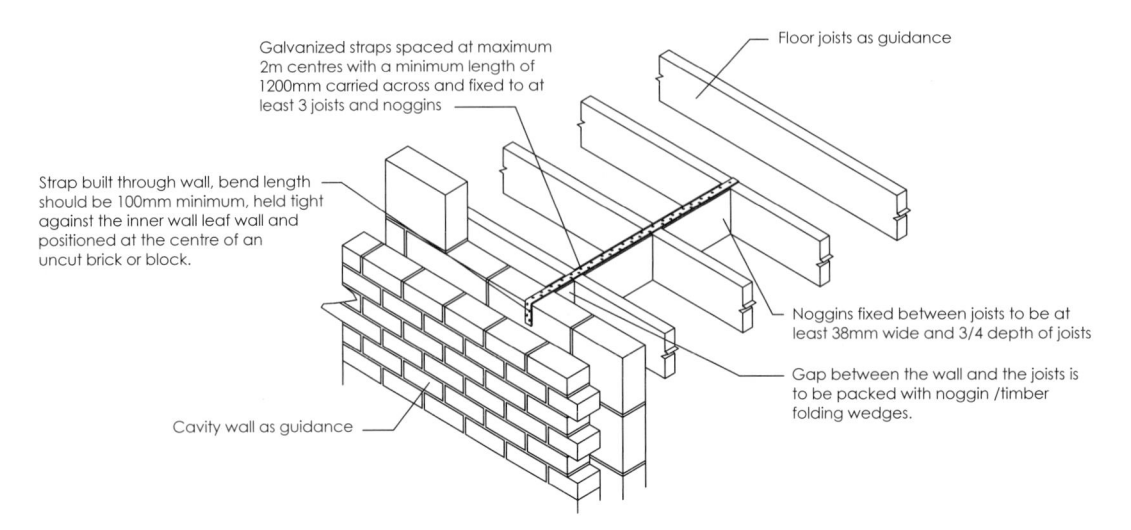

**Figure 2.27:** Strapping of floor joists parallel to walls *(not to scale)*

Labels:
- Galvanized straps spaced at maximum 2m centres with a minimum length of 1200mm carried across and fixed to at least 3 joists and noggins
- Floor joists as guidance
- Strap built through wall, bend length should be 100mm minimum, held tight against the inner wall leaf wall and positioned at the centre of an uncut brick or block.
- Noggins fixed between joists to be at least 38mm wide and 3/4 depth of joists
- Gap between the wall and the joists is to be packed with noggin /timber folding wedges.
- Cavity wall as guidance

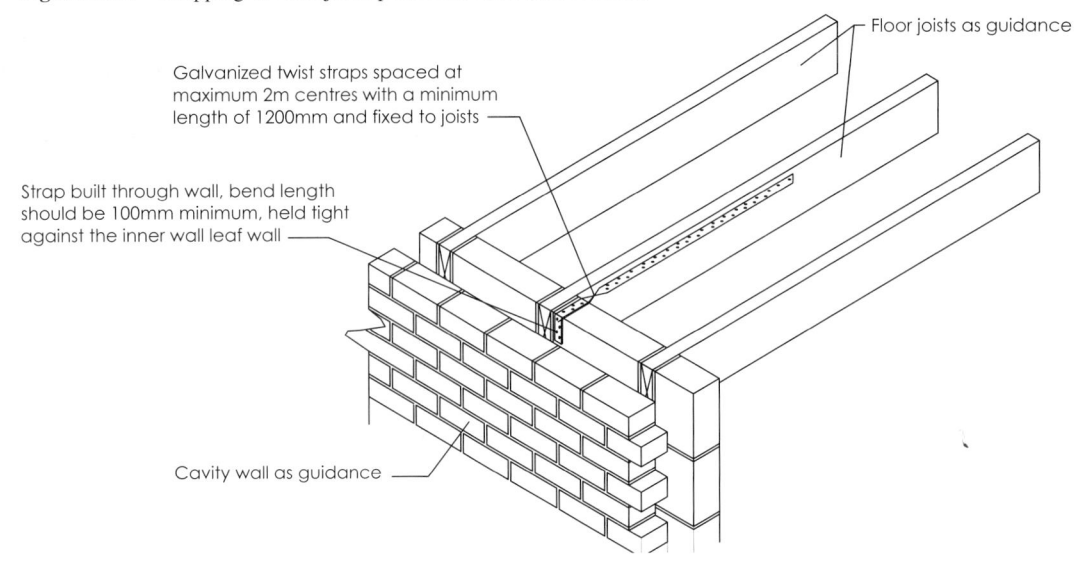

**Figure 2.28:** Strapping of floor joists at right angles to walls *(not to scale)*

Labels:
- Galvanized twist straps spaced at maximum 2m centres with a minimum length of 1200mm and fixed to joists
- Floor joists as guidance
- Strap built through wall, bend length should be 100mm minimum, held tight against the inner wall leaf wall
- Cavity wall as guidance

100 mm minimum and should be held tight against the masonry wall and positioned at the centre of an uncut brick or block. Straps can be omitted from internal walls (but not from inner leaf of external cavity walls) in houses with no more than two storeys as follows: where floors are at or about the same level on each side of a supported wall, and contact between the floors and wall is either continuous or at intervals not exceeding 2 m. Where contact is intermittent, the points of contact should be in line or nearly in line on plan as Diagram 15(e) of ADA.

### (ii) Strapping of floor joists at right angles to walls (see Figure 2.28)

Straps should be spaced at maximum 2 m centres with a minimum length of 1200 mm, fixed to joists by the use of 4 × 4 mm × 75 mm round nails or screws. The bend length should be 100 mm minimum and should be held tight against the masonry wall and positioned at the centre of an uncut brick or block. Straps can be omitted in houses with no more than two storeys as follows:

- if the joists are not more than 1200 mm centres and have at least 90 mm bearing on the supported wall or 75 mm bearing on a timber wall plate at each end;
- if the joists are supported by joist hangers built into walls at not more than 2 m centres;
- where a concrete floor has at least 90 mm bearing on the supported wall.

## Lateral restraint strapping of roofs to walls

### Strapping of roofs to gable-end walls (see Figure 2.29)

Roofs should be connected to walls with lateral restraint straps (in accordance with Paras 2C32–2C37 and Diagrams 16 of ADA), fixed horizontally to stiffen and stabilise the walls by restraining their movement in a direction at right angles to the wall length by the provision of pre-galvanised and edge-coated, heavy-duty, horizontal lateral restraint straps. Straps are to have a minimum cross-sectional size of $30 \times 5$ mm and a tensile strength of 8 kN, in compliance with BS 5268 Part 3 and BS EN 845-1. Straps should be spaced at maximum 2 m centres (strap at highest point must provide a secure connection), with a minimum length of 1200 mm carried across and fixed to at least three rafters by the use of $4 \times 4$ mm $\times 75$ mm round nails or screws into noggins (noggins are to be at least 38 mm wide and 3/4 the depth of the joist or rafter). Any gap between the wall and the timber member is to be packed with timber folding wedges. The bend length should be 100 mm minimum and should be positioned at the centre of the uncut brick or block.

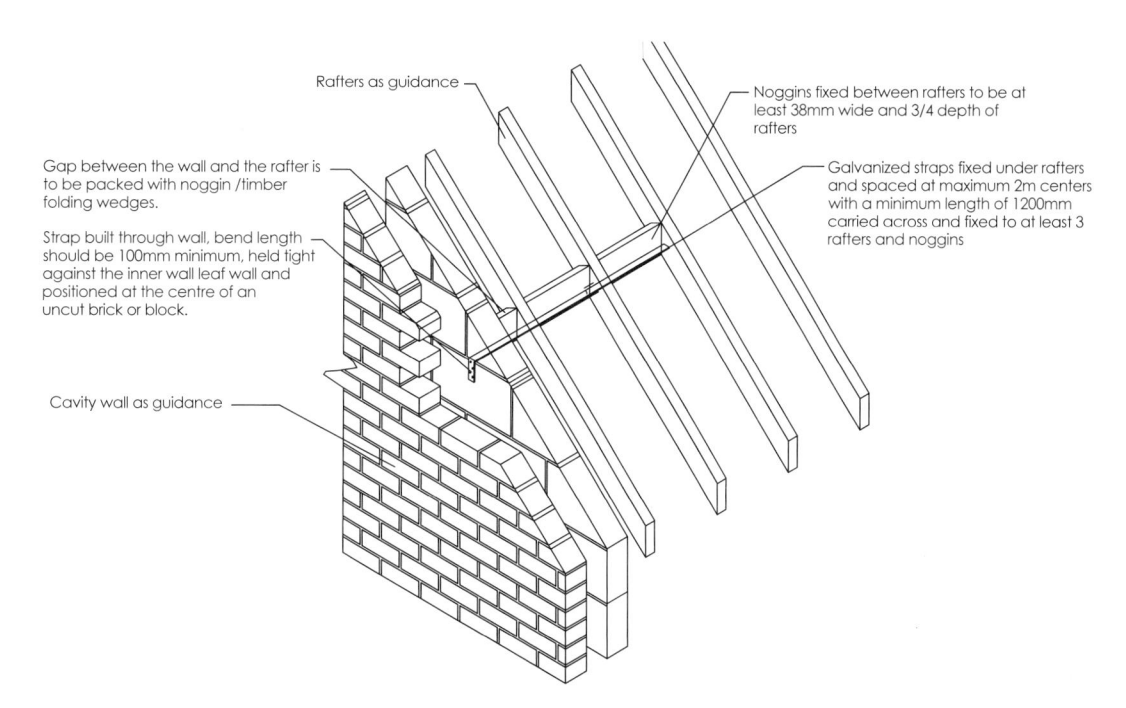

**Figure 2.29:** Strapping of roofs to gable-end walls *(not to scale)*

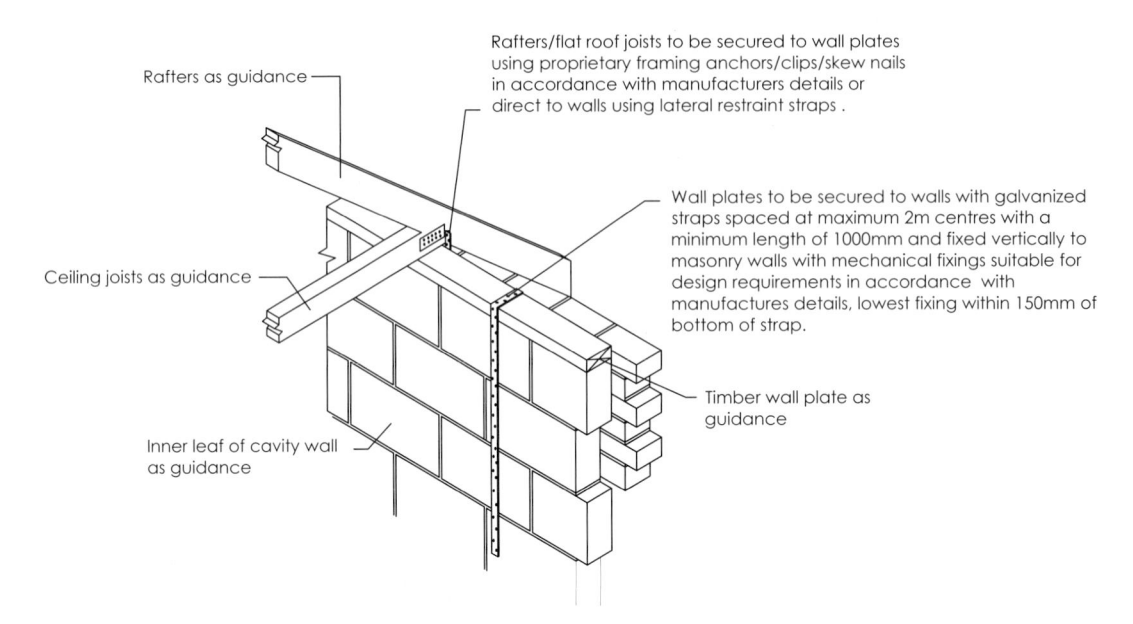

**Figure 2.30:** Strapping of wall plates and roofs at eaves level *(not to scale)*

Where the straps cannot be fixed into a cavity wall (e.g. single-skin garage walls), the bend should be fixed to masonry walls using fixings in accordance with the manufacturer's details (e.g. proprietary stainless-steel expansion bolts fixed to block/brick work as agreed with building control, typically 3 number × M6 expansion fixings per strap bend, fixed 75 mm into masonry walls).

### Strapping of wall plates and roofs at eaves level (see Figure 2.30)

Wall plates are to be secured to walls by the provision of pre-galvanised and edge-coated horizontal lateral restraint straps (in accordance with Paras 2C32–2C37 and Diagram 16 of ADA), with a minimum cross-sectional size of 30 × 5 mm or light strap in compliance with BS 5268 Part 3 and BS EN 845-1. Straps should be spaced at maximum 2 m centres, with a minimum length of 1000 mm and fixed vertically to masonry walls with mechanical fixings suitable for design requirements, in accordance with the manufacturer's details – with the lowest fixing within 150 mm of the bottom of the strap. Rafters/flat roof joists are to be secured to wall plates using proprietary framing anchors/clips/skew nails, in accordance with the manufacturer's details, or secured directly to walls using lateral restraint straps as detailed above.

Vertical strapping can be omitted if the roof has a pitch of 15° or more, is tiled or slated, is of a type known by local experience to be resistant to damage by wind gusts, and has main timber members spanning onto the supported wall at not more than 1.2 m centres.

## Lateral restraint strapping of walls at ceiling level (see Figure 2.31)

Where the height of the gable-end wall exceeds 16 × thicknesses of the external wall leaves + 10 mm (excluding cavity width), measured from the top of the floor to the centre of the gable end wall above ceiling level, lateral straps are to be provided at ceiling joist level (in accordance with Paras 2C32–2C37 and Diagram 16 of ADA), as detailed in this guidance for intermediate floors.

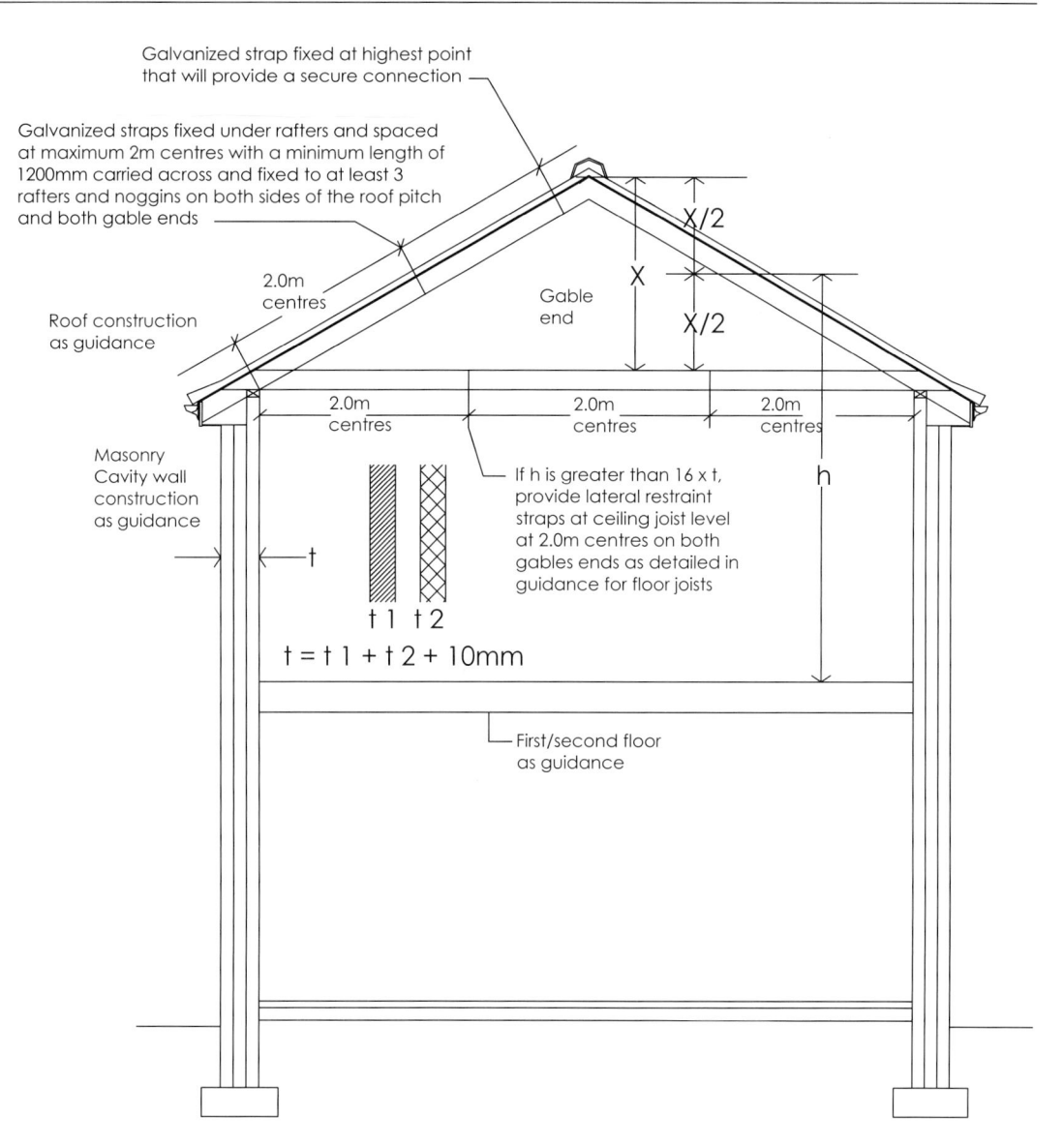

**Figure 2.31:** Strapping of walls at ceiling level *(not to scale)*

## A3: Separating walls and floors

### Masonry party walls separating dwellings (see Figure 2.32)

(U-value $0.2\,W/m^2.K$)

Party walls separating dwellings need to achieve a minimum of 60 minutes' fire resistance from both sides, and a sound insulation value of 45 dB value for airborne sound insulation (reduced to 43 dB for conversions). The walls are typically constructed using two skins of 100 mm minimum thickness dense concrete blocks (density 1990 kg/m$^3$) in 225 mm coursings, with a clear 50 mm

## 2.54 Domestic extensions

**Figure 2.32:** Section detail of masonry separating wall as Wall type 2.1 of ADE *(not to scale)*

Labels in figure:
- 2 skins of 100mm dense concrete blocks (density 1990kg/m3) in 225mm coursing heights
- 50mm minimum clear cavity
- Each wall leaf tied together with BBA approved wall ties spaced as external walls
- 13mm plaster (min mass 10kg/m2) applied to both faces.

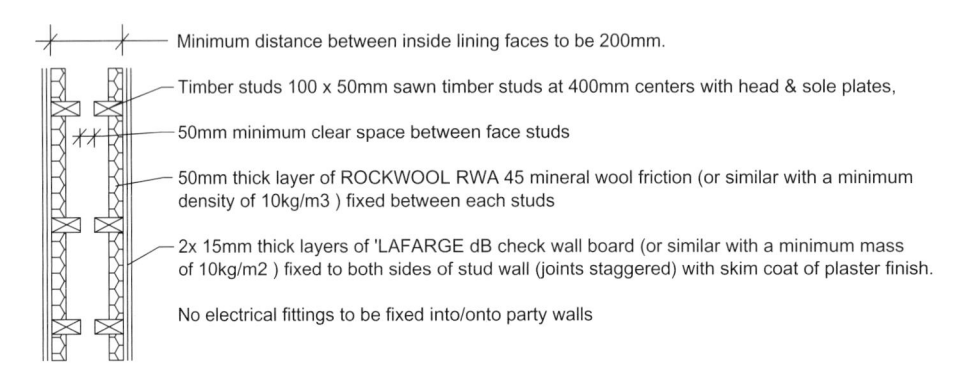

**Figure 2.33:** Plan detail of timber stud separating wall as Wall type 4.1 (new buildings) of ADE *(not to scale)*

Labels in figure:
- Minimum distance between inside lining faces to be 200mm.
- Timber studs 100 x 50mm sawn timber studs at 400mm centers with head & sole plates,
- 50mm minimum clear space between face studs
- 50mm thick layer of ROCKWOOL RWA 45 mineral wool friction (or similar with a minimum density of 10kg/m3 ) fixed between each studs
- 2x 15mm thick layers of 'LAFARGE dB check wall board (or similar with a minimum mass of 10kg/m2 ) fixed to both sides of stud wall (joints staggered) with skim coat of plaster finish.
- No electrical fittings to be fixed into/onto party walls

minimum cavity and tied together with wall ties spaced as external walls with 13 mm plaster (minimum mass 10 kg/m$^2$) applied to both faces. Walls are to be built up to the underside of the roof coverings and fire-stopped with mineral wool or an approved proprietary intumescent product. The party wall is to be built off a foundation, bonded/tied to the inner leaf, and the junction of cavities is to be fire-stopped throughout its length with a proprietary acoustic/insulated fire-stop cavity closer; all other vertical and horizontal cavities are to be closed in a similar manner to provide effective edge sealing and a U-value of 0.2 W/m$^2$.K. Additional party wall solutions are available in ADE.

## Double-leaf timber-frame party walls separating dwellings (see Figure 2.33)

(U-value 0.2 W/m$^2$.K)

Timber-framed stud party walls need to achieve a minimum of 60 minutes' fire resistance from both sides, and a sound insulation value of a minimum 45 dB value (43 dB for conversions) for airborne sound insulation. They should be constructed with two independent leaves of timber-framed walls, with 50 mm minimum clear cavity and a minimum distance between inside lining faces of 200 mm. Timber studs are constructed using 100 × 50 mm sawn timber studs at 400 mm centres with head and sole plates, with a 50 mm thick layer of ROCKWOOL RWA 45 mineral wool friction (or similar with a minimum density of 10 kg/m$^3$) fixed between each pair of studs, and two 15 mm-thick layers of 'LAFARGE dB check wallboard (or similar with a minimum mass of 10 kg/m$^2$) fixed to both sides of the stud wall (joints staggered), with skim coat of plaster finish – as in the wallboard manufacturer's details. No electrical fittings are to be fixed into or onto party walls, and all gaps are to be fire-sealed and smoke-stopped to the full height and width of the party wall and up to the underside of the roof coverings, using mineral wool (not glass wool) or an approved proprietary intumescent product to provide effective edge sealing and a U-value of 0.2 W/m².K. Additional party-wall solutions are available in ADE.

### Upgrading sound insulation of existing party walls separating dwellings (see Figure 2.34)

Existing wall should achieve 60 minutes' fire resistance, be at least 100 mm thick, of masonry construction and plastered on both faces. With other types of existing wall the independent panels should be built on both sides.

Construct new, independent frame fixed at least 10 mm from one side of the existing wall, using 100 × 50 mm timber studs at 400 mm centres fixed either onto head and sole plates or to a proprietary galvanised metal frame, fixed as manufacturer's details.

Fix 50 mm thick Rockwool RWA 45 sound insulation or other approved material (minimum density 16 kg/m³), friction-fixed between studs.

Fix two layers of 15 mm-thick dB checked wallboard with staggered joint and plaster skim finish to the independent frame using mechanical fixings. Note – allow a minimum distance of 35 mm between the face of the existing wall and the inner wallboard face.

No electrical fittings are to be fixed into/onto party walls, and all gaps are to be fire-sealed and smoke-stopped to the full height and width of the party wall and up to the underside of the roof coverings, using mineral wool (not glass wool) or an approved proprietary intumescent product to provide effective edge sealing and a U-value of 0.2 W/m².K.

Additional upgrading party wall solutions are available in ADE.

### *Party floors separating buildings*

These are outside the scope of this guidance – see relevant sections in ADE.

### *Sound-testing requirements*

Pre-completion sound testing is required for all new party walls/floors and should be carried out by a sound specialist, in accordance with ADE, and a copy of the test results sent to building control.

- Existing wall should be at least 100mm thick, of masonry construction and plastered on both faces. With other types of existing wall the independent panels should be built on both sides.
- Construct new independent frame fixed at least 10mm from one side of the existing wall
- 100 x 50mm timber stud at 400mm centers fixed onto head & sole plates
- Fix 50mm thick ROCKWOOL RWA 45 sound insulation or other approved (min density 16kg/m3) friction fixed between studs
- Fix two layers of 15mm thick dB checked wall board with staggered joint and plaster skim finish to the independent frame using mechanical fixings. Note- allow a minimum distance of 35mm between face of existing wall & inner wall board face.

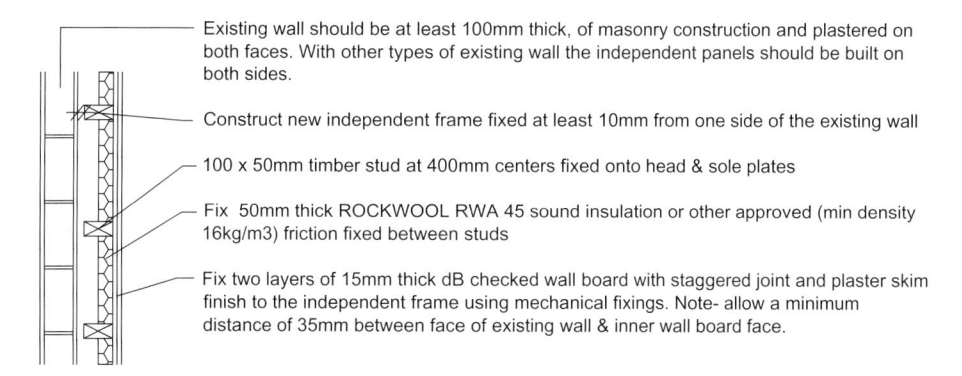

**Figure 2.34:** Plan of upgrading masonry separating wall as Wall type 4.2 (material change of use) of ADE *(not to scale)*

## A4: Internal partitions

### Internal load-bearing masonry partitions

Internal load-bearing walls are to be minimum 100 mm-thick dense concrete blocks (actual wall thickness must not be less than the wall it supports above), built off suitable concrete foundations (as guidance details above, typically 450 mm wide × 225 mm deep), with pre-cast concrete or proprietary steel lintels over openings (in compliance with the lintel manufacturer's span tables) and walls bonded/tied to external or party walls with proprietary ties on each course and restrained by floor or ceiling joists/trusses.

### Internal load-bearing timber stud partitions

Load-bearing timber stud partitions and non-proprietary lintels are to be in compliance with the structural engineer's details and calculations; they must be built off suitable concrete foundations (as guidance details above, typically 450 mm wide × 225 mm deep) and approved by building control before works commence on site. Fix a minimum of 25 mm of 10 kg/m$^3$ proprietary sound insulation quilt suspended between the studs, finished with 12.5 mm plasterboard and skim both sides. Sole/head plates are to be glued and screwed to floor joists, and where necessary additional timber members fixed to allow adequate fixing of fittings etc.

### Internal masonry non-load-bearing partitions

Internal non-load-bearing partitions are to be constructed of minimum 100 mm-thick dense concrete blocks built off a thickened floor slab (as agreed with building control) and tied/block bonded to all internal and external walls at maximum 225 mm centres with either a plaster or dry-lined finish, as the external walls.

### Internal timber studwork non-load-bearing partitions

Non-load-bearing stud partitions are to be constructed of 100 × 50 mm soft wood framing with head and sole plates and intermediate noggins fixed at 400/600 mm centres, built off a thickened floor slab (as agreed with building control), with a minimum thickness of 25 mm of 10 kg/m$^3$ proprietary sound insulation quilt suspended between the studs and finished with 12.5 mm plasterboard and skim both sides. Sole/head plates are to be glued and screwed to floor joists, and where necessary additional timber members fixed to allow adequate fixing of fittings etc.

## A5: Intermediate upper floor(s)

**Note:** Although there are no minimum headroom height requirements in the Building Regulations for habitable rooms (except for stairs and ramps – see Part K of this guidance), a minimum ceiling height of 2.3 m is recommended. (Proprietary floor systems are not covered by this guidance and should be carried out in accordance with a floor system manufacturer's details and structural calculations suitable for a particular project – approved by building control before works commence on site.)

## Floor joists

Floors are to be constructed of kiln-dried, structural-grade timber joists with sizes and spacing suitable for the proposed clear span, in compliance with Table 2.20 below. The maximum span for any

**Table 2.20:** Timber sizes and spans for domestic floor joists (strength class C24)
Supporting domestic floor loads and non-load-bearing timber stud partitions (imposed load not exceeding 1.5 kN/m$^2$; dead load (excluding self-weight of joist) not more than 0.5 kNm$^2$).

| Size of joist | | Spacing of joist (mm) | | |
|---|---|---|---|---|
| | | 400 | 450 | 600 |
| Breadth × Depth | | Maximum clear span (m) | | |
| (mm) | (mm) | | | |
| 47 | 97  | 2.10* | 1.99* | 1.74 |
| 47 | 120 | 2.67* | 2.56* | 2.31 |
| 47 | 145 | 3.21* | 3.09* | 2.80 |
| 47 | 170 | 3.76* | 3.61* | 3.28 |
| 47 | 195 | 4.30* | 4.13* | 3.75$^1$ |
| 47 | 220 | 4.83* | 4.65*$^1$ | 4.23$^1$ |
| 75 | 220 | 5.61* | 5.41* | 4.93* |

**Notes:** Where non-load-bearing partitions run at right angles to the joists, the spans in the guidance table should be reduced by 10 per cent. Two joists are to be bolted together under baths and non-load-bearing partitions running parallel with joists.
**Key:** *Increased to three joists bolted together under baths and non-load-bearing partitions running parallel with joists; $^1$80 mm minimum bearing required.
The above values have been independently compiled for the guidance table by Geomex Ltd Structural Engineers: www.geomex.co.uk
Span tables for C16- and C24-strength class solid timber members in floors, ceilings and roofs for dwellings are available from TRADA Technology at: www.trada.co.uk/bookshop.

floor supported by a wall is 6 m, measured to the centre of each bearing, in accordance with Paragraphs 2C23–2C24 of ADA or alternatively, the floor design can be in accordance with a floor specialist or structural engineer's details and calculations which must be approved by building control before works commence on site.

Joists are to have a nominal minimum bearing of 40 mm (increased to 80 mm where indicated in the guidance table below), supported by heavy-duty, proprietary, galvanised metal restraint joist hangers built into the walls or fixed to treated timber wall plates (same sizes as joists), resin-bolted (100 mm minimum) into sound walls at 600–800 mm centres using approved 12–16 mm diameter stainless-steel fixings – as agreed with building control. Alternatively, joists can be built into walls using approved proprietary sealed joist caps, or sealed with silicon sealant to provide an airtight seal for new dwellings – these require air testing, as agreed with building control. Two joists are to be bolted together under baths and non-load-bearing partitions running parallel with joists, increased to three joists under non-load-bearing partitions where indicated in the guidance table. Where non-load-bearing partitions run at right angles to the joists, the spans in Table 2.20 should be reduced by 10 per cent.

Floor void between joists is to be insulated with a minimum thickness of 100 mm of 10 kg/m$^3$ proprietary sound insulation quilt; ceiling is to be minimum 15 mm plasterboard and skim to give the required sound insulation and 30 minutes' fire resistance. Floor joists are to be provided with one row of 38 × ¾ depth solid strutting at ends between joist hangers, or proprietary galvanised struts to BS EN 10327 fixed as manufacturer's details, at mid-span for 2.5–4.5 m spans and two rows at 1/3 centres for spans over 4.5 m.

Fix 22 mm-thick, moisture-resistant, tongue-and-grooved timber floorboards laid with joints staggered, long edge fixed across the joists and all joints positioned over joists/noggins. All boards are to be glued and screwed to floor joists, with all joints glued (using waterproof glue) and pinned, in accordance with the floorboard manufacturer's details and current BS EN standards. Allow an expansion gap around wall perimeters, as manufacturer's details (typically 10–15 mm).

## Trimming and trimmer joists

Trimming joists and trimmer joist sizes supporting trimmed joists around openings should be in accordance with Figure 2.35 and Tables 2.21 and 2.22 or alternatively calculations and details are required from a suitably qualified person (i.e. a structural engineer) and should be approved by building control before works commence on site. Minimum bearing 80 mm, and double joists should be mechanically fixed together using 12 mm-diameter high-tensile bolts and $3 \times 50$ mm steel washers at each bolt end, and spaced at 600 mm minimum spacings on the centre line of the trimming/trimmer joists (minimum of two bolted connections per joist at 1/3rd span positions). Joists are to be built

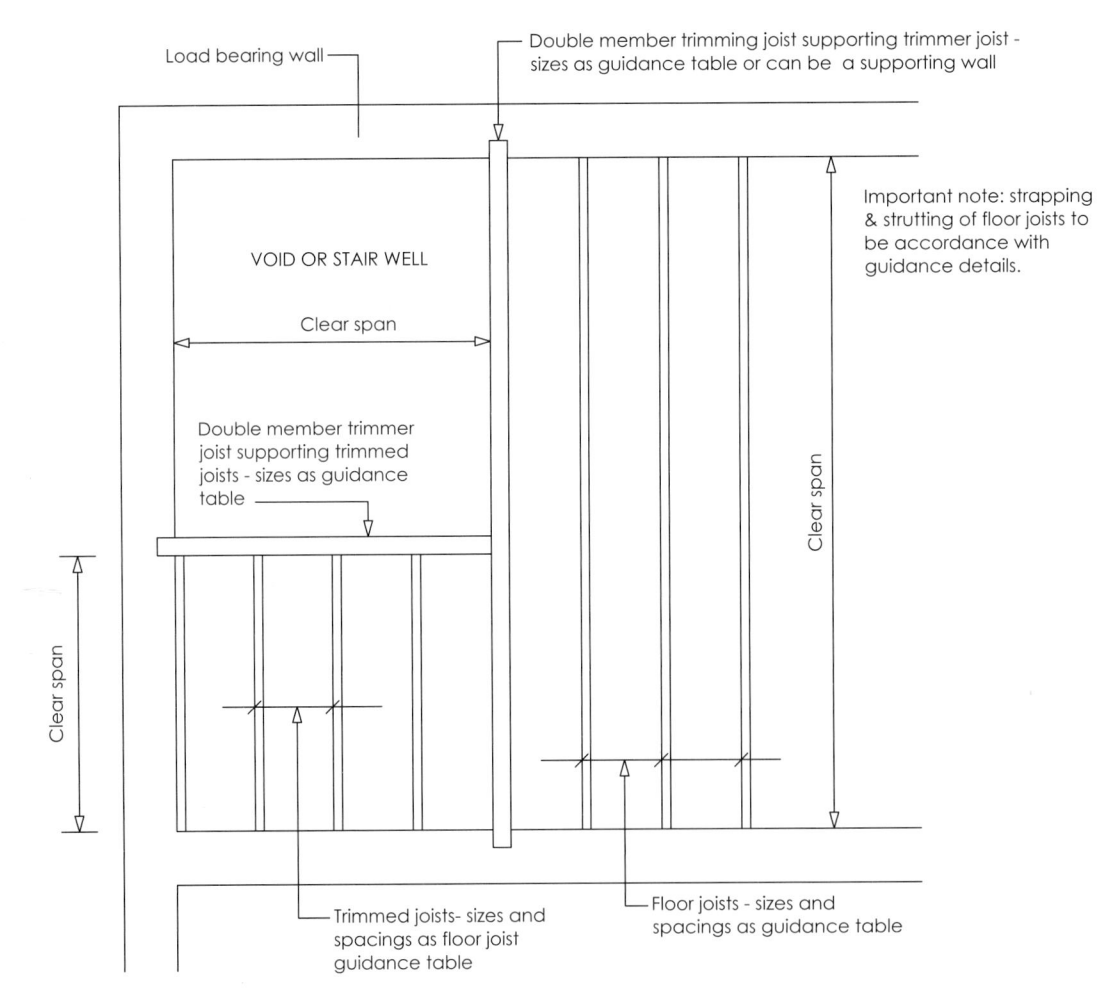

**Figure 2.35:** Typical plan layout of opening formed in suspended timber floor(s) using trimming, trimmer and trimmed joists (*not to scale*)
(*See guidance tables for joist sizes.*)

**Table 2.21:** Timber sizes and spans for trimmer joist supporting trimmed joists (strength class C24)
Supporting domestic floor loads and non-load-bearing timber stud partitions (imposed load not exceeding 1.5 kN/m$^2$; dead load (excluding self-weight of joist) not more than 0.5 kNm$^2$)

| Size of trimmer joist (mm) | Length of trimmed joists (m) | | | |
|---|---|---|---|---|
| | 1.0 | 2.0 | 3.0 | 4.0 |
| 2 no. × breadth × depth | Clear span of trimmer joist (m) supporting trimmed joists | | | |
| 2 × 47 × 145 | 2.68 | 2.27 | 1.97 | 1.68 |
| 2 × 47 × 170 | 3.21 | 2.68 | 2.33 | 1.99 |
| 2 × 47 × 195 | 3.69 | 3.09 | 2.69 | 2.31 |
| 2 × 47 × 220 | 4.17* | 3.50 | 3.06 | 2.62 |
| 2 × 75 × 220 | 4.81* | 4.13 | 3.62 | 3.21 |

See plan layout below for configuration of trimming, trimmer and trimmed joists.
**Key:** *Increased to three trimmer joists bolted together under non-load-bearing partitions running parallel with joists.
The above values have been independently compiled for the guidance table by Geomex Ltd Structural Engineers: www.geomex.co.uk. Span tables for C16- and C24-strength class solid timber members in floors, ceilings and roofs for dwellings are available from TRADA Technology at: www.trada.co.uk/bookshop.

**Table 2.22:** Timber sizes and spans for trimming joist supporting trimmer joist (strength class C24)
Supporting domestic floor loads and non-load-bearing timber stud partitions (imposed load not exceeding 1.5 kN/m$^2$; dead load (excluding self-weight of joist) not more than 0.5 kNm$^2$)

| Size of trimming joist (mm) | Length of trimmer joist (m) | | |
|---|---|---|---|
| | 1.0 | 2.0 | 3.0 max |
| 2 no. × breadth × depth | Clear span of trimming joist (m) supporting trimmer joist | | |
| 2 × 47 × 145 | 2.62 | 2.42 | 2.25 |
| 2 × 47 × 170 | 3.08 | 2.84 | 2.65 |
| 2 × 47 × 195 | 3.54 | 3.27 | 3.05 |
| 2 × 47 × 220 | 3.99 | 3.70 | 3.46 |
| 2 × 75 × 220 | 4.66 | 4.32 | 4.05 |

See plan layout below for configuration of trimming, trimmer and trimmed joists.
The above values have been independently compiled for the guidance table by Geomex Ltd Structural Engineers: www.geomex.co.uk. Span tables for C16- and C24-strength class solid timber members in floors, ceilings and roofs for dwellings are available from TRADA Technology at: www.trada.co.uk/bookshop.

into load-bearing walls or supported on heavy-duty galvanised joist hangers to current British Standards and fixed in accordance with the joist hanger manufacturer's details or supported in accordance with structural engineer's details and calculations. Notches/holes/cuts in structural timbers should be carried out in accordance with these guidance details and BS 5268: 2002.

## Notching and drilling of structural timbers

Notching and drilling in structural timbers should be in accordance with BS 5268-2:2002.

Notches should not be deeper than 0.125 times the depth of the joists and should be not closer to the support than 0.07 times the span, and not further away than 0.25 times the span. Holes drilled should have a diameter not greater than 0.25 times the depth of the joist and should be drilled at the joist centre line. They should be not less than 3 diameters (centre to centre) apart and should be located between 0.25 and 0.4 times the span from the support. Notches or holes exceeding the above requirements should be checked by a structural engineer.

### Sound insulation to floors within the dwelling

Intermediate floors are to be provided with sound insulation as described in the relevant floor section in this guidance.

### *Soil-and-vent pipe (SVP) boxing internally*

SVP pipe boxing is to consist of soft wood framing, two layers of 15 mm plasterboard and skim and the void filled with mineral wall quilt for sound insulation and fire-/smoke-stopping. Boxing is to be continuously carried up to the roof space for the soil-and-vent pipe and provided with air grilles where an air-admittance valve is used. Ensure **all gaps** and **all voids** are **sealed** to prevent any air leakage.

### *Exposed intermediate upper floors*

Semi-exposed intermediate timber floors over unheated areas such as garages, porches, walkways and canopies are to be insulated with the following minimum thickness and types of insulation to achieve a U-value of 0.22 W/m$^2$.k, in accordance with the insulation guidance table (Table 2.23) below. Where the construction is open to the environment, a vapour barrier and proprietary external mineral fibre or similar 30-minute fire- and moisture-resistant boarding are to be applied to the underside of the floor.

Table 2.23: Examples of insulation for exposed upper floors
(U-value no worse than 0.22 W/m$^2$.k, typically 50 mm wide × 200 mm deep joists at 400 mm spacings, with 22 mm floor boarding and 15 mm plaster board and skim finish to ceiling)

| Insulation product | K value | Required thickness of insulation (mm) |
|---|---|---|
| Kingspan Thermafloor TF70 | 0.022 | 110 |
| Celotex FR5000 | 0.021 | 110 |
| Rockwool Flexi | 0.038 | 200 |
| Knauf Earthwool loft roll 40 | 0.040 | 200 |

Note: Insulation is to be installed in accordance with the manufacturer's details.

## A6: Pitched roofs

**Note:** Although there are no minimum headroom height requirements in the Building Regulations for habitable rooms (except for stairs and ramps – see Part K of this guidance), a minimum ceiling height of 2.3 m is recommended.

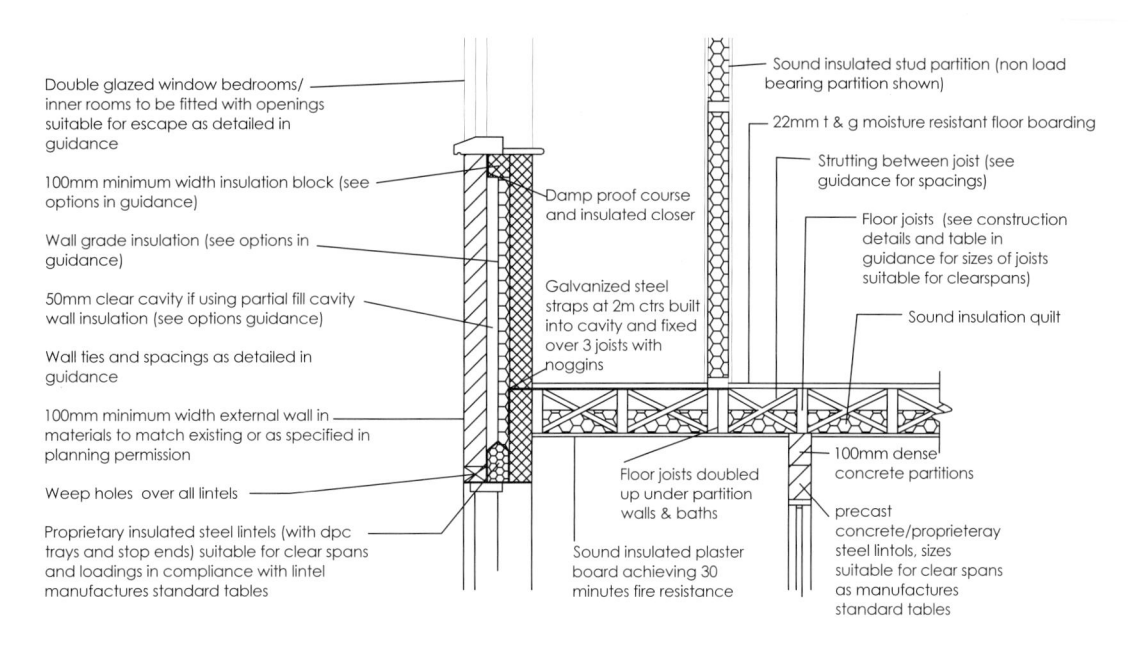

**Figure 2.36:** Typical section through an upper floor *(not to scale)*

# Pitched roof coverings

Roof covering is to consist of slates/tiles and associated ridge, verge, eaves, hip, valley, abutment and ventilation systems, etc. fitted in accordance with the tile manufacturer's details and suitable for the minimum recommended roof pitches and exposure.

Roof tiles/cladding are to be fixed in accordance with the manufacturer's details to $25 \times 50$ mm treated timber battens (battens to be at least 1.2 m long, nailed to each rafter, fixed over at least three rafters and spaced in accordance with the tile manufacturer's details). Rafters are to be overlaid with untearable underlays, using either a non-breathable/high-water-vapour-resistance underlay to BS EN 13707: 2004 (requires ventilation on opposing sides, as detailed in guidance) or a British Board of Agreement (BBA or other third-party accredited) vapour-permeable breathable/low-water-resistance-type underlay. Both types are to be fixed, ventilated and lapped in accordance with the manufacturer's details.

Where roof coverings cannot be fixed to the tile/slate manufacturer's required pitch, roof coverings can be fixed below the manufacturer's minimum recommended roof pitch by using a proprietary British Board of Agreement (BBA or other third-party accredited) corrugated roof sheet system below roof coverings so as to create an independent, secondary, weatherproof roof, which must be installed to minimum roof pitches and ventilated in accordance with the manufacturer's details – for example, 'Ondutile' under tile and slate under-sheeting system manufactured by Onduline Building Products Ltd: www.onduline.net. (Typical minimum roof pitches are: $12.5°$ for concrete interlocking tiles; $17.5°$ for clay pan tiles/natural and fibre cement slates; $22.5°$ for plain double-lap tiles –contact the manufacturer for minimum roof pitches achievable.)

## Pitched roof structure

Roofs are to be constructed using either manufactured roof trusses or a cut roof as follows (proprietary roof systems are not covered by this guidance and should be carried out in accordance with a roof system manufacturer's details and structural calculations suitable for a particular project – approved by building control before works commence on site):

### (i) Roof trusses (including attic and girder trusses)

The roof is to be constructed using specialist-designed and -manufactured trusses (or attic trusses where forming room(s) in the roof or used for storage) at 400 mm (or 600 mm maximum) spacings to BS 5268:3 or PD 6693-2. Trusses are to be fixed and braced strictly in accordance with BS 5268:3 or PD 6693-2 and the truss manufacturer's details, mechanically fixed to 100 × 50 mm treated soft wood wall plates using proprietary galvanised steel truss clips. Reinforced concrete pad stones are required to support girder trusses, to details and calculations by a suitably qualified person.

The person carrying out the building work is to check and confirm the actual roof pitch to the truss manufacturer prior to placing an order. Details of trusses and a bracing diagram are to be prepared by the specialist designer/truss manufacturer, and they must be submitted and approved by building control prior to commencing roof construction.

### (ii) Cut roof construction

The roof is to be constructed using kiln-dried, stress-graded timber. Rafters, ceiling joists, purlins, hanger and binder sizes should be as stated in the independent guidance tables below (Tables 2.24 to 2.27) – or see TRADA Technology span tables, available from: www.trada.co.uk – suitable for the proposed clear spans and all properly fixed together using approved mechanical fixings. Where the ceiling joists are raised above wall-plate level, they must be fixed within the bottom third of the rafter using 12 mm-diameter high-tensile bolts and proprietary steel-toothed connectors to connect each rafter and ceiling joist, so as to prevent possible roof spread. Joists raised above this level are to be designed by a suitably qualified person and approved by building control before works commence.

Typical minimum sizes of roof timbers are: struts and braces to be 100 × 50 mm; hip sizes to be the splayed rafter depth + 25 mm × 50 mm thick (under 30-degree pitch the hips are to be designed by a suitably qualified person); lay-boards to be the splayed rafter depth + 25 mm × 32 mm thick; ridges to be splayed rafter depth + 25 mm × 38 mm thick; all valley beams are to be designed by a suitably qualified person; wall plates are to be 100 × 50 mm fixed to inner skin of cavity wall using galvanised strapping, as detailed below. Angle ties should be used on hipped roof corners to prevent spreading. Hip rafters over 150 mm deep are to be supported on 100 × 75 mm angle ties mechanically connected across wall plates, and hip rafter notched to fit over angle tie at corners of roof. Proprietary hip irons are to be screwed to base of hip rafters to support ridge tiles unless using a proprietary ridge tile/capping system, mechanically fixed in accordance with manufacturer's details.

Soffits, fascias and barge boards, etc. should be constructed in painted/stained soft/hardwood or uPVC to BS 4576. Allow for all necessary alteration/modification of any existing adjoining roof as required to enable the proper completion of the works and in agreement with building control.

Allow for building in as work proceeds, or insertion of proprietary stepped/cavity tray dpc to follow line of new roof 150 mm above all roof/wall abutments as necessary, using proprietary dpc trays and code 5 lead flashings. Tie the new roof into the existing one, alter/modify/renew existing roof coverings and form a weather-tight structure. Fix 12.5 mm foil-backed plasterboard (joints staggered) and 3 mm skim coat of finishing plaster to the underside of all ceilings using galvanised plasterboard nails.

Roof pitch to (single-storey) single-skin buildings with walls 100 mm thick should not exceed 40° without structural engineer's details and calculations to confirm the stability of the structure. Cut roofs over 40° are to be diagonally/laterally braced in accordance with BS 5268.

### Notching and drilling of structural timbers

Notching and drilling in structural timbers should be in accordance with the guidance details above.

### Roof restraint

The roof and walls are to be provided with lateral restraint straps, as guidance details above

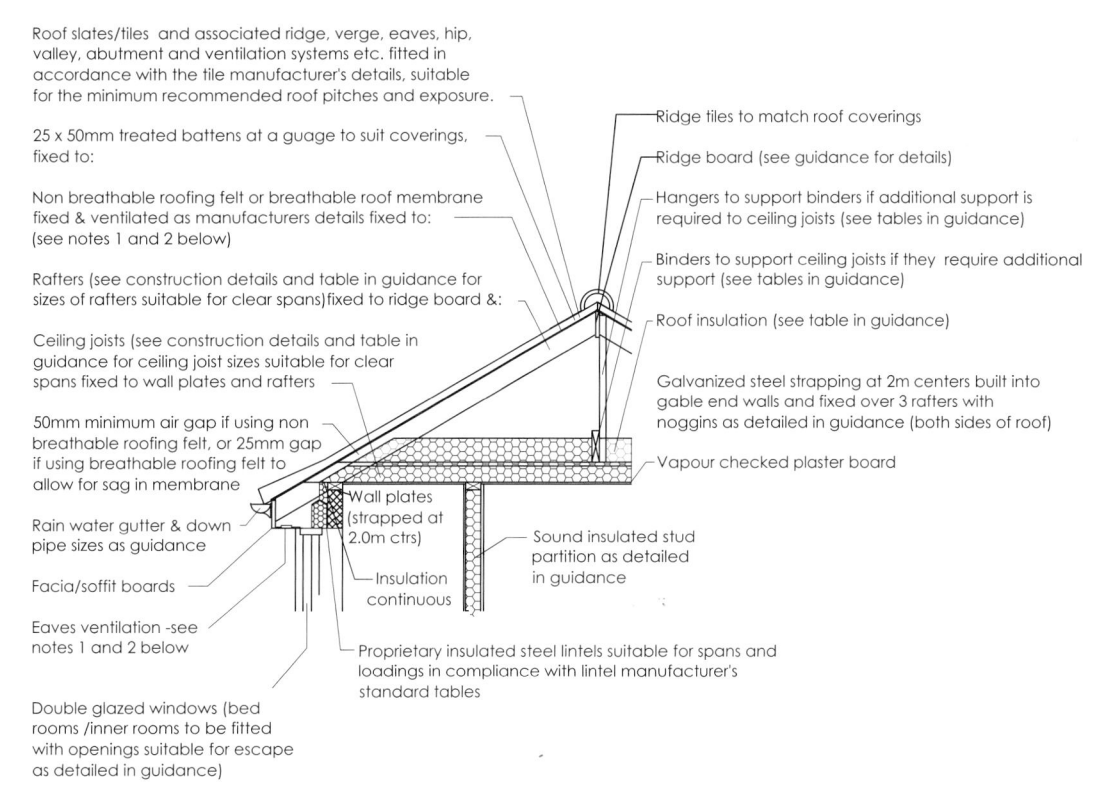

Roof slates/tiles and associated ridge, verge, eaves, hip, valley, abutment and ventilation systems etc. fitted in accordance with the tile manufacturer's details, suitable for the minimum recommended roof pitches and exposure.

25 x 50mm treated battens at a guage to suit coverings, fixed to:

Non breathable roofing felt or breathable roof membrane fixed & ventilated as manufacturers details fixed to: (see notes 1 and 2 below)

Rafters (see construction details and table in guidance for sizes of rafters suitable for clear spans) fixed to ridge board &:

Ceiling joists (see construction details and table in guidance for ceiling joist sizes suitable for clear spans fixed to wall plates and rafters

50mm minimum air gap if using non breathable roofing felt, or 25mm gap if using breathable roofing felt to allow for sag in membrane

Rain water gutter & down pipe sizes as guidance

Facia/soffit boards

Eaves ventilation -see notes 1 and 2 below

Double glazed windows (bed rooms /inner rooms to be fitted with openings suitable for escape as detailed in guidance)

Ridge tiles to match roof coverings

Ridge board (see guidance for details)

Hangers to support binders if additional support is required to ceiling joists (see tables in guidance)

Binders to support ceiling joists if they require additional support (see tables in guidance)

Roof insulation (see table in guidance)

Galvanized steel strapping at 2m centers built into gable end walls and fixed over 3 rafters with noggins as detailed in guidance (both sides of roof)

Vapour checked plaster board

Wall plates (strapped at 2.0m ctrs)

Sound insulated stud partition as detailed in guidance

Insulation continuous

Proprietary insulated steel lintels suitable for spans and loadings in compliance with lintel manufacturer's standard tables

Note1: when using non breathable roofing felt, cross ventilation is to be provided by either proprietary facia ventilation strips or soffit vents to opposing sides of roof at eaves level and fitted with an insect grill with a ventilation area equivalent to a 25mm continuous gap for roof pitches below 15° or a 10mm gap for roof pitches above 15°.

Note 2: When using non breathable roofing felt and the roof span is more than 10 metres or when the pitch is more than 35°, provide additional high level ventilated openings equivalent to a continuous 5mm air gap at ridge level to cross ventilate roofs using proprietary dry ridge systems or vent tiles spaced and fixed in accordance with tile manufacturer's details.

**Figure 2.37:** Typical section through pitched roof with ceiling joists at wall-plate level *(not to scale)* (U-value no worse than 0.16 W/m$^2$.k)

## 2.64 Domestic extensions

Roof slates/tiles and associated ridge, verge, eaves, hip, valley, abutment and ventilation systems etc fitted in accordance with the tile manufacturer's details, suitable for the minimum recommended roof pitches and exposure.

25 x 50mm treated battens at a guage to suit coverings, fixed to:

Non breathable roofing felt (or breathable roof membrane fixed & ventilated as manufacturers details)fixed to:

Rafters birds mouthed over & mechanically fixed to purlins, wall plates & ridge(see construction details and table in guidance for sizes of rafters suitable for clear spans)

Ceiling joists fixed to wall plates and rafters(see construction details and table in guidance for ceiling joist sizes suitable for clear spans

50mm minimum air gap if using non breathable roofing felt, or 25mm gap if using breathable roofing felt to allow for sag in membrane

Rain water gutter & down pipe sizes as guidance details

Facia/soffit boards

Eaves ventilation equal to a continuous 25mm air gap with insect screen both sides of roof (may not be required with certain breathable roof membranes)

Vapour checked plaster board & skim

Cavities closed at eaves level

NOTE: Proprietary high level roof vents to be installed where insulation follows slope of roof- equal to a continuous 5mm air gap with insect screen (may not be required with certain breathable roof membranes)

Ridge tiles to match roof coverings

Ridge board (see guidance for details)

Hangers to support hangers if additional support is required to ceiling joists (see tables in guidance)

Binders to support ceiling joists if they require additional support (see tables in guidance)

Roof insulation at ceiling level (see table in guidance)

Roof insulation friction fixed between rafters & under rafters to sloping part of roof. (see table in guidance

Cavities closed at eaves level

Purlins supporting rafters and preventing roof spread. (see construction details and table in guidance for sizes of purlins)or alternatives as follows:.

Alternative1: High level ceiling joists used instead of purlins which must be located within the bottom third of the rafter to prevent roof spread, each ceiling joist must be connected to each rafter with minmum 12mm diameter bolts & steel toothed connectors. Celing joist sizes as guidance tables.

Alternative 2: Ridge beam used instead of purlins to support rafters and prevent broof spread to stuctural engineers details and calculations

Galvanized steel strapping at 2m centers built into gable end walls and fixed over 3 rafters with noggins as detailed in guidance (both sides of roof)

Wall plates strapped at 2.0m centers

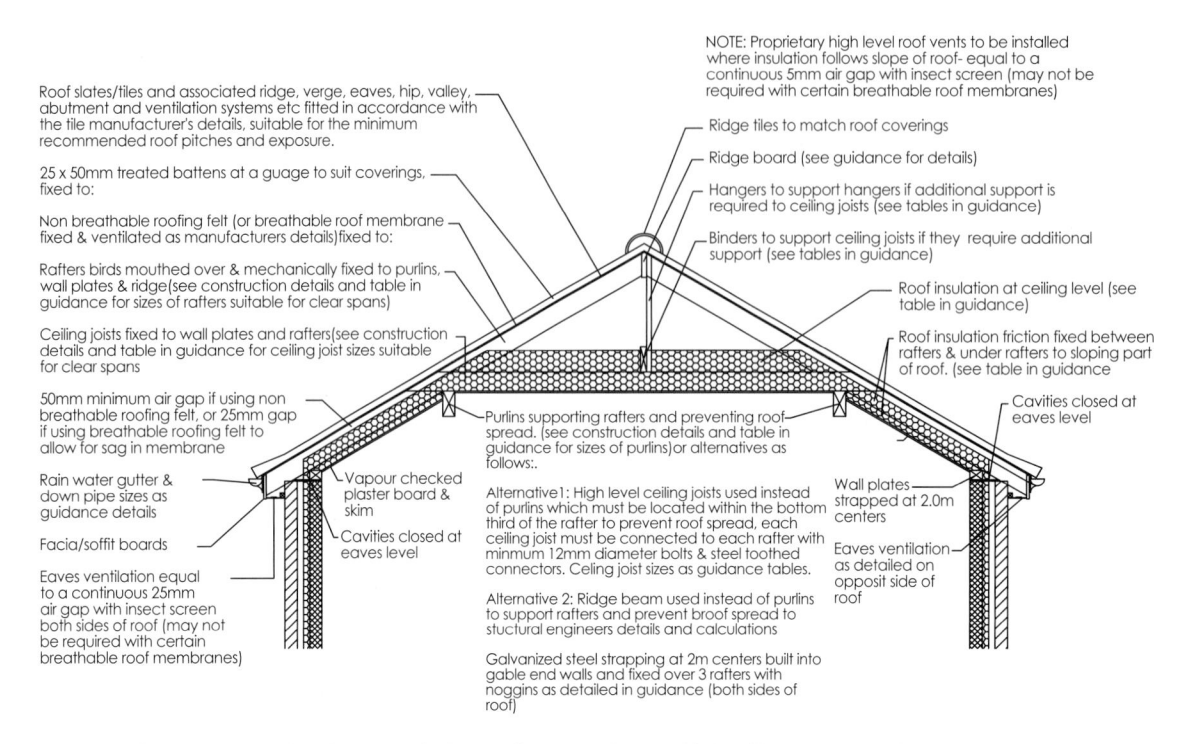

Eaves ventilation as detailed on opposit side of roof

**Figure 2.38:** Typical section through a pitched roof with purlins and high collars (*not to scale*) (U-value no worse than 0.18 W/m$^2$.k)

**Table 2.24:** Timber sizes and permissible clear spans for single-span common jack rafters at 400 mm spacing (strength class C24)

| Size of rafter | Slope of roof (degrees) | | |
|---|---|---|---|
| | 15–22.5° | 22.5–30° | 30–45° |
| Breadth × Depth (mm) | Maximum clear span (m) | | |
| 47 × 100 | 2.08 | 2.12 | 2.18 |
| 47 × 125 | 2.74 | 2.79 | 2.87 |
| 47 × 150 | 3.40 | 3.47 | 3.56 |
| 47 × 195 | 4.59 | 4.68 | 4.81 |

Minimum rafter bearing 35 mm.

Imposed load: 1.02 kN/m$^2$ (high snow load – altitudes not exceeding 100 m).

Dead load: not more than 0.75 kN/m$^2$ (concentrated load 0.9 kN) excluding self-weight of rafter.

The above values have been independently compiled for guidance table by Geomex Ltd Structural Engineers: www.geomex.co.uk. Span tables for C16- and C24-strength class solid timber members in floors, ceilings and roofs for dwellings are available from TRADA Technology at: www.trada.co.uk/bookshop.

**Table 2.25:** Timber sizes and permissible clear spans for purlins (strength class C24)

| Size of purlin (mm) | Slope of roof (degrees) | | | | | | | | | | | |
|---|---|---|---|---|---|---|---|---|---|---|---|---|
| | 15–22.5° | | | | 22.5–30° | | | | 30–45° | | | |
| | Spacing of Purlins (mm) | | | | | | | | | | | |
| | 1500 | 1800 | 2100 | 2400 | 1500 | 1800 | 2100 | 2400 | 1500 | 1800 | 2100 | 2400 |
| B × D | Maximum clear spans (m) | | | | | | | | | | | |
| 75 × 125 | 2.01 | 1.88 | 1.77 | 1.65 | 2.06 | 1.92 | 1.82 | 1.73 | 2.12 | 1.99 | 1.88 | 1.79 |
| 75 × 150 | 2.41 | 2.25 | 2.13 | 1.98 | 2.46 | 2.31 | 2.18 | 2.07 | 2.54 | 2.38 | 2.25 | 2.15 |
| 75 × 175 | 2.81 | 2.63 | 2.48 | 2.31 | 2.87 | 2.69 | 2.54 | 2.42 | 2.97 | 2.78 | 2.63 | 2.50 |
| 75 × 200 | 3.20 | 3.00 | 2.83 | 2.63 | 3.28 | 3.07 | 2.90 | 2.76 | 3.39 | 3.17 | 3.00 | 2.86 |
| 75 × 225 | 3.60 | 3.37 | 3.19 | 2.96 | 3.68 | 3.45 | 3.26 | 3.10 | 3.81 | 3.57 | 3.35 | – |

Minimum purlin bearing 80 mm.
Imposed load: 1.02 kN/m² (high snow load – altitudes not exceeding 100 m).
Dead load: not more than 0.75 kN/m² (concentrated load 0.9 kN) excluding self-weight of purlin.
The above values have been independently compiled for guidance table by Geomex Ltd Structural Engineers: www.geomex.co.uk.
Span tables for C16- and C24-strength class solid timber members in floors, ceilings and roofs for dwellings are available from TRADA Technology at: www.trada.co.uk/bookshop.

**Table 2.26:** Timber size and permissible clear spans for ceiling joists at 400 mm spacing (strength class C24)

| Size of ceiling joist | Maximum clear span (m) |
|---|---|
| Breadth × Depth (mm) | |
| 47 × 97 | 2.00 |
| 47 × 120 | 2.61 |
| 47 × 145 | 3.29 |
| 47 × 170 | 3.69 |
| 47 × 195 | 4.64 |
| 47 × 220 | 5.32 |

Minimum ceiling joist bearing 35 mm.
Imposed load: 0.25 kN/m² (concentrated load 0.9 kN).
Dead load: 0.50 kN/m² excluding self-weight of joist.
The above values have been compiled for guidance table by Geomex Ltd Structural Engineers: www.geomex.co.uk.
Span tables for C16- and C24-strength class solid timber members in floors, ceilings and roofs for dwellings are available from TRADA Technology at: www.trada.co.uk/bookshop.

### *Roof insulation and ventilation gaps*

Insulation is to be fixed in accordance with the manufacturer's details and must be continuous with the wall insulation, but stopped back at eaves or at junctions with rafters to allow for a continuous 50 mm minimum ventilated air gap above the insulation to underside of the roofing felt where a non breathable roofing felt is used, or 15–25 mm air space to allow for sag in felt if using a breathable roofing membrane, in accordance with the manufacturer's details. All guidance diagrams and details assume rafters at 400 mm centres and 12.5 mm vapour-checked plasterboard ceilings with skim finish.

**Table 2.27:** Timber sizes and permissible clear spans for ceiling binders (strength class C24)

| Size of binder (mm) | Spacing of binders (mm) | | | | | |
|---|---|---|---|---|---|---|
| | 1200 | 1500 | 1800 | 2100 | 2400 | 2700 |
| Breadth × Depth | Maximum clear span or hanger spacing (m) | | | | | |
| 47 × 175 | 2.88 | 2.69 | 2.54 | 2.42[1] | 2.32[1] | 2.23[1] |
| 47 × 200 | 3.33 | 3.11 | 2.93[1] | 2.29[1] | 2.67[1] | 2.56[1] |
| 75 × 175 | 3.43 | 3.21 | 3.04 | 2.90 | 2.78 | 2.67 |
| 75 × 200 | 3.95 | 3.70 | 3.50 | 3.33 | 3.19 | 3.07 |
| 75 × 225 | 4.47 | 4.18 | 3.95 | 3.76 | 3.60[1] | 3.47[1] |

Minimum ceiling binder bearing 60 mm.
**Key:** [1] 120 mm minimum bearing required.
Imposed load: 0.25 kN/m² (concentrated load 0.9 kN).
Dead load: 0.50 kN/m² excluding self-weight of binder.
The above values have been independently compiled for guidance table by Geomex Ltd Structural Engineers: www.geomex.co.uk. Span tables for C16- and C24-strength class solid timber members in floors, ceilings and roofs for dwellings are available from TRADA Technology at: www.trada.co.uk/bookshop.

## Pitched roof ventilation requirements when using a non-breathable roof membrane

*(i)* ***Duo-pitched roof with horizontal ceilings and insulation at ceiling level***
   Roof insulation is to be continuous with the wall insulation, but stopped back at eaves or at junctions with rafters to allow a 50 mm minimum air gap. Cross-ventilation is to be provided by either proprietary facia ventilation strips or soffit vents to opposing sides of roof at eaves level, and fitted with an insect grille with a ventilation area equivalent to a 25 mm continuous gap for roof pitches below 15°, or a 10 mm gap for roof pitches above 15°.
   When the roof span is more than 10 metres or when the pitch is more than 35°, provide additional high-level ventilated openings equivalent to a continuous 5 mm air gap at ridge level to cross-ventilate roofs, using proprietary dry ridge systems or vent tiles spaced and fixed in accordance with the tile manufacturer's details.

*(ii)* ***Mono-pitched roofs with horizontal ceilings and insulation at ceiling level***
   Roof insulation is to be continuous with the wall insulation, but stopped back at eaves or at junctions with rafters to allow a 50 mm minimum air gap. Cross-ventilation is to be provided by either proprietary facia ventilation strips or soffit vents at eaves level, and fitted with an insect grille with a ventilation area equivalent to a 25 mm continuous gap for roof pitches below 15°, or a 10 mm gap for roof pitches above 15°.
   Provide high-level ventilated openings fitted with an insect grille equivalent to a continuous 5 mm air gap to cross-ventilate roofs, using proprietary ventilation systems or vent tiles spaced and fixed in accordance with the tile manufacturer's details.

*(iii)* ***Duo-pitched roof with insulation following slope of rafters (rooms in the roof)***
   Roof insulation is to be continuous with the wall insulation, but stopped back at eaves or at junctions with rafters to allow a continuous 50 mm air gap between the top of the insulation and the underside of the roof membrane. Cross-ventilation is to be provided by proprietary eaves-ventilation strips equivalent to a 25 mm continuous air gap to opposing sides of roof at eaves level, fitted with an insect grille and at ridge/high level to provide ventilation equivalent

**Table 2.28:** Examples of roof insulation fixed between/under rafters
(Vented cold roof achieving a U-value of 0.18 W/m².k)

| Product | K value | Position in roof |
|---|---|---|
| Kingspan Kooltherm K7 Pitched Roof Board and Kingspan Kooltherm K18 Insulated Plasterboard | 0.020 0.021 | 100 mm friction fixed between rafters and 42.5 mm fixed under rafters, with integral vapour-checked plasterboard and skim finish |
| Kingspan Kooltherm K7 Pitched Roof Board and Kingspan Kooltherm K18 Insulated Plasterboard | 0.020 0.021 | 100 mm friction-fixed between rafters and 37.5 mm fixed under rafters, with integral vapour-checked plaster-board and skim finish* |
| Celotex GA4000 | 0.022 | 100 mm friction-fixed between rafters and 35 mm fixed under rafters, with vapour-checked plasterboard and skim finish* |
| Celotex GA4000 | 0.022 | 165 mm friction-fixed between rafters with vapour-checked plasterboard and skim finish fixed to underside of rafters* |
| **Multi foils** | | |
| Web Dynamics TLX Silver FB multi foil and Insulation with a K value of 0.022 or better | R-value 1.69 0.022 | One layer of multi foil fixed under rafters with vapour-checked plasterboard fixed to 25 mm-deep battens to create air space and 75 mm Kingspan or Celotex (or other approved foil-faced rigid insulation) fixed between rafters, allowing a 25 mm cavity between the multi foil and rigid insulation* |
| YBS SuperQuilt multi foil and Insulation with a K value of 0.023 or better | R-value 2.71 (including both air spaces) 0.023 | One layer of multi foil fixed under rafters with vapour-checked plasterboard fixed to 25 mm-deep battens to create air space and 65 mm Kingspan or Celotex (or other approved foil-faced rigid insulation) fixed between rafters, allowing a 25 mm cavity between the multi foil and rigid insulation |

**Key:** *All unvented roofs using vapour-permeable underlay.
**Note:** Insulation is to be installed in accordance with the manufacturer's details.
Source: a representative selection of values taken from *Technical Note 10, U-Values of Elements (Approved Document L1B 2010)*, produced by Hertfordshire Technical Forum for Building Control. Reproduced by permission of Hertfordshire Technical Forum for Building Control.

**Table 2.29:** Examples of roof insulation laid horizontally between and over ceiling joists
(Vented cold roof achieving a U-value of 0.16 W/m².k)

| Product | K value | Position in roof |
|---|---|---|
| Earthwool Loft Roll 44 | 0.044 | 100 mm between joists and 170 mm laid over joists |
| Rockwool Roll | 0.044 | 100 mm between joists and 170 mm laid over joists |
| Earthwool Loft Roll 44 and Polyfoam Space Boards | 0.044 0.029 | 100 mm between joists and 2 layers 52.5 mm Space Boards fixed over joists and overlaid with 18 mm floorboards |

**Note:** Insulation is to be installed in accordance with the manufacturer's details.
Source: a representative selection of values taken from *Technical Note 10, U-Values of Elements (Approved Document L1B 2010)*, produced by Hertfordshire Technical Forum for Building Control. Reproduced by permission of Hertfordshire Technical Forum for Building Control.

### 2.68 Domestic extensions

**Table 2.30:** Examples of roof insulation fixed over/between rafters
(Warm roof achieving a U-value of $0.18\,W/m^2.k$)

| Product | K value | Position in roof |
|---|---|---|
| Kingspan Kooltherm K7 Pitched Roof Board | 0.020 | 100 mm fixed over rafters with breathable membrane: for example, Kingspan Nilvent fixed beneath counter battens* **or** 90 mm fixed over rafters with breathable membrane fixed over counter battens* |
| Kingspan Kooltherm K7 Pitched Roof Board | 0.020 | 55 mm fixed over rafters and 50 mm fixed between rafters with breathable membrane: for example, Kingspan Nilvent fixed beneath counter battens* **or** 50 mm fixed over rafters and 50 mm fixed between rafters with breathable membrane fixed over counter battens* |
| Celotex GA4000 | 0.022 | 100 mm fixed over rafters with breathable membrane* |
| Celotex GA4000 | 0.022 | 60 mm fixed over rafters and 60 mm fixed between rafters with breathable membrane* |

**Key:** *All unvented roofs using vapour-permeable underlay.
**Note:** Insulation fixed over the roof should be carried out in accordance with the insulation manufacturer's details, and may require specialist fixings for the build-up of insulation; battens/counter battens and breathable membrane positions will be required.
Source: a representative selection of values taken from *Technical Note 10, U-Values of Elements (Approved Document L1B 2010)*, produced by Hertfordshire Technical Forum for Building Control. Reproduced by permission of Hertfordshire Technical Forum for Building Control.

to a 5 mm air gap in the form of a proprietary dry ridge system or vent tiles spaced and fixed in accordance with the tile manufacturer's details.

#### *Proprietary vapour-permeable roof membrane*

Ventilation to the roof space may be omitted only if a proprietary British Board of Agreement (BBA or other third-party accredited) vapour-permeable, breathable roof membrane is used. Vapour-permeable, breathable roof membranes must always be installed in strict accordance with the manufacturer's details (note. some breathable membranes may also require additional roof ventilation, in accordance with the manufacturer's details).

#### *Valleys and lead work* (see Figure 2.40.1)

Lead work, flashing, soakers, valleys and gutters, etc. are to be formed from Code 5 lead sheet and fully supported on treated valley boards etc. They are to have minimum 150 mm lap joints, dressed 200 mm under tiles, to have suitable gradients for drainage etc., and are not to be fixed in lengths exceeding 1.5 m; but they are to be fixed in accordance with the roof cladding manufacturer's and the Lead Sheet Association's recommendations. Proprietary fibreglass valleys are to be fixed in accordance with manufacturer's details.

#### *Roof abutment with cavity wall* (see Figure 2.40.2)

Proprietary dpc trays/stepped dpc trays, stop ends and weep holes etc. for new works or retrospective fitting into existing walls etc. at roof/cavity wall abutments etc. should have a British Board of

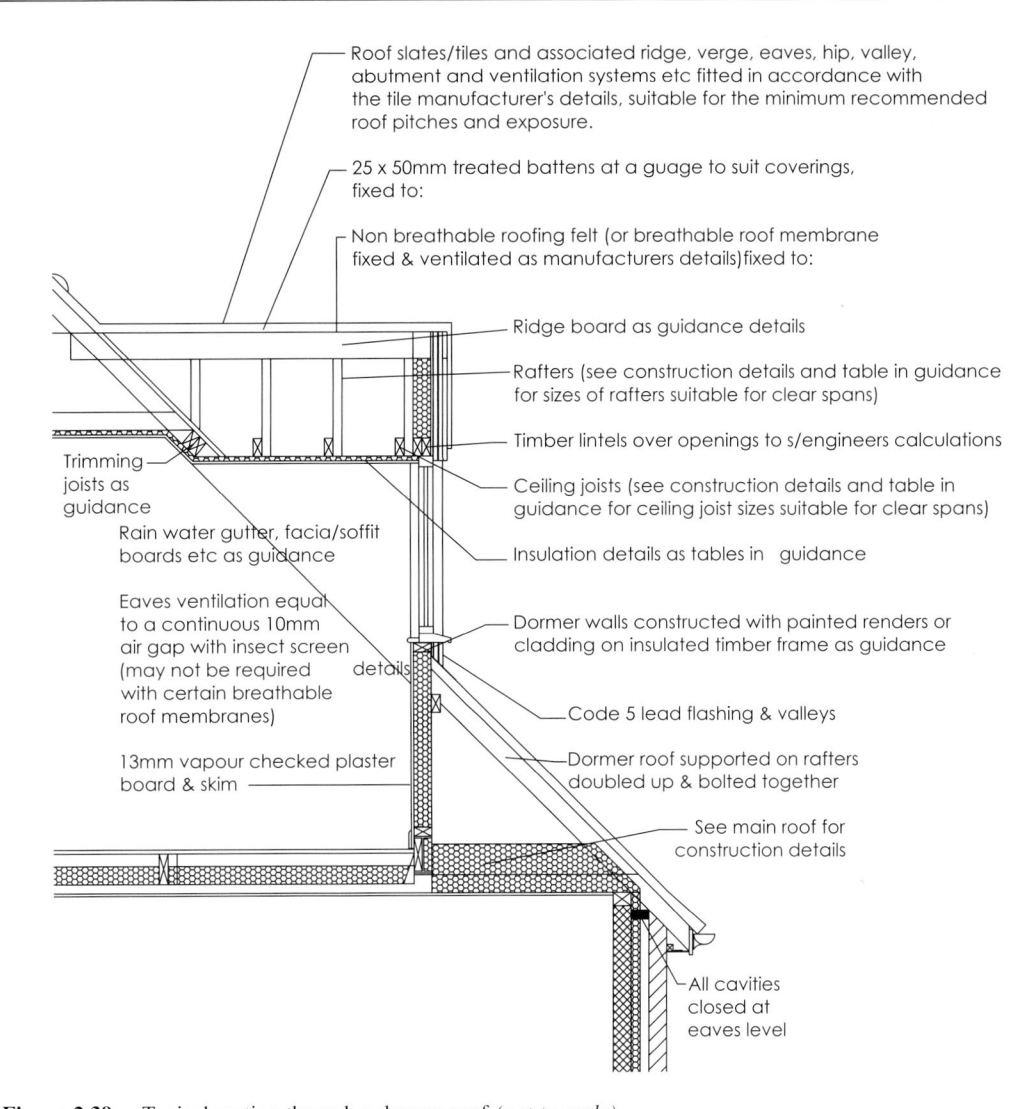

**Figure 2.39:** Typical section through a dormer roof *(not to scale)*

Agreement (BBA or other third-party accredited) certification, and are to be fixed in accordance with the manufacturer's details.

### Lofts hatches, doors and light wells to roof spaces

All hatches, doors and light wells in the roof space are to be insulated to the same standard as the roof, draft stripped and positively fixed.

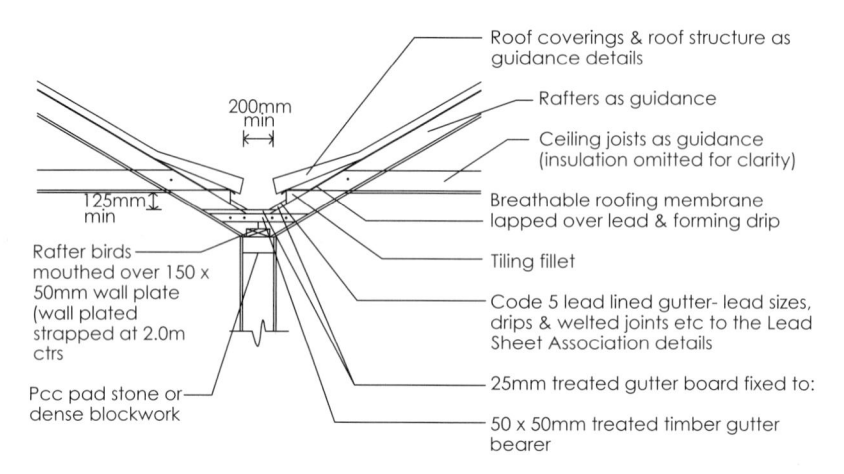

**Figure 2.40.1:** Typical roof valley detail *(not to scale)*

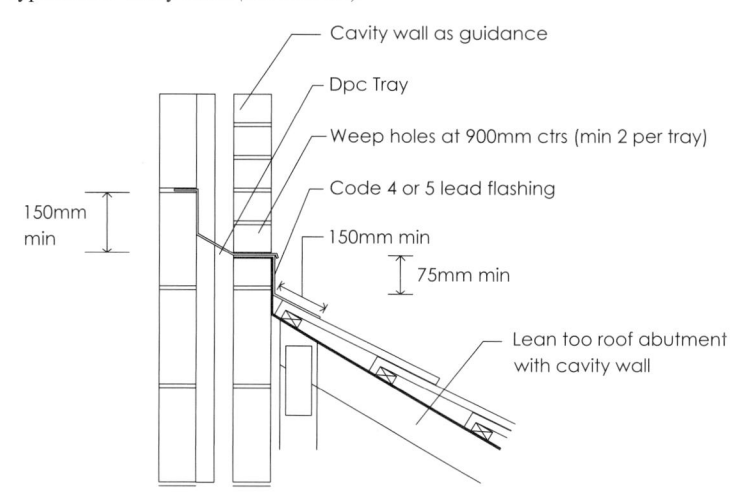

**Figure 2.40.2:** Typical lean-to roof abutment with cavity wall detail (section detail – *not to scale*)

## A7: Flat-roof construction

There are four options, as follows:

## Option 1: Flat roof with 'cold deck'

The insulation layer is placed at ceiling level with an air space between the top of the insulation and the underside of the deck, ventilated to external air on opposing sides; this option should be considered only for timber structural decks, as detailed in Figure 2.41. (Note: owing to the technical difficulties in achieving the required levels of insulation between roof timbers and the associated risks of condensation within the 'cold roof' and at thermal bridges, a flat roof with a 'warm deck' is the preferred option.)

Waterproof coverings are normally one of the following:
- three layers of high-performance felt (hot-bonded together with bitumen) to a current BBA Certificate in compliance with BS 8217;

- a single-layer system with a current BBA or WIMLAS Certificate;
- a glass-reinforced plastic (GRP) system with a current BBA or other approved accreditation;
- rolled lead sheet, fixed in compliance with the Lead Sheet Association's publication 'Rolled Lead Sheet – The Complete Manual', obtainable from: www.leadsheet.co.uk;
- mastic asphalt, fixed in accordance with the Mastic Asphalt Council's technical guides and specifications, obtainable from: www.masticasphaltcouncil.co.uk.

Proprietary flat roof systems should have a British Board of Agreement (BBA or other third-party accredited) certification, and are to be fixed in accordance with the manufacturer's details and approved by building control before works commence.

Waterproof covering is to be fixed by a flat-roofing specialist in accordance with the manufacturer's details, typically onto a timber deck as follows:

- 22 mm external-quality plywood decking or similar approved material, laid to 1:60/80 minimum gradient using firing strips at spacing to match joists, fixed onto:
- Timber flat-roof joists constructed of kiln-dried, structural-grade timber with sizes and spacing suitable for the proposed clear span, in compliance with independent span table below, or see TRADA Span Tables.
- Flat-roof covering (excluding lead and areas used for habitable use) is to have a surface finish of bitumen-bedded stone chippings covering the whole surface to a depth of 12.5 mm, to achieve a class AA (or B (t4) European class) fire-rated designation for surface spread of flame.
- Restrain flat roof to external walls by the provision of $30 \times 5 \times 1000$ mm lateral restraint straps at maximum 2000 mm centres, fixed to $100 \times 50$ mm wall plates and internal wall faces.
- Insulation is to be fixed between/under joists in compliance with Table 2.31 below and is to be continuous with the wall insulation. Fix 12.5 mm vapour-checked plasterboard (unless plasterboard is integral with the insulation) and 3 mm skim to underside of joists.
- Eaves ventilation: provide cross-ventilation to each and every void of the flat roof by installing eaves ventilation on opposing sides (fitted with insect grilles) equivalent to a continuous 25 mm gap up to 5 m span, or 30 mm gap for 5–10 m span.
- Ventilated air space over roof: provide an unrestricted ventilated air space between the top of the insulation and the underside of the complete roof deck at not less than 50 mm for up to 5 m spans, and not less than 60 mm for 5–10 m spans.

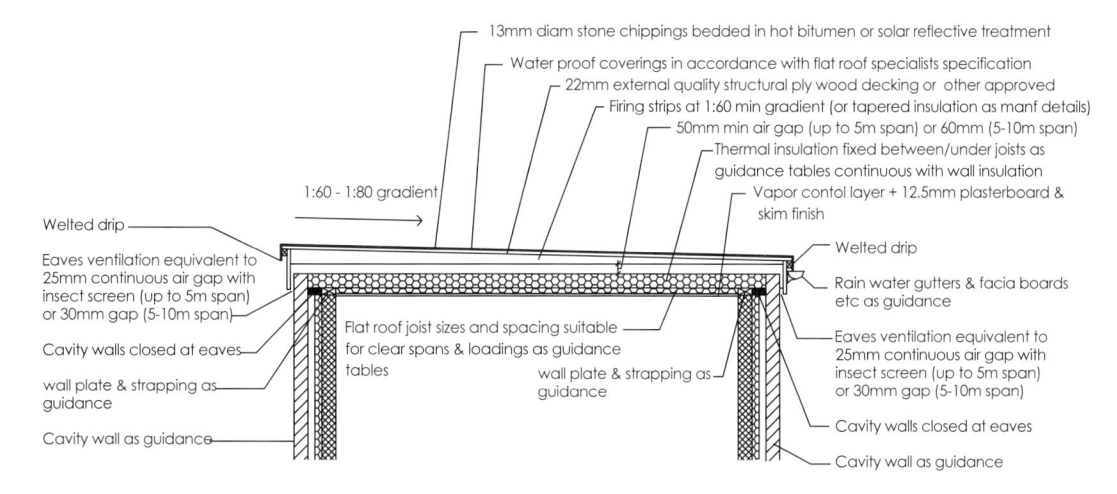

**Figure 2.41:** Typical section through a flat roof with 'cold deck' *(not to scale)*

**Table 2.31:** Examples of 'cold roof' insulation fixed between/under flat-roof joists
(Vented 'cold roof' achieving a U-value of $0.18\,W/m^2.k$)

| Product | K value | Position in roof |
|---|---|---|
| Kingspan Thermapitch TP10 | 0.022 | 185 mm (105 + 80 mm) friction-fixed between joists |
| Kingspan Thermapitch TP10 and Kingspan Kooltherm K18 Insulated Plasterboard | 0.022 0.021 | 120 mm friction-fixed between joists and 37.5 mm fixed under joists, with integral vapour-checked plasterboard and skim finish |
| Celotex XR4000 | 0.022 | 185 mm friction-fixed between joists **or** 125 mm friction-fixed between joists and 25 mm fixed under joists, with 12.5 mm vapour-checked plasterboard and 3 mm skim finish |
| Jablite Premium Board | 0.030 | 220 mm friction-fixed between joists **or** 150 mm between joists and 50 mm fixed under joists with 12.5 mm vapour-checked plasterboard and 3 mm skim finish |

**Note 1:** The joist depth must be sufficient to maintain a 50 mm air gap above the insulation and cross-ventilation to be provided on opposing sides by a proprietary ventilation strip, equivalent to a 25 mm continuous gap at eaves level with insect grille for ventilation of the roof space.
**Note 2:** All specifications assume 50 mm-wide joists at 400 mm centres with 12.5 mm vapour-checked plasterboard and 3 mm skim finish fixed to underside of joists. Vapour-control layer to be fixed in accordance with the flat roof specialist's/manufacturer's details.
**Note 3:** Insulation is to be installed in accordance with the manufacturer's details.
Source: a representative selection of values taken from *Technical Note 10, U-Values of Elements (Approved Document L1B 2010)*, produced by Hertfordshire Technical Forum for Building Control. Reproduced by permission of Hertfordshire Technical Forum for Building Control.

### Flat roofs with unlimited access/habitable use

Flat roofs with unlimited access are to have proprietary non-slip finishes/tiles in accordance with the manufacturer's details, and suitable protection from falling in accordance with guidance details and Approved Document K.

## Option 2: Flat roof with 'warm deck'

The insulation layer is placed above the roof deck, but below the weatherproof membrane and normally there should be no insulation below the deck unless it is British Board of Agreement (BBA or other third party) accredited. Ventilation of the roof is not required, as the insulation is fixed/bedded on an approved vapour-control layer. Warm roofs can be used above timber structural decks, as detailed in Figure 2.42 below, and are also suitable for concrete and metal structural decks, in accordance with a flat-roof specialist's design for the particular project.

Waterproof coverings normally one of the following:

- three layers of high-performance felt (hot-bonded together with bitumen) to a current BBA Certificate, in compliance with BS 8217;
- a single-layer system with a current BBA or WIMLAS Certificate;
- a glass-reinforced plastic (GRP) system with a current BBA or other approved accreditation;
- rolled lead sheet, fixed in compliance with the Lead Sheet Association's publication 'Rolled Lead Sheet – The Complete Manual', obtainable from: www.leadsheet.co.uk;
- mastic asphalt, fixed in accordance with the Mastic Asphalt Council's technical guides and specifications, obtainable from: www.masticasphaltcouncil.co.uk.

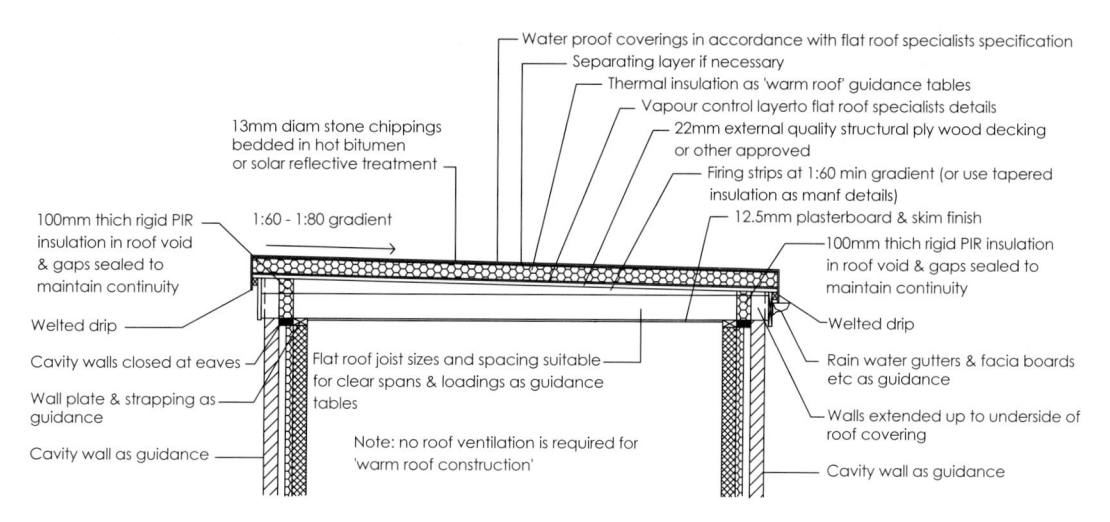

**Figure 2.42:** Typical section through a flat roof with 'warm deck' *(not to scale)*

**Table 2.32:** Examples of 'warm roof' insulation fixed above flat-roof joists
(Non-vented 'warm roof' achieving a U-value of 0.18 W/m².k)

| Product | K-value | Insulation fixed above deck |
|---|---|---|
| Kingspan Thermaroof TR26 LPC/FM (use with mechanically fixed single-ply membranes) | 0.022 | 110 mm fixed using telescopic tube fixings |
| Kingspan Thermaroof TR27 LPC/FM (use with bonded or mechanically fixed to substrate – finish with 3 layers of partially bonded built-up felt, mastic asphalt or single-ply membrane and approved liquid applied systems) | 0.024 | 120 mm |
| Celotex TD4000 (use with mechanically fixed single-ply membrane or 3 layers of built-up felt. Note: 12 mm additional ply layer required for single-ply membranes) | 0.022 | 126 mm |
| Knauf Polyfoam Roofboard Standard (use with single-ply membrane only) with timber deck | 0.035 | 180 mm |
| Knauf Rocksilk Krimpact Flat Roof Slab (use with bonded fixing over a plywood deck – finished with 3 layers of built-up felt, mastic asphalt or single-ply membrane) | 0.038 | 185 mm |
| Jablite Jabdec (use with bonded fixing over a plywood deck – finished with 3 layers of built-up felt, mastic asphalt or single-ply membrane | 0.035 | 183 mm (with mechanical fixings), 163 mm (without mechanical fixings) |

**Note 1:** Where composite deck insulation is to be used with a single-ply membrane, ensure the conditions of use of the membrane are met. It may be necessary to use an additional layer of 12 mm external-quality structural plywood above the insulation to meet the conditions of use.
**Note 2:** All specifications assume 50 mm-wide joists at 400 mm centres with 12.5 mm vapour-checked plasterboard and 3 mm skim finish fixed to underside of joists. Vapour-control layer to be fixed in accordance with the flat roof specialist's/manufacturer's details.
**Note 3:** Insulation is to be installed in accordance with the manufacturer's details.
Source: a representative selection of values taken from *Technical Note 10, U-Values of Elements (Approved Document L1B 2010)* produced by Hertfordshire Technical Forum for Building Control. Reproduced by permission of Hertfordshire Technical Forum for Building Control.

Proprietary flat roof systems should have a British Board of Agreement (BBA or other third party accredited) Certification, and are to be fixed in accordance with the manufacturer's details and approved by building control before works commence.

Waterproof covering is to be fixed by a flat-roofing specialist in accordance with the manufacturer's details, typically onto a timber deck as follows:

- Waterproof coverings are to be fixed onto a separating layer where necessary and onto:
- a roof insulation layer, in compliance with Table 2.32 below, fixed/bonded to an approved vapour-control layer (fully supported by the roof deck, in accordance with the manufacturer's details) onto:
- 22 mm external-quality plywood decking or similar approved material, laid to 1:60/80 minimum gradient using firing strips at spacing to match joists, fixed onto:
- timber flat-roof joists constructed of kiln-dried, structural-grade timber with sizes and spacing suitable for the proposed clear span, as annotated on the drawing or in compliance with the independent span table below – or see TRADA span tables.
- Flat-roof covering (excluding lead and areas accessible for habitable use) is to have a surface finish of bitumen-bedded stone chippings covering the whole surface to a depth of 12.5 mm, to achieve a class AA (or B (t4) European class) fire-rated designation for surface spread of flame.
- Restrain flat roof to external walls by the provision of $30 \times 5 \times 1000$ mm lateral restraint straps at maximum 2000 mm centres fixed to $100 \times 50$ mm wall plates and internal wall faces.
- Fix 12.5 mm vapour-checked plasterboard and 3 mm skim to underside of joists.

### Flat roofs with unlimited access/habitable use

Flat roofs with unlimited access are to have proprietary non-slip finishes/tiles in accordance with the manufacturer's details, and suitable protection from falling in accordance with guidance details and Approved Document K.

## Option 3: Flat roof with inverted 'warm deck' (insulation on top of waterproof coverings)

This would be typically constructed as a ballast layer, over a filter layer, over an insulation layer, over waterproof coverings/vapour-control layer and structural deck, designed and constructed by a flat-roofing specialist. Not covered by this guidance.

## Option 4: Flat roof with green roof on 'warm deck' (either intensive or extensive)

### *Intensive green roof*

Typically constructed with vegetation in 1.0 m-deep soil layer, over a filter layer, over a drainage/reservoir layer, over a protective layer, over root barrier, over waterproof coverings/ vapour-control layer and structural deck, ALL designed and constructed by a flat-roofing specialist. Not covered by this guidance.

### Extensive green roof

Typically constructed with a sedum blanket, over a filter layer, over a root barrier, over water proof coverings/a vapour-control layer and structural deck, all designed and constructed by a flat-roofing specialist. Not covered by this guidance.

## The design, workmanship and selection of materials for flat roofs

The design, workmanship and selection of materials should comply with Model Specification Sheet P.L.1 Built-Up Roofing: Plywood Deck, published by The British Flat Roofing Council. Metallic roof trims are to be of non-corrodible material, resistant to sunlight and not fixed through the water-proof covering. All timber is to be treated using CCA vacuum/pressure or O/S double vacuum to BS 5268:5, including all cut ends of timber etc. within 300 mm of any joint.

All flat-roofing works are to be carried out by a specialist flat-roofing contractor and all materials etc. are to be fitted in compliance with the manufacturer's details. Work should not be carried out during wet weather or when the deck has not fully dried out. A vapour-control barrier is required on the underside of the roof below the insulation level (typically 500 g polythene or foil-backed plasterboard). Fix 12.5 mm plasterboard (joints staggered) and 5 mm skim coat of finishing plaster to the underside of all ceilings, using galvanised plasterboard nails.

Proprietary flat-roof systems should have a British Board of Agreement (BBA or other third-party accredited) certification, and are to be fixed in accordance with the manufacturer's details and approved by building control before works commence. The moisture content of timber should not exceed 20 per cent and it is to be kiln-dried and of grade C24. Workmanship is to comply with BS 8000:4. All fixings are to be proprietary stainless steel or galvanised steel.

### Cavity closers

Proprietary British Board of Agreement (BBA or other third-party accredited) acoustic/thermally insulated/fire-resistant cavity closers, or similar, are to be provided to all cavity openings/closings, tops of walls and junctions, with other properties in accordance with the manufacturer's details.

Tops of cavity walls can be closed to prevent the passage of fire using a proprietary British Board of Agreement (BBA or other third-party accredited) 30 minutes' fire-resistant proprietary closer or rigid board, fixed in accordance with the manufacturer's details.

## A8: Mortars, renders and gypsum plasters

## Cement mortars and renders

There are various types of cement for particular uses, Ordinary Portland cement (OPC) is the most widely used for general building works, including binder for mortars, screeds, concrete and renders for use in modern buildings. Cements are factory-made by calcining natural clay, silica, alumina and iron ore with limestone. Cement is hydraulic and hardens by chemical reaction with water; no air is needed in the process.

Cement has a strong, rapid set and the strength is related to the water: cement ratio.

**Table 2.33:** Timber sizes and permissible clear spans for joists for flat roofs – with access and use for maintenance and repairs only (strength class C24)

| Size of joist (mm) | Spacing of joist (mm) | | |
|---|---|---|---|
| | 400 | 450 | 600 |
| Breadth × Depth | Maximum clear span (m) | | |
| 47 × 97 | 1.97 | 1.93 | 1.83 |
| 47 × 120 | 2.57 | 2.51 | 2.37 |
| 47 × 145 | 3.22 | 3.15 | 2.95 |
| 47 × 170 | 3.88 | 3.78 | 3.45 |
| 47 × 195 | 4.51 | 4.34 | 3.95 |

Minimum flat-roof joist bearing 40 mm.
Imposed load: $1.02\,kN/m^2$ (high snow load – altitudes not exceeding 100 m).
Dead load: not more than $0.75\,kN/m^2$ (concentrated load 0.9 kN), excluding self-weight of joists.
The above values have been independently compiled for the guidance table by Geomex Ltd Structural Engineers: www.geomex.co.uk. Span tables for C16- and C24-strength class solid timber members in floors, ceilings and roofs for dwellings are available from TRADA Technology at: www.trada.co.uk/bookshop.

**Table 2.34:** Timber sizes and permissible clear spans for joists for flat roofs – with unlimited access (strength class C24)

| Size of joist (mm) | Spacing of joist (mm) | | |
|---|---|---|---|
| | 400 | 450 | 600 |
| Breadth × Depth | Maximum clear span (m) | | |
| 75 × 120 | 2.50 | 2.46 | 2.37 |
| 75 × 145 | 3.20 | 3.15 | 3.02 |
| 75 × 170 | 3.91 | 3.85 | 3.65 |
| 75 × 195 | 4.63 | 4.55 | 4.17 |

Minimum flat roof joist bearing 40 mm.
Imposed load: $1.5\,kN/m^2$ (high snow load – altitudes not exceeding 100 m).
Dead load: not more than $0.75\,kN/m^2$ (concentrated load 2 kN), excluding self-weight of joists.
The above values have been independently compiled for guidance table by Geomex Ltd Structural Engineers: www.geomex.co.uk. Span tables for C16 and C24 strength class solid timber members in floors, ceilings and roofs for dwellings are available from TRADA Technology at: www.trada.co.uk/bookshop.

**Table 2.35:** Cement mortar – general mix ratio

| Wall construction | Cement: sand with plasticiser (mix ratio by volume) | |
|---|---|---|
| | Sheltered | Exposed |
| Hard durable materials[1 and 2] | 1:5–6 | 1:3–4 |
| Moderately hard materials[1 and 2] | 1:7–8 | 1:5–6 |
| Soft weak materials | See notes 1, 2 and 3 | See notes 1, 2 and 3 |

**Note 1:** If sulfates are present in the ground, use sulfate-resisting cement.
**Note 2:** Retaining walls and 3-storey mortar mix are to be designed by a suitably qualified person i.e. a structural engineer.
**Note 3:** Mortar mix is to be specified by a suitably qualified and experienced specialist – it must be suitable for the type of wall material and degree of exposure.

Table 2.36: Cement render – general mix ratio

| Wall construction | Exposure | Cement: sand with plasticiser (mix ratio by volume) | |
|---|---|---|---|
| | | Undercoat | Top coat with smooth float finish |
| Hard, durable materials | Severe | 1:3–4 | 1:5–6 |
| | Sheltered/moderate | 1:3–4 | 1:5–6 |
| Moderately hard materials | Severe | 1:5–6 | 1:5–6 |
| | Sheltered/moderate | 1:5–6 | 1:7–8 |
| Soft, weak materials | Severe | See note 1 | See Note |
| | Sheltered/moderate | See note 1 | See Note |

Note: The mortar mix is to be specified by a suitably qualified and experienced specialist; it must be suitable for the type of backing wall material and degree of exposure, in accordance with the manufacturer's details.

Table 2.37: Thickness of render coats

| Application | Type of coat | Thickness | Notes |
|---|---|---|---|
| **External render** | Dubbing out | 16 mm max | |
| | Undercoat | 10 mm | Comb-scratched for key |
| | Top coat | 10 mm | Smooth float finish |

Note: Thickness of render coats is to be specified by a suitably qualified and experienced specialist, in accordance with the manufacturer's details.

Table 2.38: Gypsum plaster for internal finishes – general mix ratio

| Background[1] | Gypsum plaster[1] | |
|---|---|---|
| | Undercoat[1] | Final coat (smooth finish)[1] |
| New masonry walls | Bonding coat – 11 mm thick | Finish coat – 3 mm thick |
| Existing masonry walls | Renovation plaster bonding coat – 11 mm thick | Renovating plaster finish coat – 3 mm thick |
| New plaster board | New plasterboard | 3 mm-thick board finish plaster |

Note: [1]Preparation of walls, product suitability, application, and protection of gypsum plaster are to be specified by a suitably qualified and experienced specialist, in accordance with the manufacturer's details. They must be suitable for the type of backing wall material.

## Gypsum plasters

Gypsum is a naturally occurring mineral found in limestone strata and is factory-manufactured into a commercial plaster. Gypsum plaster sets and hardens by chemical reaction with water and no air is needed in the process. It has a strong, rapid set suitable for modern buildings, but has poor permeability and flexibility and is unsuitable for older 'breathing' buildings and damp conditions.

## PART B: FIRE SAFETY AND MEANS OF ESCAPE

Please refer fully to the relevant Approved Document.

## Fire detection and fire alarm systems

Self-contained, mains-operated smoke alarms, with battery backup to BS 5446, and fire alarm systems are to be installed in accordance with BS 5839 to the following standards:

### New houses and extensions

Grade D Category to an LD3 standard. Self-contained, mains-operated smoke and heat alarms with battery backup to BS 5446 are to be installed in accordance with the relevant recommendations of BS 5839 Part 6: 2004, as follows:

- in all circulation areas at each storey level that forms part of the escape route from the dwelling
- within 7.5 m of all doors to habitable rooms
- sited at ceiling level (or other as stated in BS 5839 for other locations), at least 300 mm from walls and light fittings
- heat alarms are to be installed in kitchens if the kitchen is open to the stairway or circulation area (in two-storey dwellings)
- all interconnected together
- smoke alarms should not be fixed in positions or by appliances etc. that are likely to give false alarms.

### Large houses

Requirements are as follows if houses have more than one storey and any of those storeys exceeds $200\,m^2$:

#### *Large two-storey house (excluding basement)*

Grade B Category to an LD3 standard.

- Fire detection and alarm system comprising fire detectors (other than smoke/heat alarms), fire alarm sounders and control and indicating equipment to either BS EN 54-2 (and power supply to BS EN 54-4), or to Annex C of BS 5839: Pt.6.
- Detection system to be installed in all circulation areas at each storey level that forms part of the escape route from the dwelling.

#### *Large three-storey house (excluding basements)*

Grade A Category to an LD2 standard (smoke/heat alarms sited in accordance with BS 5839-1 for a Category L2 system).

- Fire detection and alarm system incorporating control and indicating equipment to BS EN 54-2, and power supply to BS EN 54-4, installed to BS 5839: Pt.1 with some minor exceptions.
- Detection system to be installed in all circulation spaces that form part of the escape routes from the dwelling, and in all rooms or areas that present a high risk of fire.

## Means of escape

### 1. Means of escape from the ground storey in dwellings

Except for kitchens, all habitable rooms in the ground storey should either:

(i) open directly onto a hall leading to the entrance or other suitable exit; or
(ii) be provided with a means of escape window (or door) in compliance with Figures 2.43 and 2.44 and the guidance details below.

### 2. Means of escape from a two-storey dwelling with a floor not more than 4.5 m above ground level

Habitable rooms in the first-floor storey served by only one set of stairs should either:

(i) be provided with means-of-escape windows (or doors) in compliance with the guidance diagrams and details below; or
(ii) be provided with direct access to protected stairs, as described in three-storey buildings guidance below.

### Option 1

Habitable rooms in ground storey open directly onto a hall leading to the entrance or other suitable exit, or;

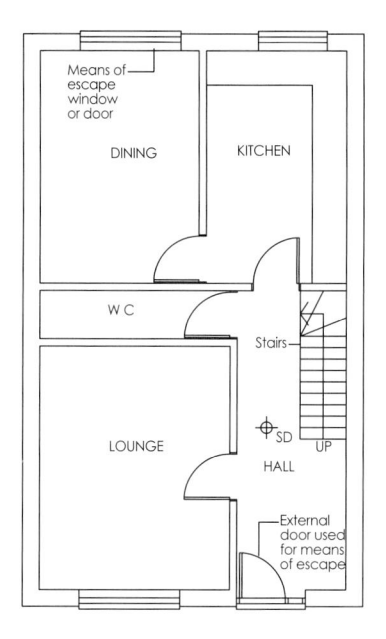

### Option 2

Inner habitable rooms (rooms which pass through another room) to be provided with a means of escape window or door

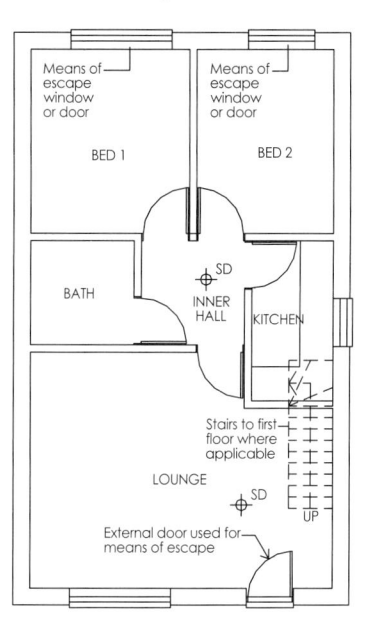

**KEY TO ITEMS INDICATED ON LAYOUT** (to be read in conjunction with guidance details)
SD  Interconnected mains operated smoke alarm with battery back up fitted at ceiling level. Note: Where the kitchen is not separated from the stairway or circulation space by a door, there should be a compatible Interconnected mains operated smoke alarm with battery back up at ceiling level in the kitchen, in addition to the smoke alarm(s).

**Figure 2.43:** Means of escape from the ground storey (typical plan layouts – *not to scale*)

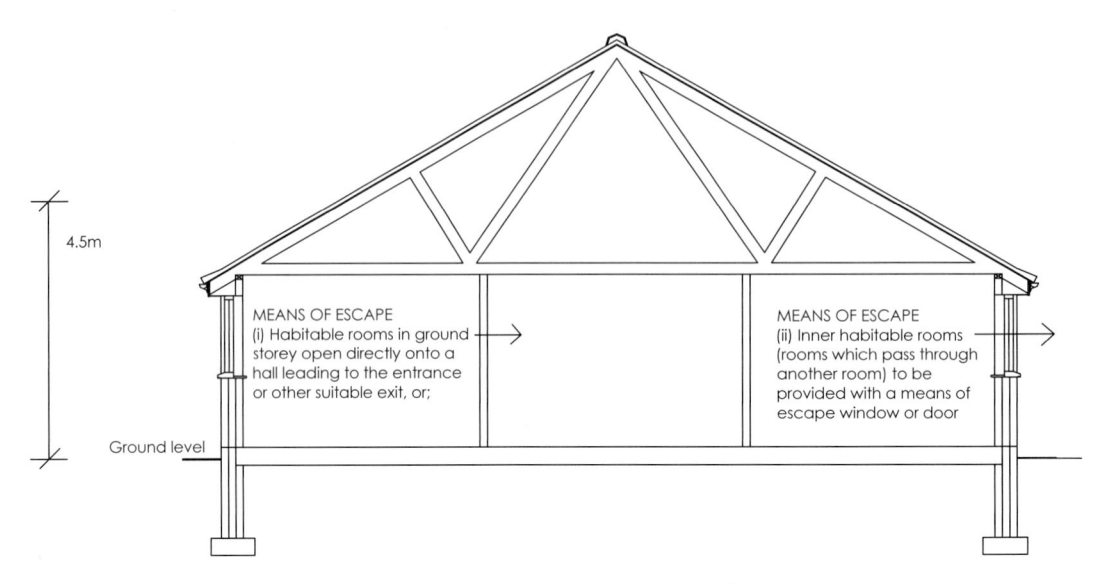

## 2.80 Domestic extensions

**Figure 2.44:** Means of escape from the ground storey (typical section detail – *not to scale*)

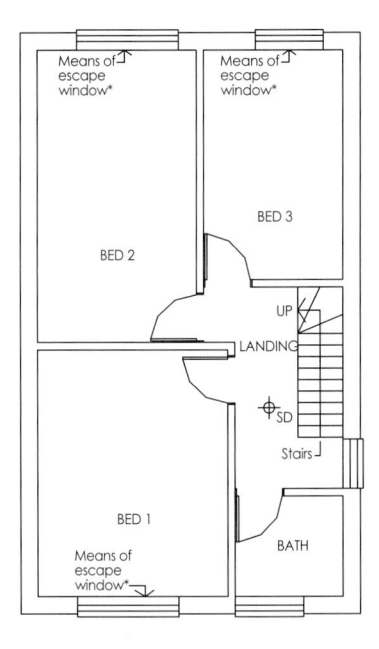

Labels on figure:
- 4.5m
- Ground level
- MEANS OF ESCAPE (i) Habitable rooms in ground storey open directly onto a hall leading to the entrance or other suitable exit, or;
- MEANS OF ESCAPE (ii) Inner habitable rooms (rooms which pass through another room) to be provided with a means of escape window or door

**KEY TO ITEMS INDICATED ON LAYOUT** (to be read in conjunction with guidance details)

SD    Interconnected mains operated smoke alarm with battery back up fitted at ceiling level.

\*    A single means of escape window is acceptable to serve two habitable rooms providing a communicating door is provided between rooms and both rooms have their own access to the stairs.

Notes: (i) 30 minutes fire resistance is required to underside of first floor.

**Figure 2.45:** Means of escape from the first floor (typical plan layout – *not to scale*)

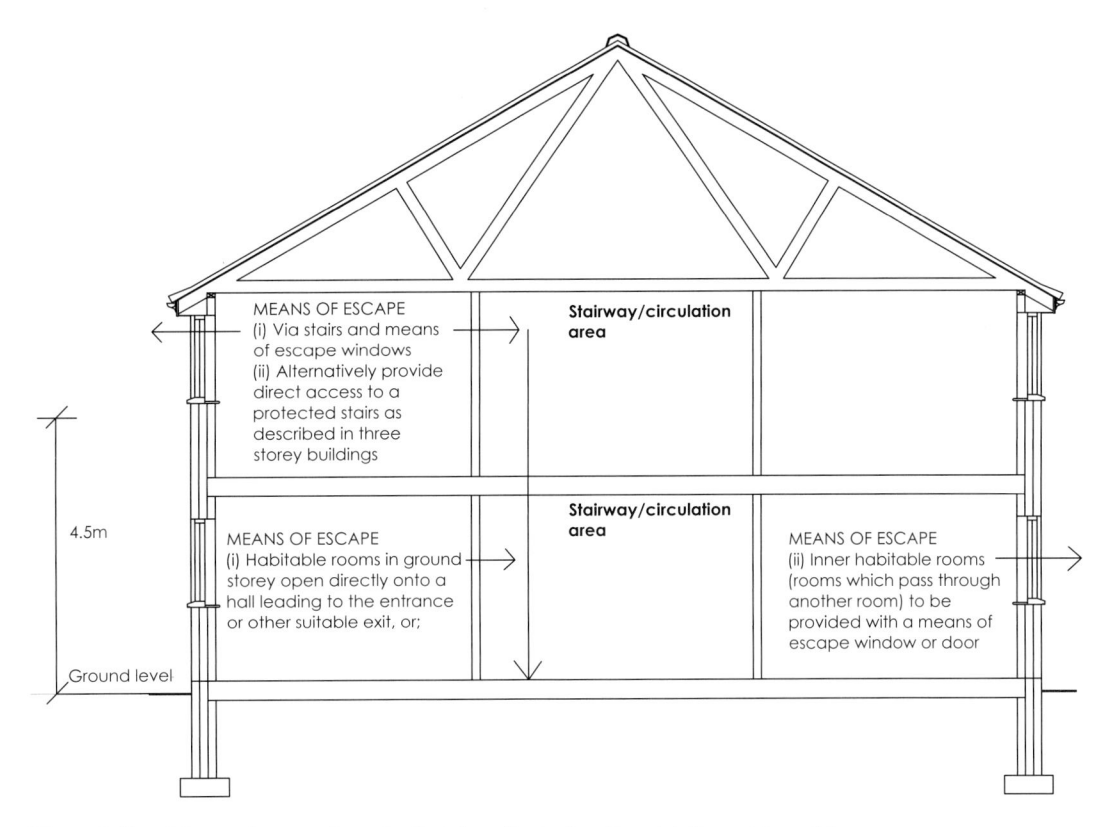

**Figure 2.46:** Means of escape from the first floor (typical section detail – *not to scale*)

## 3. Means of escape from three-storey dwellings with one upper floor more than 4.5 m above ground level

The dwelling house may either have protected stairs as described in Option 1 below, or the top floor can be separated and given its own alternative escape route as described in Option 2 below. Alternatively, where the fire safety requirements of the building regulations cannot be met, a sprinkler system may be allowed against the requirements of Approved Document B, as described in the guidance below. **Important note:** The options below need not be followed if the dwelling house has more than one internal stairway, which afford an effective alternative means of escape in two directions and are physically separated from each other (to be approved by building control).

### Option 1: Protected stairway

The stairs, landings and hallway from the top storey down to the ground floor must be protected and enclosed in 30-minute fire-resisting construction and the protected stairs must discharge directly to an external door, as described in guidance diagram (i) below; or the protected stairs can give access to at least two escape routes and final exits separated by 30 minutes' fire-resisting construction at ground level, as described in guidance diagram (ii) below.

## 2.82 Domestic extensions

# Diagram (i) Protected stairs extending to a final exit

**TYPICAL GND FLOOR LAYOUT**

- Means of escape window
- DINING
- KITCHEN
- WC
- Stairs
- FD20
- All partitions enclosing the stairs is to have 30 minutes fire resistance including glazing*
- LOUNGE
- HALL
- FD20
- SD
- UP
- External door used for means of escape

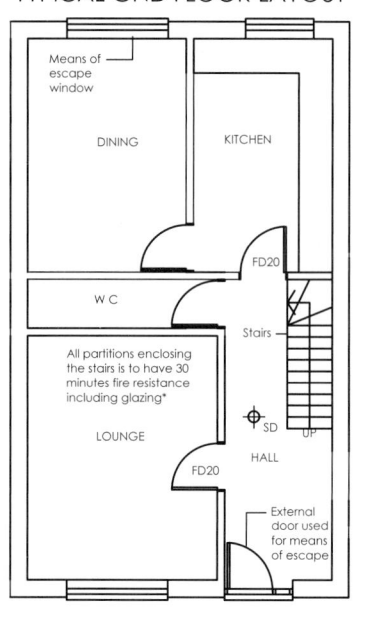

**TYPICAL 1ST FLOOR LAYOUT**

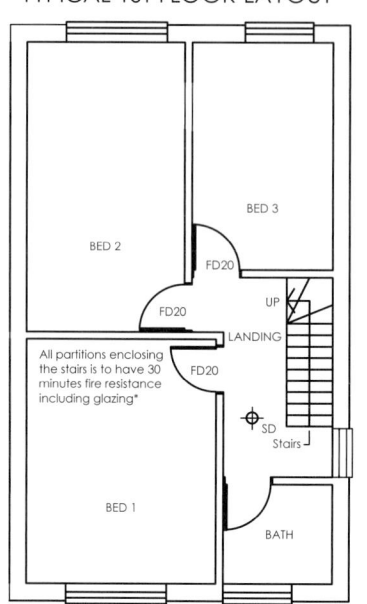

- BED 2
- BED 3
- FD20
- FD20
- LANDING
- All partitions enclosing the stairs is to have 30 minutes fire resistance including glazing*
- FD20
- SD
- Stairs
- UP
- BED 1
- BATH

**TYPICAL 2ND FLOOR LAYOUT**

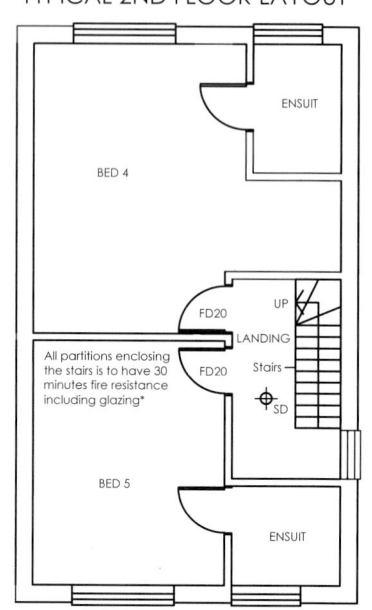

- ENSUIT
- BED 4
- FD20
- UP
- LANDING
- All partitions enclosing the stairs is to have 30 minutes fire resistance including glazing*
- FD20
- Stairs
- SD
- BED 5
- ENSUIT

# Diagram (ii) Alternative access to two escape routes at ground level

**TYPICAL GND FLOOR LAYOUT**

- External door used for means of escape
- HD
- KITCHEN/DINING ROOM
- Underside of stairs outside of protected stairs to have 30 minutes fire resistance
- cupb
- FD20
- All partitions enclosing the stairs is to have 30 minutes fire resistance including glazing*
- SD
- UP
- LOUNGE
- FD20
- SD
- External door used for means of escape

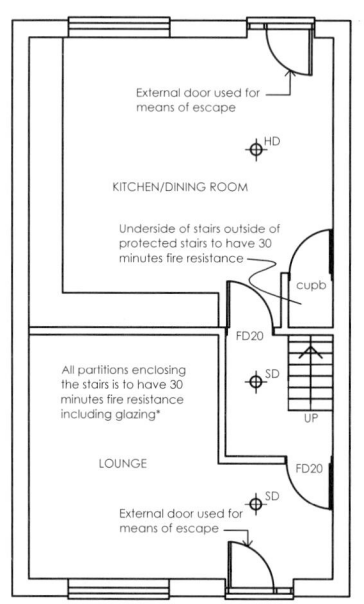

**KEY TO ITEMS INDICATED ON LAYOUTS** (to be read in conjunction with guidance details)

FD20  20 minute fire resisting door and frame fitted with intumescent strips and 3 fire resistant hinges excludes toilet/ensuit and bathroom doors- providing the partitions between the stairway and habitable room has 30 minutes fire resistance (both sides).

SD  Interconnected mains operated smoke alarm with battery back up fitted at ceiling level

HD  Interconnected mains operated heat alarm with battery back up fitted at ceiling level

*  Fire resistant glazing to be 1.1m minimum above floor level in walls - (unlimited in doors - 100mm minimum above floor level in basement doors, no glazing is permitted in a fire door between an attached garage and the dwelling). These glazing limitations do not apply to glazed elements which satisfy the relevant fire insulation criterion in Table A1 of ADB1

Notes:
(i) 30 minutes fire resistance is required to underside of upper floors.
(ii) Cupboards within the protected stairway to be fitted with FD 20 fire doors as detailed above.
(iii) If there is a doorway between the dwelling and an attached garage it should be fitted with a self closing FD 30s fire door with intumescent strips, smoke seals and 100mm high fire resistant threshold as guidance (no glazing is permitted other than fire insulated glazing)

## Option 2: Fire-separated third storey with alternative external/internal fire exit

The top, third storey should be separated from the lower storeys by 30-minute fire-resisting construction and provided with an alternative escape route leading to its own final exit, as described in the diagram below. Fire resistance of areas adjacent to external stairs is detailed in the guidance below.

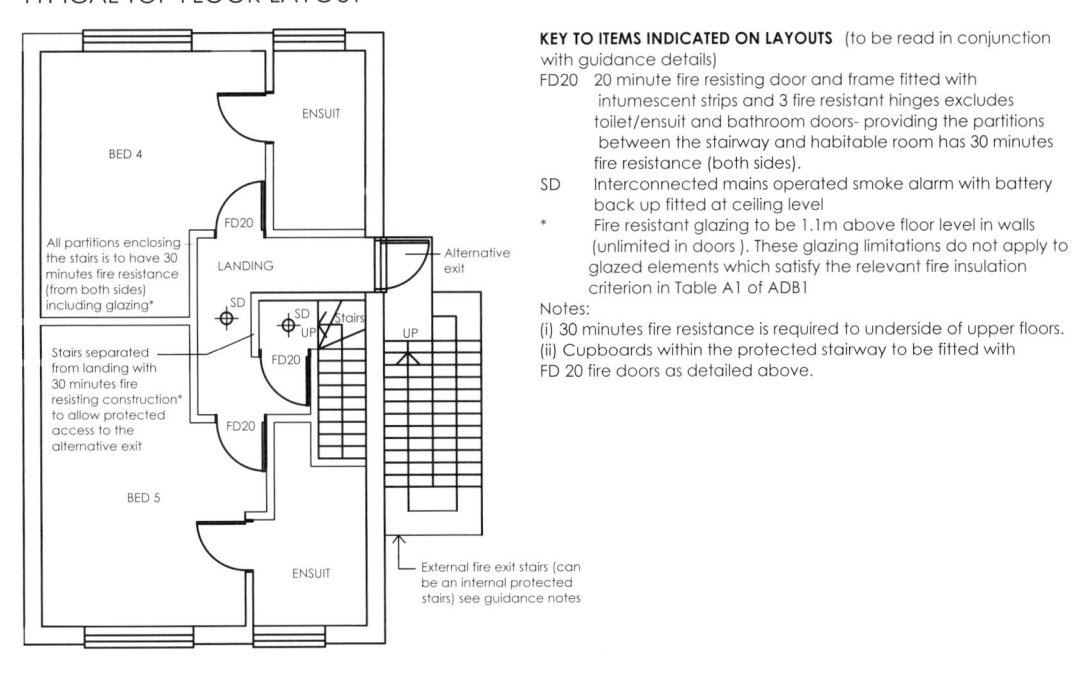

TYPICAL TOP FLOOR LAYOUT

**KEY TO ITEMS INDICATED ON LAYOUTS** (to be read in conjunction with guidance details)
FD20   20 minute fire resisting door and frame fitted with intumescent strips and 3 fire resistant hinges excludes toilet/ensuit and bathroom doors- providing the partitions between the stairway and habitable room has 30 minutes fire resistance (both sides).
SD   Interconnected mains operated smoke alarm with battery back up fitted at ceiling level
\*   Fire resistant glazing to be 1.1m above floor level in walls (unlimited in doors ). These glazing limitations do not apply to glazed elements which satisfy the relevant fire insulation criterion in Table A1 of ADB1

Notes:
(i) 30 minutes fire resistance is required to underside of upper floors.
(ii) Cupboards within the protected stairway to be fitted with FD 20 fire doors as detailed above.

## General provisions for means of escape

### Fire doors to protected stairway enclosures

Fire doors to protected stairway enclosures are to be FD 20 fire-resisting doors having 20 minutes' fire resistance to BS 476-22:1987, fitted with intumescent strips rebated around sides and top of door or frame; this excludes toilets/bathrooms/en-suite, providing the partitions protecting the stairs have 30 minutes' fire resistance from both sides. (Note: a self-closing FD30 fire door is required between the dwelling and the garage, in accordance with the guidance details below.)

Existing or new solid hardwood/timber doors may achieve 20 minutes' fire resistance, or the doors may be suitable for upgrading to achieve 20 minutes' fire resistance (as agreed with building control) with a proprietary intumescent paint/paper system in accordance with the manufacturer's details. More details are available from: www.fireproof.co.uk, who can supply (and apply where required) an intumescent paint/paper system, which must be applied in accordance with the manufacturer's details. A copy of the manufacturer's certificate of purchase/ application must be provided for building control on completion. Upgraded doors/frames are to be fitted with intumescent strips as detailed in the guidance above.

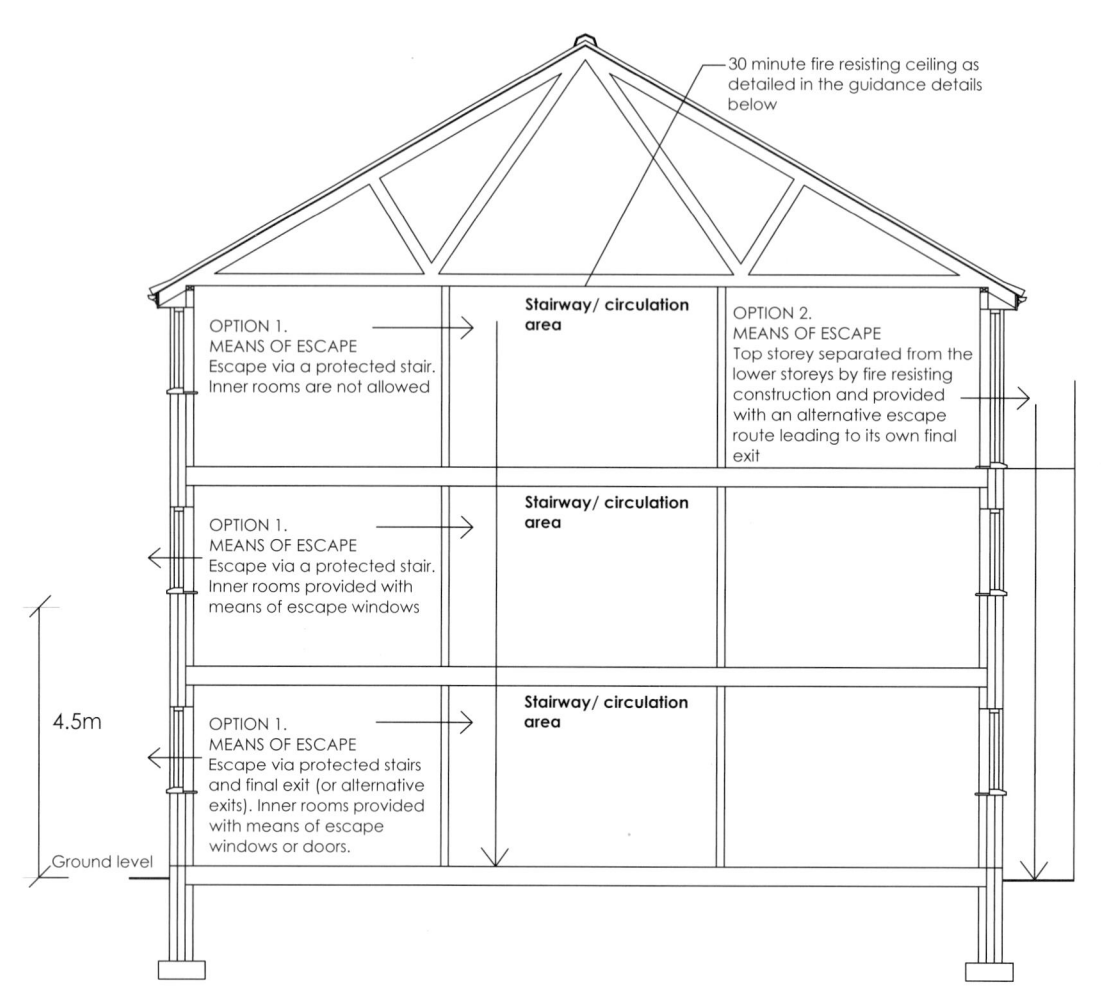

**Figure 2.47:** Means of escape from three-storey dwellings with one upper floor more than 4.5 m above ground level

### *Protected stairway enclosures*

These are to have 30 minutes' fire-resisting construction from both sides, constructed in accordance with partition wall details given in Part A of this guidance.

### *Limitations on the use of uninsulated glazed elements*

Limitations on the use of uninsulated glazed elements on escape routes are to be in compliance with Table A4 of Approved Document B: Volume 1 – Dwelling Houses ('uninsulated' refers to the fire insulation value of the glazing (normally 30 minutes) and not the thermal insulation value).

### *Fire resistance to upper floors and elements of structure*

Upper floors are to have 30 minutes' fire-resisting construction from the underside, constructed in accordance with upper-floor details in Part A of this guidance. Other elements of the structure are to have fire resistance in compliance with Table 2.40 below.

### Means-of-escape windows and external doors

Means-of-escape windows are to be fitted with proprietary hinges to open the window to the minimum required clear width of 450 mm. Escape windows must have minimum clear opening casement dimensions of $0.33\,m^2$ (typically 450 mm wide × 750 mm high), with the opening located between 800 and 1100 mm above floor level to all bedrooms and habitable rooms at first-floor level, and inner habitable rooms on the ground-floor level. (Roof window openings may be acceptable 600 mm above floor level, subject to approval by building control.)

The means-of-escape window or door should lead to a place of safety away from the fire. A courtyard or back garden from which there is no exit other than through other buildings would have to be at least as deep as the dwelling house is high to be acceptable, as detailed in Para 2.8 (b) and Diagram 4 of ADB1.

Openings (windows/doors etc.) above the ground-floor storey and within 800 mm of floor level are to be provided with containment/ guarding/proprietary catches that should be removable (but childproof) in the event of a fire. Where an escape window cannot be achieved, direct access to protected stairs (or a protected route to inner rooms) is acceptable, in compliance with the guidance details above, for three-storey buildings and ADB1 para 2.6 (a) or (b).

Windows should be designed to remain in the open position while an escape is made

Locks (with or without removable keys) and stays may be fitted to escape windows, subject to the stay being fitted with a release catch that is child-resistant.

### Galleries

A gallery floor providing a raised area or platform around the sides or at the back of a room (to provide extra space) should be provided with:

- an alternative exit, or
- where the gallery floor is not more than 4.5 m above ground level, a means-of-escape window in accordance with the guidance details above.

Where a gallery is not provided with an alternative exit or means-of-escape window, it should comply with the following requirements (see Diagram 5 of ADB1 for full details):

- the gallery should overlook at least 50 per cent of the room below;
- the distance between the foot of the access stairs to the gallery and the door to the room containing the gallery should not exceed 3 m;
- the distance from the head of the access stairs to any point on the gallery should not exceed 7.5 m; and
- any cooking facilities within a room containing a gallery should either:
  (i)  be enclosed in fire-resisting construction; or
  (ii) be remote from the stair to the gallery and positioned such that they do not prejudice an escape from the gallery.

### Basements

If the basement storey is served by a single stairway and contains a habitable room, the basement should be fitted either with a means-of-escape window or door, in compliance with this guidance, or protected stairs leading from the basement to the final exit, in compliance with guidance details for three-storey buildings. Fire-resistant glazing in protected routes is to be in compliance with Table A4 of ADB1.

### *Passenger lifts*

Lifts installed in dwellings with a floor more than 4.5 m above ground level should either be located within the enclosures to the protected stairway or contained in a 30-minute fire-resisting lift shaft, in accordance with the lift manufacturer's details.

### *Replacement windows (excludes repairs)*

The replacement window opening should be sized to provide at least the same potential for means of escape as the window it replaces, or where the original window is larger than necessary for purposes of means of escape, the window opening could be reduced down to the minimum specified in this guidance for means-of-escape windows.

Cavity barriers should be provided around windows where necessary and the window should also comply with the requirements of Parts L and N in this guidance.

### *Fire separation between an integral garage and dwelling*

The wall and any floor between an integral garage and the dwelling house are to be constructed as a compartment wall/floor to give 30 minutes' fire resistance from both sides of the wall, and taken up to the ceiling/roof level and fire-stopped with mineral wool. Any door between the house and garage is to be fitted with: an FD30s fire door, in compliance with BS476-22:1987; proprietary mechanical self-closers; intumescent strips; and smoke seals. The garage floor should be laid to falls to allow fuel spills to flow away from the fire door to the outside; alternatively, the door opening should be positioned at least 100 mm above the garage floor level, as detailed in the guidance diagram below. Fire-resistant glazing is to be in compliance with Table A4 of ADB1

### *Protection of openings, fire stopping and cavity barriers*

uPVC pipes passing through compartment walls/floors should not exceed an internal diameter of 110 mm and should either be fitted a proprietary intumescent collar at the wall/floor junction, or enclosed throughout the pipe length with 30 minutes' fire-resisting construction (typically soft wood framing fixed around pipework, packed with acoustic quilt with 12.5 mm plasterboard and skim finish, or two layers of plasterboard with staggered joists). Pipes with a diameter of 40 mm or less do not require fire protection in accordance with the guidance above. Other pipe materials and pipe sizes are to be in compliance with Table 3 of ADB1.

A 30-minute fire-resisting ceiling should be provided between a protected stairway and roof void in a dwelling house with a floor more than 4.5 m above ground level. Alternatively, if the stairway extends through the roof void up to the roof level, a 30-minute fire-resisting cavity barrier or wall should separate the stairway from the roof space.

### *Fire resistance of areas adjacent to external fire exit stairs*

The external stairs must not be within 1.8 m of any unprotected opening at the side of the stairs, and no openings are permitted below the stairs – unless the opening is fitted with 30-minute fire-resisting

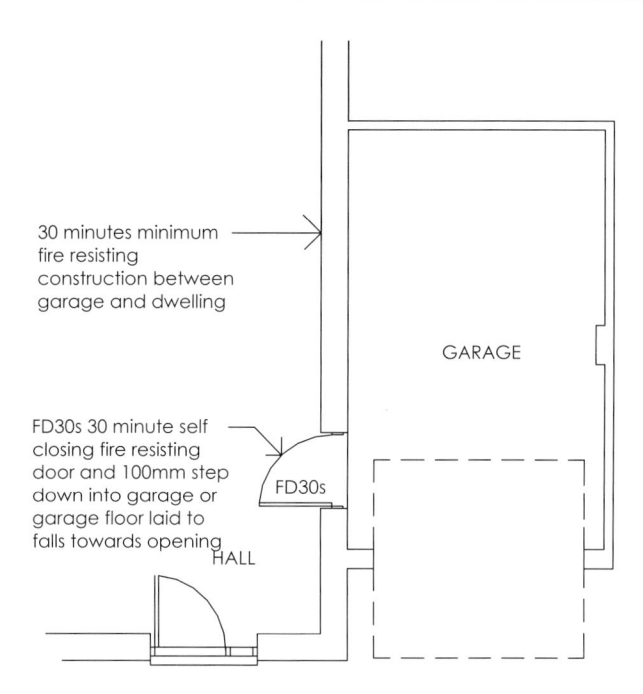

**Figure 2.48:** Fire separation between an integral garage and dwelling (plan detail – *not to scale*) (See Diagram 10 of ADB1 for full details.)

glass and a proprietary bead system and is permanently sealed shut (subject to adequate ventilation requirements for the room), as detailed in Figure 2.49 below.

### Circulation systems in houses with a floor more than 4.5 m above ground level

Where ventilation ducts pass through compartment walls into another building, follow the guidance in ADB2.

Air circulation systems that circulate air within an individual dwelling house should be designed to prevent smoke and fire spread into a protected stairs as follows:

- No air-transfer grilles are to be fitted in any walls, floors or ceilings to the protected stairs.
- Ducts passing through protected stairs or entrance hall are to be constructed of rigid steel and all joints between the duct work and the enclosure must be fire-stopped.
- Ventilation ducts serving protected stairs should not serve other areas.
- Ventilation systems serving protected stairs and other areas should be designed to shut down on detection of smoke within the system.
- A room thermostat for a ducted warm-air system should be mounted in a living room 1370–1830 mm above the floor and set at 27 degrees maximum.

### Residential sprinkler systems for means of escape

Where fire safety requirements of the building regulations cannot be met, the proposals for fire-engineered solutions that may incorporate a sprinkler suppression system as part of the solution can

**2.88** Domestic extensions

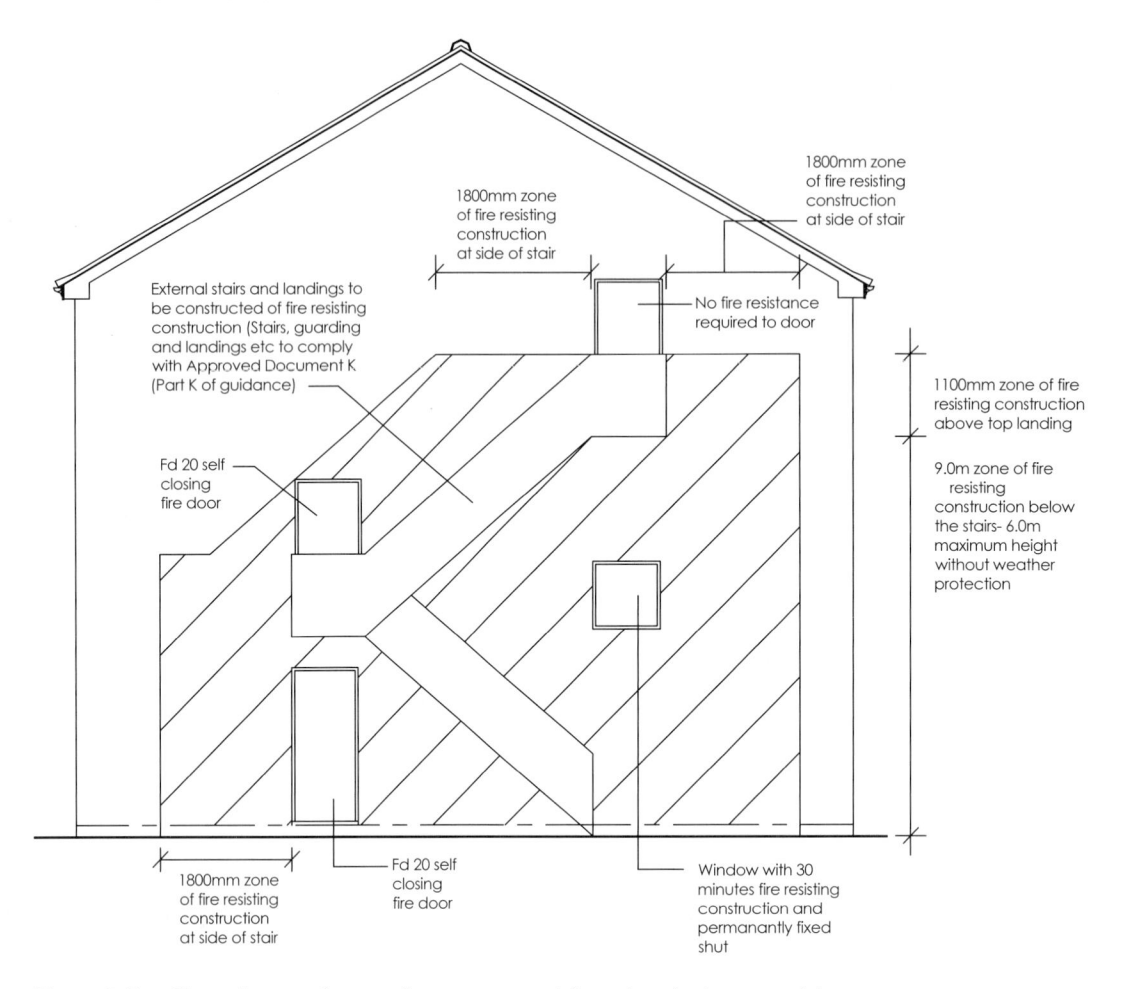

**Figure 2.49:** Fire resistance of areas adjacent to external fire exit stairs *(not to scale)* (See Diagram 7 of ADB1 for full details.)

Labels on figure:
- 1800mm zone of fire resisting construction at side of stair
- 1800mm zone of fire resisting construction at side of stair
- No fire resistance required to door
- 1100mm zone of fire resisting construction above top landing
- 9.0m zone of fire resisting construction below the stairs- 6.0m maximum height without weather protection
- External stairs and landings to be constructed of fire resisting construction (Stairs, guarding and landings etc to comply with Approved Document K (Part K of guidance)
- Fd 20 self closing fire door
- Fd 20 self closing fire door
- 1800mm zone of fire resisting construction at side of stair
- Window with 30 minutes fire resisting construction and permanantly fixed shut

be allowed, against the requirements of ADB, where a risk assessment has been carried out by a suitably qualified and experienced fire engineer and approved by building control before works commence on site. The residential sprinkler system is to be designed and installed by a suitably qualified specialist to BS 9251:2005, incorporating BAFSA technical guidance note No.1 of June 2008, and must be approved by Building Control before works commence on site.

In three-storey dwellings where the stairs discharge into a habitable open-plan area, a partial sprinkler installation to the whole of all connected open-plan areas may be used. Fire separation of the route will be required from the upper floor from this open-plan area with a 30-minute fire-resisting partition and FD20 fire door fitted with intumescent strips. Instead of the separation it may be possible to fully sprinkler the whole dwelling and retain the open-plan arrangement – with the agreement of building control.

(An alternative solution to the open-plan arrangement effecting the means of escape as detailed above is to link an automatic opening vent (AOV) designed to be opened by an electronic sprinkler control panel.)

With the agreement of building control, it should be possible to reduce fire protection throughout the dwelling by 30 minutes with the introduction of a full sprinkler installation

Where dwellings are unable to meet the requirements for access and facilities for the fire service under Section 5 of ADB1, it should be possible to install a full sprinkler installation as a compensatory measure, with the agreement of building control.

For further information on sprinkler systems, contact Keith Rhodes of Nationwide Fire Sprinklers at: www.Nationwide-Fire.co.uk, or contact the British Automatic Sprinkler Association (BAFSA)- *Sprinklers for Safety: Use and Benefits of Incorporating Sprinklers in Buildings and Structures* (2006) ISBN: 0 95526 280 1. See also:
- www.bafsa.org.uk
- Technical guidance note no.1, ISBN 0-9552628-3-6
- www.firesprinklers.org.uk.

## Surface spread of flame: internal wall and ceiling linings including roof lights

Surface spread of flame over internal wall or ceiling finishes is to be in compliance with the product manufacturer's details and in compliance with BS 476–7:1997 (as amended).

Please refer to Section 3: Wall and ceiling linings of ADB1, for full details.

### Fire resistance to elements of structure

Load-bearing elements of structure are to have the minimum standard of fire resistance for buildings up to three storeys, as stated in Table 2.40 below, to prevent premature failure of the structure and to minimise the risk to occupants. This will also reduce the risk to fire-fighters and reduce the danger to people in the vicinity of the building should failure of the building occur. See Table A1: Appendix A of ADB1 for full details.

### External wall construction in relation to a boundary

External walls with less than 1.0 m to a relevant boundary should have 30 minutes' fire resistance from each side separately, and external walls with more than 1.0 m to a relevant boundary should have 30 minutes' fire resistance from inside the building (maximum 3 storeys high). Typical construction details are detailed in Part A of this guidance. External walls within 1.0 m of a boundary

**Table 2.39:** Internal surface spread of flame: classification of wall and ceiling linings
(See Table 1 of ADB1 for full details.)

| Location | National class | European class |
| --- | --- | --- |
| Small rooms up to 4 m$^2$ and domestic garages up to 40 m$^2$ | 3 | D–s3,d2 |
| Other rooms over 4 m$^2$ incl. garages over 40 m$^2$ and circulation spaces within dwellings (e.g. hall, stairs and landings) | 1 | C–s3,d2 |

**Note:** Plaster on masonry walls and plasterboard and skim linings in this guidance will achieve class 1. Exposed timber linings should be treated with a proprietary treatment to achieve the above classifications.

**Table 2.40:** Fire resistance to common elements of structure etc.[1]

| Building element | Fire resistance in minutes | | | Meth Method of protection |
| --- | --- | --- | --- | --- |
| | Load-bearing capacity | Integrity | Insulation | |
| Structural beam, column or frame | 30 | Na | Na | All exposed faces |
| Load-bearing wall (which is not also a wall described in any of the following items) | 30 | Na | Na | Each side separately |
| Upper floors (not above a garage or basement) | 30 | 15 | 15 | From underside |
| Roof (only if forming part of an escape route) | 30 | 30 | 30 | From underside |
| External walls:[2 and 3] (i) Less than 1.0 m to relevant boundary (max 3 storey building) | 30 | 30 | 30 | Each side separately |
| (ii) More than 1.0 m to relevant boundary (max three-storey building) | 30 | 30 | 30 | From inside building |
| Walls and upper floors separating an integral garage from the dwelling | 30 | 30 | 30 | From garage side |
| Compartment walls separating dwellings | 60 | 60 | 60 | Each side separately |
| Compartment floors separating dwellings | 30 | 30 | 30 | Each side separately |
| Protected stairs and lift shaft (not forming part of a compartment wall) | 30 | 30 | 30 | Each side separately |
| Cavity barriers etc. (i.e. junctions between roof and compartment/ separating walls or in cavities between separating walls and floors) | Na | 30 | 15 | Each side separately |
| Ceiling above protected stairs (three-storey) | Na | 30 | 30 | From underside |
| Ducts in cavity barriers | Na | 30 | No provision | From outside |
| Casing around soil pipes etc. | Na | 30 | No provision | From outside |

**Notes:**
1. 12.5 mm plasterboard with a plaster skim finish applied to 100 × 50 mm timber stud partitions/ ceiling/ floor joists as detailed in this guidance will achieve 30 minutes' fire resistance; two layers of 12.5 mm plasterboard (joints staggered) with a plaster skim finish will achieve 60 minutes' fire resistance. Masonry walls detailed in this specification will achieve 60 minutes' minimum fire resistance.
2. External walls within 1.0 m of a boundary should achieve class 0 surface spread of flame.
3. Combustible materials used on external surfaces should comply with Diagram 10 of Approved Document B: Volume 1.

should achieve class 0 external surface spread of flame. Please refer to Section 8 of ADB1 for full details. Combustible materials used on external surfaces should comply with Diagram 19 of ADB1.

## Compartment walls and floors separating buildings

Compartment walls (party walls) and compartment floors (party floors) separating buildings should have 60 minutes' fire resistance (including load-bearing capacity, integrity and insulation) from each side separately. Typical construction details for party walls are detailed in Part A of this guidance. Party floors are beyond the scope of this guidance and reference should be made to ADE.

## Permitted building openings in relation to a boundary

### Openings within 1.0 m of a boundary

An unprotected opening of $1\,m^2$ (e.g. a window) is permitted every 4.0 m on the same building face (openings must be separated by at least 4 m vertically and horizontally). This unprotected opening can consist of two or more smaller openings within an area of $1\,m^2$ (openings less than $0.1\,m^2$ are permitted every 1500 mm on the same building face – openings must be separated by at least 1500 mm vertically and horizontally). There are no restrictions on dimensions between openings separated by compartment walls and floors. Please refer to Diagram 20 of ADB1 for full details.

### Openings more than 1.0 m from a boundary

Permitted unprotected openings are to be in compliance with Table 2.41 below for buildings not exceeding three storeys in height (excludes basements) or more than 24 m long.

## Designation of roof covering and minimum distance to boundary

Roof coverings (not roof structure) near a boundary should give adequate protection against the spread of fire when exposed to fire from outside, in accordance with Table 2.42 below.
  Please refer to Section 10 of ADB1 for full details.

**Table 2.41:** Permitted unprotected areas in relation to a relevant boundary[1, 2]
(See Diagram 22 (and Table 4 of ADB1 for full details.)

| Minimum distance between side of building and relevant boundary | Maximum total area of unprotected openings |
| --- | --- |
| 1.0 m | $5.6\,m^2$ |
| 2.0 m | $12\,m^2$ |
| 3.0 m | $18\,m^2$ |
| 4.0 m | $24\,m^2$ |
| 5.0 m | $30\,m^2$ |
| 6.0 m | no limit |

**Notes:**

1. Refer to Section 9 of ADB1 for full details relating to space separation and other methods of calculating unprotected areas.
2. If sprinklers are fitted throughout the building to BS 9251, the above distances can be reduced by 50 per cent (minimum 1.0 m) or the unprotected opening area doubled.

**Table 2.42:** Limitations on designation of roof coverings* and minimum distance to boundary (See Table 5 of ADB1 for full details.)

| Designation# of coverings of a roof or part of a roof | | Minimum distance from any point on relevant boundary | | | |
|---|---|---|---|---|---|
| National Class | European Class | Less than 6 m | At least 6 m | At least 12 m | At least 20 m |
| AA.AB or AC | $B_{roof}(t4)$ | Acceptable | Acceptable | Acceptable | Acceptable |
| BA.BB or BC | $C_{roof}(t4)$ | Not acceptable | Acceptable | Acceptable | Acceptable |
| CA.CB or CC | $D_{roof}(t4)$ | Not acceptable | Acceptable [1][2] | Acceptable [1] | Acceptable |
| AD.BD or CD | $E_{roof}(t4)$ | Not acceptable | Acceptable [1][2] | Acceptable [1] | Acceptable [1] |
| DA.DB, DC or DD | $F_{roof}(t4)$ | Not acceptable | Not acceptable | Not acceptable | Acceptable [1][2] |

**Notes:** *See Table 5 of ADB1 for limitations on glass, thatch, wood shingles and plastic roof lights.
#For explanation of the designation of external roof surfaces and separation distances, see Table 5 of ADB1. Separation distances do not apply to the boundary between roofs of a pair of semi-detached houses or enclosed/covered walkways, but they do apply over the top of compartment walls – see Diagram 11 of ADB1.
[1] Not acceptable in any of the following buildings:
  a. Houses in terraces of three or more houses
  b. Any other building with a cubic capacity of more than 1500 m³.
[2] Acceptable on buildings not listed in Note[1], providing that part of the roof is no more than 3 m² in area and is at least 1500 mm from any similar part, with the roof between the parts covered with a material of limited combustibility.

### Typical fire and rescue service vehicle access route specification for dwellings

See Section 11 of ADB1 for full details or new dwellings in this guidance.

## PART C: SITE PREPARATION AND RESISTANCE TO CONTAMINANTS AND MOISTURE

Please refer fully to the relevant Approved Document.

### C1: Resistance to contaminants

The site should be prepared and the building constructed in accordance with Approved Document C of the Building Regulations and the details in this guidance, so as to prevent and resist contaminates (and moisture) from causing damage to the building and affecting the health of its occupants.

An initial desk study, walk-over assessment and evaluation of the site and surrounding area should be carried out in accordance with BS10175 and BS5930 by a suitable person before works commence to ensure it does not contain any hazard that is, or may become, harmful to persons, controlled waters or buildings, including substances that are corrosive, explosive, flammable, radioactive or toxic (radon gas, methane and other gases are covered in more detail below). All results should be recorded.

For sites where hazards are not suspected, basic site investigations i.e. trial pits, may be required by building control to establish if any hazards exist. For sites where hazards are known or suspected, detailed specialist geotechnical and/or geo-environmental site investigations may be required by building control and, where necessary, remedial works may be needed to manage or remove the hazard. For more information on contaminates, contact building control or your local authority environmental health department.

# Radon gas

## What is radon gas?

Radon is a colourless, odourless gas that is radioactive. It is formed within the ground where uranium and radium are present and moves through cracks and fissures within the subsoil into the atmosphere or spaces under and in dwellings. Where radon occurs in high concentrations it can be a risk to health, and exposure to particularly high levels of radon may increase the risk of developing lung cancer.

## Protective measures

To reduce the radon risk, all new buildings, extensions and changes of use may need to incorporate precautions against radon; you can determine the level of radon protection your development requires (at a cost) either through the British Research Establishment website at: www.ukradon.org for individual properties, or www.bre.co.uk/radon or the British Geological Society at: www.bgs.ac.uk, or contact building control for more information.

The information acquired from the maps will show the highest radon potential within 1-km grid squares. These 1-km squares are based on the National Grid used by Ordnance Survey and are colour-coded to identify the level of radon precaution required – either full or basic protection, or areas that do not require any protection at all.

Protective measures against radon gas should be incorporated into the building in compliance with the following guidance details and approved by building control before works commence:

## Basic radon protection

Ground-supported concrete floors are acceptable only for basic radon protection (and are not suitable for full radon protection as detailed below in full radon protection details). Ground-supported floors with a continuous 1200 g (300 micrometre) polythene damp-proof membrane (dpm) are acceptable as a radon barrier over the footprint of the building, continuing through the external/cavity walls – subject to building control approval (recycled products may not be suitable and proprietary radon gas-proof membranes are available) which must also be suitable for use as a damp-proof membrane with British Board of Agreement (BBA or other third-party accredited) certification, positioned and fixed in accordance with manufacturer's details), positioned within the floor as illustrated in the guidance diagrams in Part A of the guidance above. One-piece membranes manufactured by specialists to fit the footprint of the building, with preformed welded barriers, are available.

To prevent damage to the radon barrier, it should be installed at a later stage of construction and sealed with gas-proof tape to strips of membrane already built into the walls, or else a proprietary reinforced radon barrier/damp-proof membrane can be used. Any damaged areas are to be repaired with radon membrane and sealed with two strips of gas-proof tape with 150 mm minimum laps.

## Full radon protection

Full radon protection is achieved by the provision of a continuous 1200 g (300 micrometer) polythene damp-proof membrane or proprietary gas-proof membrane which must also be suitable for use as a damp-proof membrane (as detailed above for basic radon protection) over the footprint of the building, continuing through the external/cavity walls, supported by suspended beam-and-block floors with vented sub-floor void or alternatively a cast in-situ reinforced concrete slab/raft

## 2.94 Domestic extensions

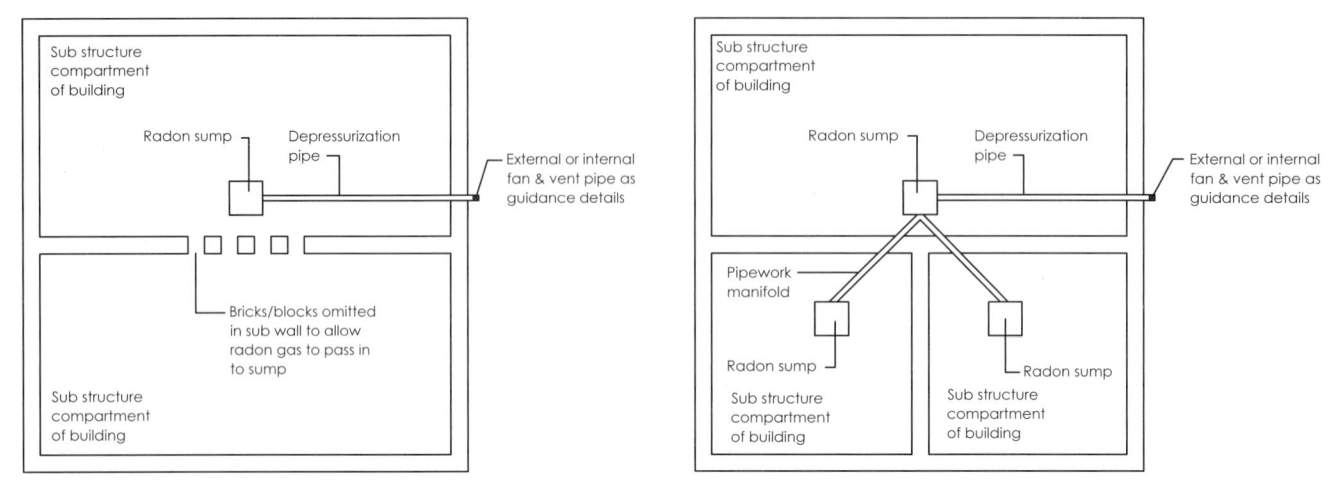

**Figure 2.50:** Plan layout of sump and depressurisation pipe positions in two different building sub-structures *(not to scale)*

foundation with sumps and sub floor depressurisation pipes as illustrated in the guidance diagrams in Part A of the guidance above, and ground floor diagrams below. In areas requiring full radon protection the floor needs to be suspended/supported on the cavity wall to prevent settlement of the ground-floor and rupture of the radon-proof barrier at the external wall junction. Also see guidance details above for prevention of damage to, and repair of, radon membranes.

### Number and position of sumps

Where clean permeable fill has been used in the sub-floor make-up, a single radon sump is suitable for a single dwelling over an area of approximately $250\,m^2$ or for a distance of 15 m from the sump. Sumps can be connected together in multi-compartmented sub-floor areas using a pipework manifold and connected to an external or internal fan, or else vent openings/ducts can be formed through sub walls.

### Sump construction

(i) *Site-constructed sumps*

Site-constructed sumps are typically 600 mm × 600 mm square × 400 mm deep, constructed using bricks laid in a honeycomb bond so as to form a box around the end of the pipe. The top of the box is covered with a paving slab. To avoid subsequent collapse when compacting fill around the sump, mortar should be used for horizontal joints. However, it is essential that all vertical joints are left open. Further details are available on the BRE website at: www.bre.co.uk/radon.

(ii) *Proprietary prefabricated sumps*

Prefabricated sumps used as an alternative to brick construction should be installed in accordance with the manufacturer's details.

(iii) *Depressurisation pipes*

The pipe from the sump needs to be 110 mm-diameter uPVC, with joints using standard couplings that are sealed and airtight. The pipe needs to leave the building so that it could be coupled to a fan mounted on the external wall. The pipe should terminate about 100 mm from the external wall, and should be located at the rear of the house or at a re-entrant corner where subsequent installation of a boxed-in fan and vertical stack will be least obtrusive. Until such time as a fan is installed, the pipe should be capped off 300 mm above ground level to prevent vermin and rain penetration, and capped off with an access plug and sign identifying radon pipework fixed to the wall above the capping.

(iv) *Geo-textile drainage matting (as an alternative to sumps)*
Geo-textile drainage matting is to be laid beneath the slab and connected to an extract pipe to provide a sump, in accordance with the manufacturer's details. The matting is likely to prove more expensive than a sump. The sump and pipework are installed only as a fallback measure and do not provide any radon removal until such time as a fan is installed or until the sump is connected to a passive stack system.

(v) *Edge-located sumps (mainly used for retrospective fitting or conversion work)*
Edge-located mini-sumps can be used instead of a centrally located sump, in accordance with details available on the BRE website at: www.bre.co.uk/radon, but must be agreed with building control before works commence on site. Edge-located sumps are typically constructed by excavating a hole 400 × 400 × 400 mm in the hardcore or fill beneath the dwelling, alongside the perimeter wall, to form an open area around the end of the extract pipe. Fix 600 × 600 mm paving slab or similar over the sump to provide a permanent formwork to support the floor slab (or make good existing cast in-situ floor slabs where sumps are fitted retrospectively). Seal the pipe where it passes through the wall. Prefabricated sumps or site-constructed sumps as detailed above can also be used as edge-located sumps. (Concrete floor slabs are to be reinforced over sumps if required by building control.)

### Radon fan locations

When required, the fan should be positioned with the outlet well away from windows, doors and ventilation grilles and should discharge just above eaves level. Low-level discharge is permitted if there are no openings or vents close by. To avoid penetrating the radon-proof floor membrane, the pipe should be taken through the wall, not up through the floor. The pipework can be installed in ducts inside the house and connected as close as possible to the roof-space fan and outlet terminating out through the roof, using a proprietary roof vent and flashing system 900 mm above any roof opening or vent – that is, within 3.0 m of the terminal.

### Stepped foundations and retaining walls

Where possible stepped foundations should be avoided, as they complicate the achievement of radon protection using only sealing techniques. It may prove less expensive to excavate around the house to provide a ventilated space, rather than trying to build into the hillside and seal all the faces of the building that are below ground level.

## Further details

Guidance on protective measures against radon gas is available on the BRE website at: www.bre.co.uk/radon.

## Methane and other ground gas protection

Where necessary, protection is to be provided against methane and other ground gases entering the building voids and compartments in compliance with a specialist's design. Methane is explosive in air at 5–15%, carbon dioxide is highly toxic, and both are asphyxiants.

Sources of ground gas include: farmland, made-up ground, landfill, sewers, river/pond sediments, coal mining, peat bogs, limestone/chalk and moss lands. The method of protection depends on the type of development, gas type, source, and volume of gas and rate of emission. Protection measures typically include ground barriers, gas-alarm systems, gas-proof membranes and passive/active ventilation of the building substructure. If within 250 m of a landfill site the Environment Agency's policy on building development on or near landfill sites should be followed.

Information regarding landfill and land contamination (including biodegradable substances, or land that may have been subject to use that could have petrol oil or solvent spillages; or naturally occurring methane, carbon dioxide and other hazardous gases, e.g. hydrogen sulfide) may be available from your local authority environmental health department. Risk assessments, ground investigations, necessary remedial works and substructure designs should be carried out by a specialist in accordance with ADC which should be approved by Building Control/Environmental Health Department before works commence on site.

Further information on protective measures for methane and other ground gases is available in BRE publication 'Protective measures for housing on gas-contaminated land' (2001) which is available to purchase from: www.brebookshop.com. and 'Protecting development from methane' CIRIA Report 149 (1995) which is available to purchase from: www.ciria.org.

### Landfill gas and radon

There may be cases where the dwelling being constructed is located on or adjacent to a landfill site or old coalfield. In such cases additional precautions that exceed those required for radon, to a specialist's design, may be needed to deal with methane; so where both methane and radon are present, methane-protective measures should be applied, and only intrinsically safe (non-sparking) fans and switchgear should be used.

## C2: Resistance to moisture

The walls, floor and roof of the building should be constructed in accordance with ADC and the details/diagrams in this guidance to prevent and resist the passage of moisture into the building.

### Horizontal damp-proof courses (dpcs)

A horizontal damp-proof course (dpc) and dpc trays with weep holes at 900 mm centres are to be provided 150 mm above external ground level, continuous with and sealed to the floor damp-proof membrane (dpm) and radon/dpc tray to prevent the ingress of moisture into the building.

### Vertical damp-proof courses (dpcs) and damp-proof course trays, etc.

Stepped and horizontal dpc/cavity trays are to be provided over all openings, roof abutments/ projections and over existing walls with different construction or materials. Install vertical dpc or proprietary insulated cavity closers at all closings, returns, abutments to cavity work and openings, etc. to prevent the ingress of moisture into the building.

### External cavity walls

50 mm-wide minimum, clear, continuous cavity should extend the full height and width between the internal and external wall leafs, bridged only by wall ties, cavity trays, cavity barriers, fire stops and cavity closures. (Where a cavity is to be partially filled, the residual cavity should not be less than 50 mm wide – unless the product has a British Board of Agreement (BBA or other approved third-party) accreditation for use and is approved by building control.) The cavity should be carried down at least 225 mm below damp-course level at ground-floor level, to protect the inner wall leaf, and damp-proof (cavity) trays should be at least 150 mm deep, as diagrams/details in this guidance.

### Tanking systems

These, providing either barrier, structural or drained protection to the building, must be assessed, designed and installed for the particular project in compliance with BS 8102: 2009 Code of Practice for Protection of Below Ground Structures Against Water from The Ground. Tanking systems can be installed internally or externally, in accordance with a tanking specialist's details.

The illustrated tanking section details in this guidance are suggested details only and actual details must be approved by building control before works commence on site. Forms of tanking include: bonded sheet materials, liquid applied membranes, mastic asphalt, drained cavity membranes, and cementitious crystallisation and cementitious multi-coat renders.

Suitable tanking systems need to have British Board of Agreement (BBA or other approved third-party) accreditation and be individually assessed by a tanking specialist as suitable for the proposed situation. Tanking systems above ground should be vapour-permeable, to prevent condensation problems within the building and also to prevent mould growth.

Tanking systems must be designed/installed/applied by a tanking specialist for the particular project, in compliance with the tanking manufacturer's details, and where necessary additional measures must be taken to prevent radon gas and other such ground gases and contaminates from entering the building.

Tanking systems are to be properly connected to, and made continuous with, wall damp-proof courses/radon dpc trays. Perforation of the tanking system by service entry pipes etc. should be avoided or carried out strictly in accordance with the tanking manufacturer's details.

### Flood risk

Flood risk should be assessed and precautions carried out in compliance with paragraph 0.8 of ADC

### Condensation risks

The technical details and diagrams in this guidance document should be read in conjunction with the BRE publication 'Thermal Insulation Avoiding Risks', which explains the technical risks and condensation risks that may be associated with meeting the Building Regulations requirements for thermal insulation for the major elements of the building. A copy of the publication can be obtained from: www.brebookshop.com.

## PART D: CAVITY WALL FILLING WITH INSULATION

Please refer fully to the relevant Approved Document.

If insulating material is inserted into a cavity in a cavity wall, precautions must be taken to prevent the subsequent permeation of any toxic fumes from that material into any part of the building occupied by people.

The suitability of the cavity wall for filling must be assessed before the works are carried out by an insulation specialist, in accordance with BS 8208: Part 1: 1985, and the insulation system must be British Board of Agreement (BBA or other third-party) accredited.

The insulation specialist carrying out the work must hold or operate under a current BSI Certificate of Registration of Assessed Capability for the work being carried out. The insulation material must be in accordance with BS 5617: 1985 and the installation must be in accordance with BS 5618: 1985

The Installation of urea–formaldehyde (UF) in cavity walls is to be carried out in compliance with paragraphs 1.1–1.2 of ADD1

## PART E: RESISTANCE TO THE PASSAGE OF SOUND

Please refer fully to the relevant Approved Document.

### New party walls and floors in new extension

Sound insulation details for new party walls are to be carried out in accordance with the relevant details in ADE Section 2, and for floors are to be carried out in accordance with the relevant details contained within Section 2 of this guidance and ADE Section 3.

Please note that the Party Wall Act may be applicable to works to party walls. Further information is contained in Section 1 of this guidance.

### New internal walls and floors in new extension

Sound insulation details between internal walls and floors separating bedrooms, or a room containing a WC and other rooms, is to be carried out in accordance with the relevant details contained in Section 2 of this guidance and ADE Section 5.

### Pre-completion sound testing

Where new party walls or party floors are constructed, pre-completion sound testing is to be carried out to demonstrate compliance with ADE Section 1 and as follows3

Pre-completion sound testing is to be carried out by a suitably qualified person or specialist with appropriate third-party accreditation (UKAS or ANC registration) to demonstrate compliance with ADE1, and a copy of test results is to be sent to building control.

#### *Remedial works and retesting*

Remedial works and retesting will be required where the test has failed, in compliance with Section 1 of ADE

### Exemptions and relaxations

If the requirements of the Building Regulations will unacceptably alter the character or appearance of a historic/listed building/ancient monument or building within a conservation area, then the requirements may be exempt or relaxed to what is reasonably practical or acceptable, ensuring that any exemption or relaxation would not increase the risk of deterioration of the building fabric or fittings, in consultation with the local planning authority's conservation officer (before works commence). For further information see ADE and contact your local authority planning department

## PART F: VENTILATION

Please refer fully to the relevant Approved Document.

## Purge (natural) ventilation

Purge (natural) ventilation equal to 1/20 (5 per cent) of the floor area is to be provided to all habitable rooms. The 1/20 applies where the external windows/doors open more than 30 degrees, and is increased to 1/10 (10 per cent) of the floor area where the windows open between 15 and 30 degrees. Window openings that open less than 15 degrees are not suitable for purge ventilation and alternative ventilation details are required, as detailed below (in compliance with Section 5 and Appendix B of ADF1). Purge ventilation openings to habitable rooms are to be typically located 1.75 m above floor level. The area of external windows, roof windows and doors should not exceed 25 per cent of the useable internal floor area, otherwise SAP calculations may be required from a suitably qualified person to confirm design flexibility.

Unprotected openings (glazed window and door openings) should not exceed the permitted areas in relation to a boundary, in compliance with Part B of this guidance.

Means-of-escape windows are to be fitted with proprietary hinges to open to the minimum required clear width of 450 mm. Escape windows must have minimum clear opening casement dimensions of $0.33\,m^2$ (typically 450 mm wide × 750 mm high), and must be located within 800 to 1100 mm above floor level to all bedrooms and habitable rooms at first-floor level and inner habitable rooms on the ground floor. Windows above the ground-floor storey and within 800 mm of floor level are to be provided with containment/guarding/proprietary catches, which should be removable (but childproof), in the event of a fire. Where escape windows cannot be achieved, direct access to protected stairs (or a protected route to inner rooms) is acceptable, in compliance with ADB1 para 2.6 (a) or (b).

## Mechanical extract ventilation and fresh air inlets for rooms without purge ventilation

Mechanical extract ventilation and fresh-air inlet are required for habitable rooms without purge (natural) ventilation and must be designed by a ventilation specialist; they are to have a minimum of four air changes per hour and must be manually controlled, in compliance with Section 5 of ADF1. This system can incorporate heat recovery if required. Note: means-of-escape windows are required to all bedrooms and habitable rooms at first-floor level and inner habitable rooms on the ground floor, in accordance with the guidance details above.

## Background ventilation

Background ventilation is to be provided equivalent to 8000 mm$^2$ to habitable rooms and 2500 mm$^2$ to wet rooms via operable hit-and-miss vents into frames (or two-stage security catches fitted to operable windows if agreed with building control). Fans and background vents fitted in the same room should be a minimum of 0.5 m apart.

## Intermittent mechanical extract ventilation

Mechanical ventilation is to be provided to the rooms listed below, directly ducted to the outside air, equivalent to the following rates:

| | |
|---|---|
| Kitchen | 30 litres per second over hob or 60 litres elsewhere |
| Utility room | 30 litres per second |
| Bathroom | 15 litres per second (including shower rooms and en-suites) |
| Toilet | 6 litres per second WC (with or without a window) |

### Ventilation systems for basements

To be carried out in compliance with Paragraphs 5.11–5.13 of ADF1

### Ventilation of a habitable room through another room or conservatory

To be carried out in compliance with Paragraphs 5.14–5.16 of ADF1

### General requirements for mechanical extract ventilation

Mechanical ventilation to rooms without operable windows is to be linked to light operation, independent switch or PIR, have 15 minutes' overrun and have a 10 mm gap under the door for air supply. Fans must not be installed in rooms containing open-flue appliances, unless the interaction of mechanical ventilation and open-flue heating appliances is checked and certified by an approved method and suitably qualified person, as contained in ADJ.

Mechanical ventilation is to be ducted in proprietary insulated ducts to outside through walls to a proprietary vent or through roof space to a proprietary tile or soffit vent.

## PART G: SANITATION, HOT-WATER SAFETY AND WATER EFFICIENCY

Please refer fully to the relevant Approved Document.

## Wholesome hot and cold water supply

Sinks with wholesome hot and cold running water are to be provided in all food-preparation areas; bathrooms are to be fitted with either a bath or shower. Hot and cold water supplies to wash basins, baths, showers and sinks, including external taps, are to have water from a wholesome water supply.

Hot taps should be located on the left-hand side (traditionally – as most people are right-handed it prevents people from unwittingly running the hot tap and burning themselves).

Chemically softened, wholesome cold water should not be provided where drinking water is drawn off or to any sink where food is prepared.

Wholesome water supply is to comply with the Water Supply (Water Quality) Regulations 2000 (SI2000/3148), and in Wales the Water Supply (Water Quality) Regulations 2001 (SI2001/3911) and Annex 1 of AD. Private water supplies are to comply with the Private Water Supplies Regulations 2009 (SI 2090/3101), and in Wales the Private Water Supplies (Wales) Regulations (SI 2010/66).

## Scale of provisions

Any dwelling-house or flat must have at least one bathroom, with a fixed bath or shower, wash basin and WC in compliance with BS 6465. Hot taps should be located on the left-hand side. In new dwellings, the WC should be located in the principal entrance storey.

## Wash basins and separation of w/c from any food-preparation areas

Wash hand basins are to be provided in all rooms containing a WC (or in an adjacent room, providing the room is not used to prepare food), and a door must separate the WC and wash basin from any food-preparation area in a dwelling.

## Water tanks/cisterns base

Water tanks/cisterns must have an adequately designed flat platform base to support the proposed loads.

## Pumped small-bore foul-water drainage

Pumped small-bore foul-water drainage from a toilet is permitted only if there is also access to a gravity-draining toilet in the same dwelling. Proprietary pumped foul-water macerator systems must have BBA or other approved accreditation and be fitted in compliance with the manufacturer's details to a suitable foul-water drainage system.

## Vented and unvented hot-water storage systems

Vented and unvented hot-water storage systems are to be designed, installed, commissioned and tested by a suitably qualified heating engineer/specialist (unvented systems are to be indelibly marked with the information contained in paragraph 3.23 of ADG), in compliance with paragraphs 3.10–3.27 of ADG. A copy of commissioning certificates is to be issued to building control on completion of the works.

## Safety valves, prevention of scalding and energy cut-outs

See 'New dwellings'.

### Discharge pipes from safety devices

Discharge pipes from safety devices should be 600 mm maximum length, constructed of metal (or other material suitable for proposed temperatures to BS 7291-1:2006) and should connect to a tundish fitted with a suitable air gap, in compliance with the current Water Supply (Water Fittings) Regulations. Any discharge into the tundish must be visible (and where the dwelling is occupied by visibly or physically impaired persons, the device must be electronically operated and able to warn of discharge).

Discharge pipes from the tundish should be at least 300 mm in length and fixed vertically below the tundish, before connection to any bend or elbow and at a continuous fall of 1:200 thereafter until the point of termination. Pipes from the tundish should be at least one pipe size larger than the outlet of the safety device up to 9 m in length (2 × larger for 9–18 m and 3 × larger for 18–27 m), and constructed of metal (or other material suitable for proposed temperatures to BS 7291- 1:2006).

Point of termination from discharge pipes can be either:

(i) to a trapped gully – below grating, but above the water seal.
(ii) downward discharges at low level – up to 100 mm above external surfaces (car parks, hard-standings, grassed areas, etc.) and fitted with proprietary wire guard to prevent contact.
(iii) discharges at high level into metal hopper and metal down-pipes at least 3 m from plastic guttering collecting the discharge.

Note: Visibility of discharge must be maintained at all times and discharges of hot water and steam should not come into contact with materials that could be damaged by such discharges.

## Solar water heating

Solar water heating roof/wall panel systems are to be factory-made to BS EN 12976-1:2006, fitted with safety devices and an additional heating source to maintain an adequate water temperature, and fitted in compliance with the manufacturer's details. Solar water heating systems should comply with current European/British Standards.

## Electrical water heating

Fixed electrical immersion heaters must comply with BS EN 60335-2-73:2003, electrical instantaneous water heaters with BS EN 60335-2-35:2002, electrical storage water heaters with BS EN 60335-2-21:2003, and safety devices are to be manufactured and installed in accordance with ADG, the manufacturer's details and current European/British Standards.

## Insulation of pipework to prevent freezing

All hot and cold water service pipework, tanks and cisterns should be located within the warm envelope of the building to prevent freezing.

Where hot and cold water service pipework, tanks and cisterns are located in unheated spaces they should be insulated to prevent freezing, in compliance with BS 6700 and BS 8558, and typically as follows:

(i) All tanks and cisterns should be thermally insulated to prevent freezing, with proprietary insulated systems in compliance with the manufacturer's systems (insulation is normally omitted from below tank, where it benefits from heat in the heated area below).
(ii) Pipework should be insulated with proprietary insulated sleeves of phenolic, polyisocyanurate or polyurethane foam having a minimum wall thickness of 30 mm for 15 mm-diameter pipes and 12 mm for 22 mm-diameter pipes (or other approved material), and fixed in accordance with manufacturer's details.

Incoming cold water supply service pipes should be at least 750 mm below the ground level and other precautions should be carried out to prevent freezing and protect the pipe, in accordance with the relevant water authority's requirements. Consent from the water authority will be required before works commence.

### Commissioning certificates

Commissioning certificates for fixed building services are required on completion, with a copy sent to building control

## Supply (Water Fittings) Regulations 1999

All new water installations must be in compliance with the 'Supply (Water Fittings) Regulations 1999' for England and Wales, for protection against frost and freezing, prevention of waste, misuse, undue consumption, contamination and erroneous measurement of a water supplier's mains water supply. A free copy of regulations can be downloaded from the HMSO website, or alternatively a hard copy of the new Regulations can be purchased directly from your local HM Stationery Office. The Regulations are Statutory Instrument No 1148 and the amendments are Statutory Instrument No 1506, both dated 1999.

## PART H: DRAINAGE AND WASTE DISPOSAL

Please refer fully to the relevant Approved Document.

## H1: Foul- and storm-water drainage

### Foul-, rain- and storm-water drainage systems (single dwellings)

An adequate system of foul-water drainage shall be provided to carry foul water from appliances within the building to one of the following, listed in order of priority: public sewer, private sewer, sewage treatment system, septic tank or cesspool, as detailed in this guidance.

## 2.104 Domestic extensions

An adequate system of rainwater drainage shall be provided to carry rainwater from roofs of the building and paved areas around the building to one of the following, listed in order of priority: adequate soak-away, as detailed in this guidance (or similar approved filtration system); where that is not practicable, a watercourse; or where also that is not practicable, a sewer (note: discharge to a watercourse or sewer is subject to the relevant water authority's written approval).

Both storm and foul drainage are to consist of a proprietary underground drainage system with BBA certification (or other approved accreditation), with minimum 100 mm-diameter pipes laid at a minimum gradient of 1:40 or 1:80 where they serve one or more WCs.

uPVC pipes should be surrounded in a single-size aggregate (size 5–10 mm) at a minimum/maximum depth of 0.6/7.0 m in fields, or 0.9/7.0 m in drives and roads, in compliance with Figure 2.51 below. If minimum depths cannot be achieved, pipes can be protected with a 100 mm reinforced concrete slab with compressible material under and 300 mm minimum bearing on original ground, in compliance with Figure 2.52.

Drainage/services are to incorporate adequate precautions to prevent excessive movement due to possible ground movement in shrinkable clay subsoils, in accordance with design details from a suitably qualified specialist.

### *Bedding and backfilling requirements for rigid and flexible pipes*

(See Diagram 10 of ADH1 for full details.)

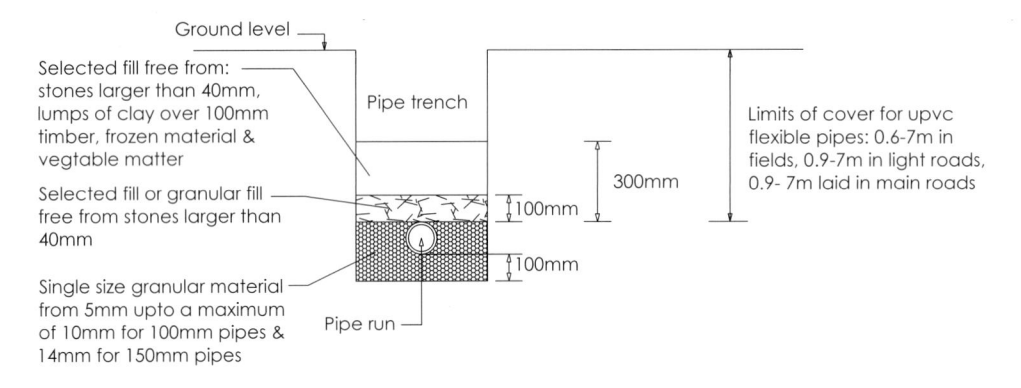

**Figure 2.51:** Typical bedding detail for flexible pipes *(not to scale)*

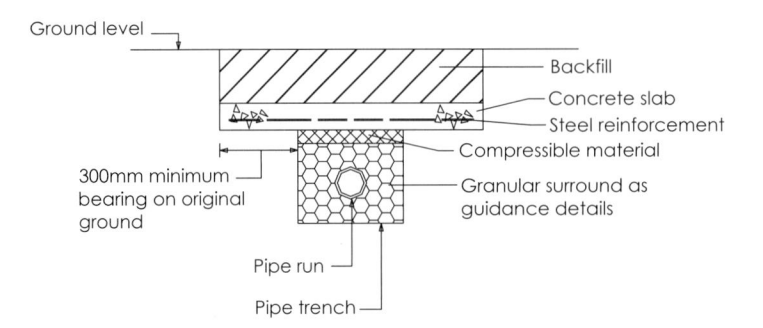

**Figure 2.52:** Typical protection detail for pipes laid at shallow depths *(not to scale)*
(See Diagram 11 of ADH1 for full details.)

## Pipes penetrating through walls

Pipes penetrating through walls should have joints formed within 150 mm of either the wall face, with 600 mm maximum length adjacent rocker pipes fitted both sides with flexible joints, or alternatively lintels provided above openings through the walls to give 50 mm clear space around pipes and openings, in-filled with inert sheet material and sealed to prevent ingress of fill, vermin and radon gas.

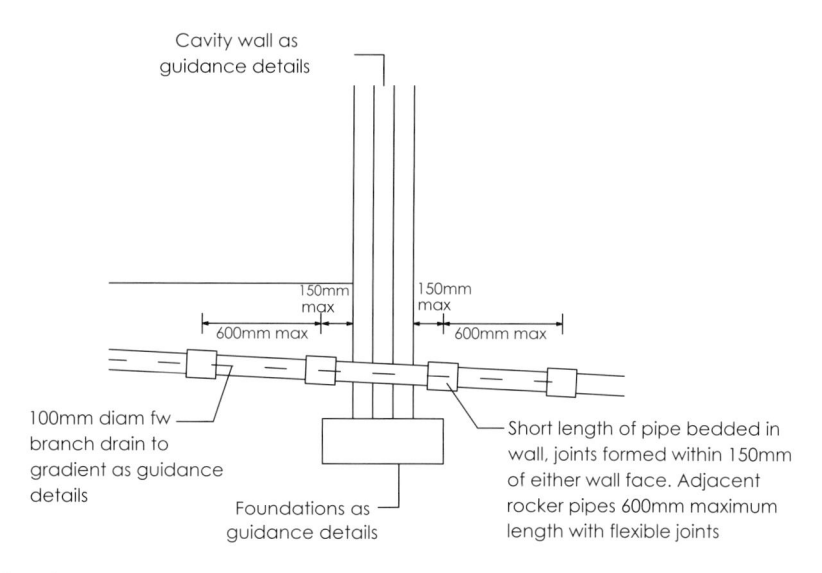

**Figure 2.53.1:** Pipes penetrating through walls using rocker pipe connections *(not to scale)* (See Diagram 7 of ADH1 for full details.)

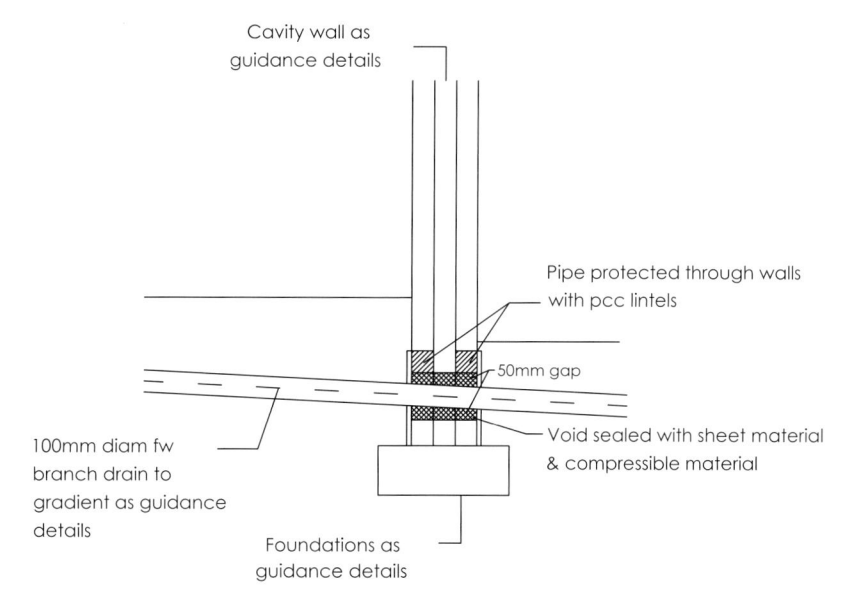

**Figure 2.53.2:** Pipes penetrating through walls with lintels over *(not to scale)* (See Diagram 7 of ADH1 for full details.)

### Drain trenches near buildings

Trench excavations for pipe runs located within 1.0 m of buildings, which extend below the level of the existing foundations, should have trenches backfilled with concrete up to the underside of the existing foundations, as detailed in Figure 2.54.1 below. Trench excavations for pipe runs located more than 1.0 m from buildings and that extend below the level of the existing foundations should have trenches backfilled with concrete up to the underside of the existing foundations less 150 mm, as detailed in Figure 2.54.2. Concrete-encased pipes should have movement joints formed at each socket or sleeve joints using proprietary compressible material (at least 25 mm thick or as pipe manufacturer's details) as detailed in Figure 2.55.

### Inspection chambers, gullies and access fittings etc.

Proprietary uPVC 450 mm-diameter inspection chambers are to be provided at all changes of level and/or direction and at 45 m maximum spacing in straight runs up to 1.2 m in depth. Minimum dimensions for access fittings and inspection chambers are to be in accordance with Table 2.43 (see Table 11 of ADH1 for full details). Maximum spacing of access points is to be in accordance with Table 2.45 (see Table 13 of ADH1 for full details). All gullies are to be trapped and must have rodding access where serving branches. Inspection chamber covers are to be mechanically fixed and suitable for vehicular loads in drives and roads, and must have double-sealed, air-tight, bolt-down covers and frames in buildings, in accordance with the manufacturer's details.

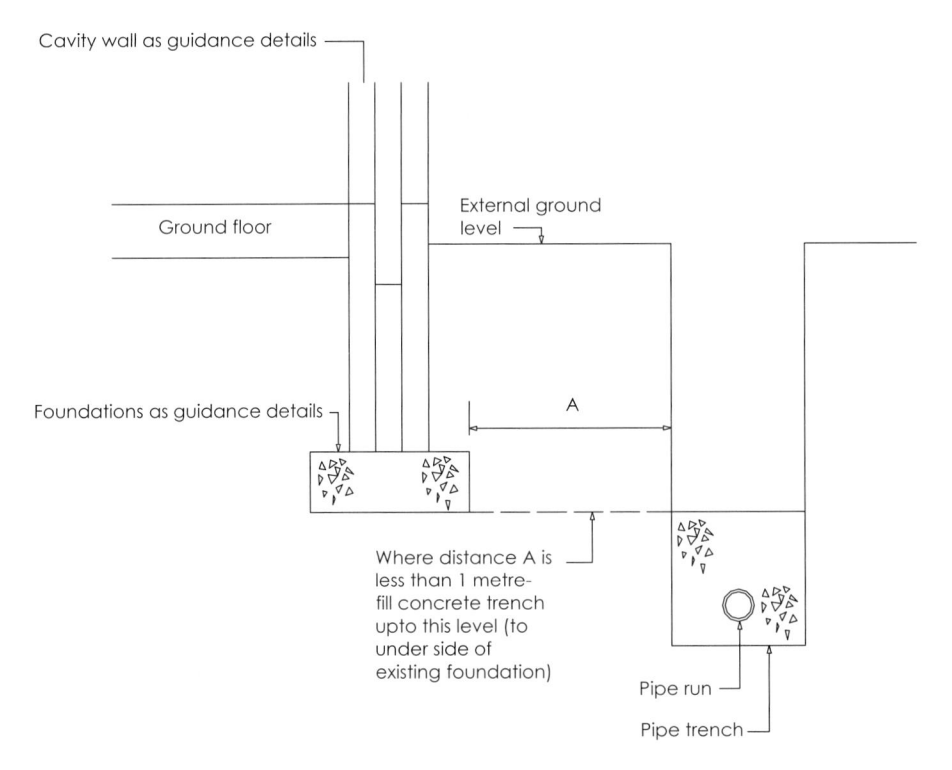

**Figure 2.54.1:** Typical detail of drain trenches within 1 m of buildings *(not to scale)* (See Diagram 8 of ADH1 for full details.)

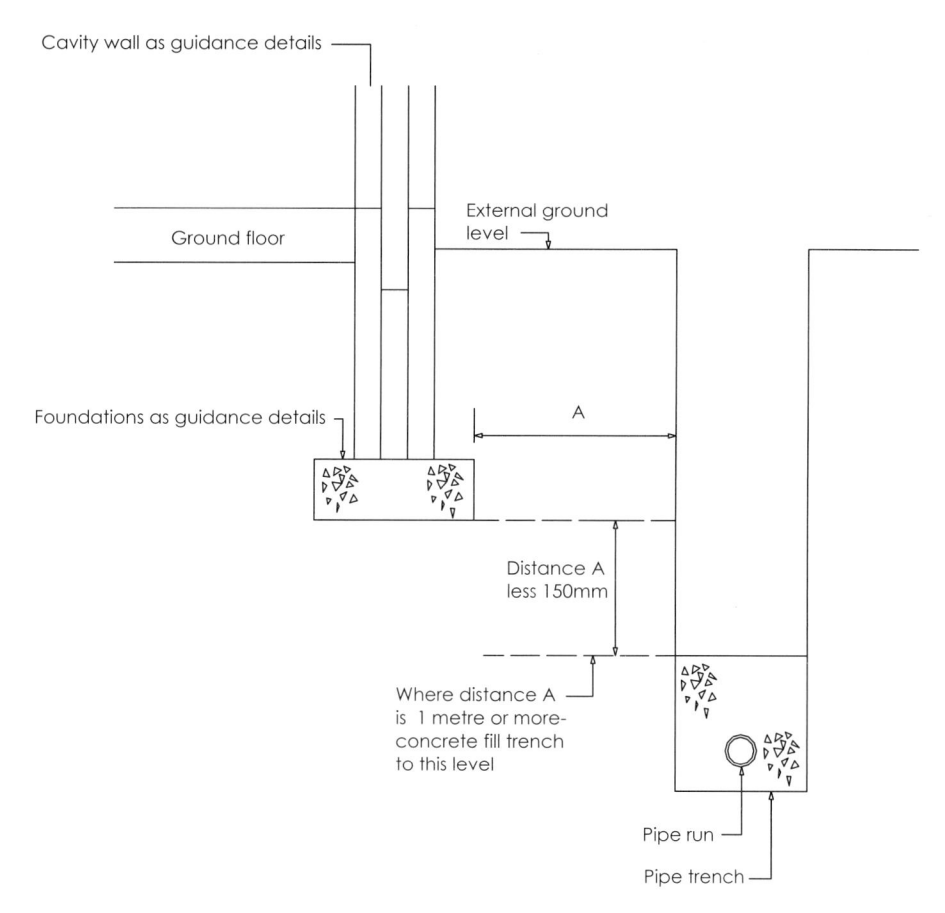

**Figure 2.54.2:** Typical detail of drain trenches more than 1 m from buildings *(not to scale)* (See Diagram 8 of ADH1 for full details.)

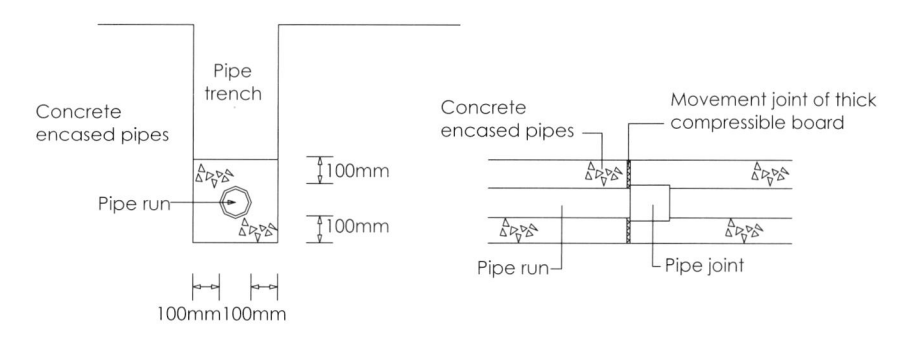

**Figure 2.55:** Typical movement joint detail for concrete-encased pipes – minimum sizes stated *(not to scale)* (See Diagram 12 of ADH1 for full details.)

**Table 2.43:** Minimum dimensions for access fittings and inspection chambers (See Table 11 of ADH1 for full details.)

| Type | Depth to invert from cover level (m) | Internal sizes | | Cover sizes | |
|---|---|---|---|---|---|
| | | Length × width (mm) | Circular (mm) | Length × width (mm) | Circular (mm) |
| **Rodding eye** | n/a | As drain but minimum 100 | n/a | n/a | Same size as pipework[1] |
| **Access fitting** | | | | | |
| Small: 150 diameter 150 × 100 | 0.6 or less except where situated in a chamber | 150 × 100 | 150 | 150 × 100[1] | Same size as access fitting |
| Large: 225 × 100 | 0.6 or less except where situated in a chamber | 225 × 100 | 225 | 225 × 100[1] | Same size as access fitting |
| **Inspection chamber** | | | | | |
| Shallow | 0.6 or less 1.2 or less | 225 × 100 450 × 450 | 190[2] 450 | Min 430 × 430 | 190[1] 430 |
| Deep | Greater than 1.2 | 450 × 450 | 450 | Max 300 × 300[3] | Access restricted to max 350[3] |

**Notes:**
[1] The clear opening may be reduced by 20 mm in order to provide proper support for the cover and frame.
[2] Drains up to 150 mm diameter.
[3] A larger clear-opening cover may be used in conjunction with a restricted access. The size is restricted for health and safety reasons to deter entry.

### *Manholes*

Manholes (either proprietary or to a specialist's design in accordance with British/European Standards) should be provided for pipe depths 1.2 m to 3.0 m and manhole shafts provided for pipe depths exceeding 3.0 m deep in accordance with Table 2.44 (see Table 12 of ADH1 for full details). Maximum spacing of access points to be in accordance with Table 2.45 (see Table 13 of ADH1 for full details). Manhole covers to be mechanically fixed and suitable for vehicular loads in drives and roads in accordance with manufacturer's details.

### *Foul-water disposal*

Foul water should be discharged into new or existing foul-water drainage facilities using existing or new inspection chamber connection, as shown on plans/specification, or as agreed with building control on site.

Foul drainage systems to low-lying buildings or basements that carry storm water or other vulnerable drainage systems should be provided with anti-flood protection, such as one-way valves etc., to prevent flooding and sewage from entering the building.

### *Waste pipes*

All WCs are to have a trapped outlet connected to 100 mm-diameter pipes (can be reduced to 75 mm if there is no risk of blockage, i.e. sanitary towels). Sanitary appliances such as wash hand basins, baths, showers, sinks, etc. are to be provided with waste pipes laid to falls and fitted with traps, with sizes as stated in Table 2.46 below. Where waste-pipe runs exceed 4 m, British Board of Agreement

**Table 2.44:** Minimum dimensions for manholes
(See Table 12 of ADH1 for full details.)

| Type | Size of largest pipe (DN) (mm) | Minimum internal dimensions[1] | | Min. clear opening size[1] | |
|---|---|---|---|---|---|
| | | Rectangular length and width (mm) | Circular diameter (mm) | Rectangular length and width (mm) | Circular diameter (mm) |
| **Manhole** | | | | | |
| Less than 1.5 m deep to soffit | Equal to or less than 150<br>225<br>300<br>Greater than 300 | 750 × 675[7]<br>1200 × 675<br>1200 × 750<br>1800 × (DN+450) | 1000[7]<br>1200<br>1200<br>The larger of 1800 or (DN+450) | 750 × 675[2]<br>1200 × 675[2] | n/a[3] |
| Greater than 1.5 m deep to soffit | Equal to or less than 225<br>300<br>375–400<br>Greater than 450 | 1200 × 1000<br>1200 × 1075<br>1350 × 1225<br>1800 × (DN+775) | 1200<br>1200<br>1200<br>The larger of 1800 or (DN+775) | 600 × 600 | 600 |
| **Manhole shaft[4]** | | | | | |
| Greater than 3.0 m deep to soffit pipe | Steps[5] | 1050 × 800 | 1050 | 600 × 600 | 600 |
| | Ladder[5] | 1200 × 800 | 1200 | – | – |
| | Winch[6] | 900 × 800 | 900 | 600 × 600 | 600 |

**Notes:**
1. Larger sizes may be required for manholes on bends or where there are junctions.
2. May be reduced to 600 × 600 where required by highway loading considerations, subject to a safe system of work being specified.
3. Not applicable due to working space needed.
4. Minimum height of chamber in shafted manhole 2 m from benching to underside of reducing slab.
5. Minimum clear space between ladder or steps and the opposite face of the shaft should be approximately 900 mm.
6. Winch only – no steps or ladders, permanent or removable.
7. The minimum size of any manhole serving a sewer (i.e. any drain serving more than one property) should be 1200 mm × 675 mm rectangular or 1200 mm diameter.

**Table 2.45:** Maximum spacing of access points in metres
(See Table 13 of ADH1 for full details.)

| From | To access fitting | | | | |
|---|---|---|---|---|---|
| | Small | Large | Junction | Inspection chamber | Manhole |
| Start of external drain[1] | 12 | 12 | – | 22 | 45 |
| Rodding eye | 22 | 22 | 22 | 45 | 45 |
| Access fitting: Small | | | | | |
| 150 diam and 150 × 100 | – | – | 12 | 22 | 22 |
| Large 225 × 100 | – | – | 45 | 22 | 45 |
| Inspection chamber shallow | 22 | 45 | 22 | 45 | 45 |
| Manhole and inspection chamber deep | – | – | – | 45 | 90[2] |

**Notes:**
1. Stack or ground-floor appliance.
2. May be up to 200 for man-entry-size drains and sewers.

**Table 2.46:** Waste pipe and trap design limits (see Tables 1 and 2 of ADH1 for full details).

| Appliance | Minimum diameter of pipe and trap (mm) | Minimum depth of trap seal (mm) | Slope of pipe (mm/m) | Maximum length of pipe to stack (m) |
|---|---|---|---|---|
| Sink | 40 | 75 | 18 to 90 | 3 (increased to 4 for 50 mm-diam pipe[1]) |
| Washing Machine[2] | 40 | 75 | As appliance manufacturer's details | As appliance manufacturer's details |
| Dishwasher[2] | 40 | 75 | As appliance manufacturer's details | As appliance manufacturer's details |
| Bath[2] | 40 | 50 | 18 to 90 | 3 (increased to 4 for 50 mm-diam pipe) |
| Shower[2] | 40 | 50 | 18 to 90 | 3 (increased to 4 for 50 mm-diam pipe) |
| WC outlet < 80 mm<br>outlet > 80 mm | 75[5]<br>100 | 50<br>50 | 18 to 90<br>18 to 90[4] | 6<br>6 |
| Washbasin[1] | 32 | 75 | 120/0.5<br>80/0.75<br>50/1.0<br>35/1.25<br>25/1.5<br>20/1.75 | 1.7 (increased to 3 for 40 mm-diam pipe) |

**Notes:**
1. Depth of seal may be reduced to 50 mm only with flush grated wastes without plugs on spray tap basins.
2. Where these appliances discharge directly to a gully the depth of seal may be reduced to 38 mm.
3. Traps used on appliances with flat bottom (trailing waste discharge) and discharging to a gully with a grating the depth of seal may be reduced to 38 mm.
4. May be reduced to 9 mm/m on long drain runs where space is restricted, but only if more than one WC is connected.
5. Not recommended where disposal of sanitary towels may cause blockages.

(BBA or other third-party accredited) air-admittance valves are to be fitted above the appliance spill-over level. Waste pipes are to discharge either below a trapped gully grating or into soil-and-vent pipes via proprietary waste manifolds or bossed junctions. Internally all waste and drainage pipes are to have rodding access/eyes at changes of direction and must be adequately clipped/supported and provided with 30 minutes' fire protection where passing through floors.

### *Soil-and-vent pipes (discharge stack)*

These are to consist of a uPVC proprietary above-ground drainage system, sized in accordance with Table 2.47 below. The discharge stack is normally installed internally through the building in sound-insulated boxing, as in the guidance details, and is fitted with a proprietary flashing system through the roof or vent tile, or alternatively soil-and-vent pipe fixed externally, in accordance with the manufacturer's details. A ventilated stack should terminate 900 mm minimum above any opening into the building that is within 3.0 m of the stack and fitted with a proprietary grilled vent cap. An open soil-and-vent pipe should always be fitted wherever possible at the head of the drainage system, particularly where a septic tank or sewage treatment system is installed.

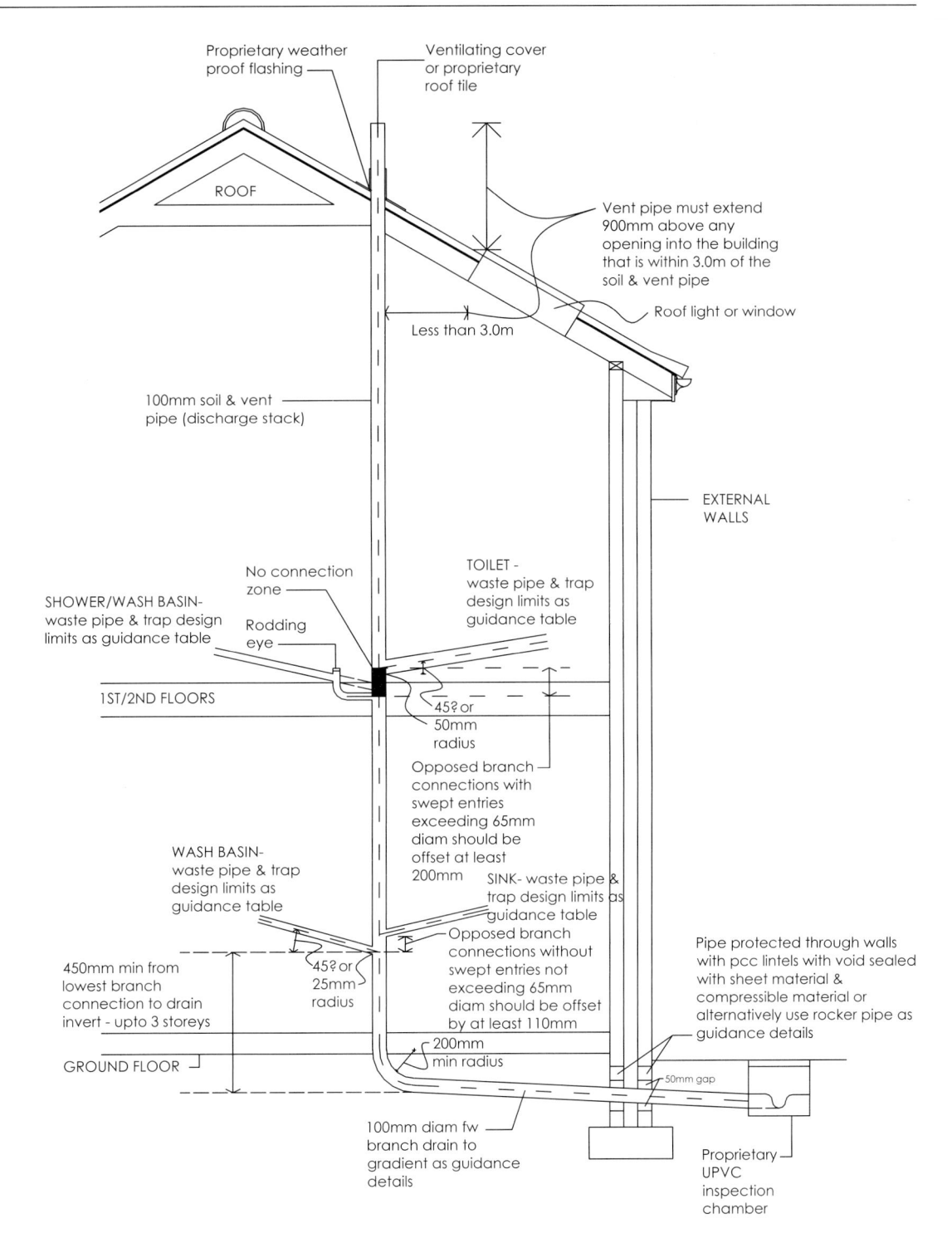

**Figure 2.56:** Typical section indicating sanitary pipework through dwelling *(not to scale)* (See Diagrams 1, 2, 3, 4, 5 and 6 of ADH1 for full details.)

**Table 2.47:** Minimum diameters for discharge stacks
(See Table 3 of ADH1 for full details.)

| Stack size (mm) | Maximum capacity (litres/sec) |
|---|---|
| 50* | 1.2 |
| 65* | 2.1 |
| 75** | 3.4 |
| 90 | 5.3 |
| 100 | 7.2 |

Key:
*No WCs.
**Not more than 1 WC with outlet size <80 mm.

### Waste-pipe connections to soil-and-vent pipes (discharge stack) – to prevent cross-flow

(i) *Waste pipes up-to 65 mm diameter* – opposed pipe connections (without swept entries) should be offset at least 110 mm on a 100 mm-diameter stack and 250 mm on a 150 mm-diameter stack, at a radius of 25 mm or angle of 45 degrees; or alternatively a proprietary manifold can be fitted in accordance with the manufacturer's details.

(ii) *Waste pipes over 65 mm diameter* – opposed pipe connections (with swept entries) should be offset at least 200 mm irrespective of stack diameter (no connections are allowed within this 200 mm zone), at a radius of 50 mm or angle of 45 degrees. Unopposed connections may be at any position.

(iii) *Lowest waste pipe connection to soil-and-vent pipe* – 450 mm minimum distance is required between centre line of waste-pipe connection to soil-and-vent pipe and invert level of below-ground drain, ensuring a 200 mm minimum radius bend connects the soil-and-vent pipe to the drain.

### Stub stacks

These are to consist of 100 mm-diameter uPVC proprietary above-ground drainage system (can be reduced to 75 mm for not more than 1 WC), with wash basins/sinks connected to the sub-stack within 2.0 m of the invert level of the drain, and the WC floor level is to be within 1.3 m of the invert level of the drain.

### Air-admittance valves

Proprietary air-admittance valves fitted to sub-stacks or soil-and-vent pipes should comply with BS EN 12380 and be installed in accordance with the manufacturer's details; the valve is to be located above the spill-over level of the highest appliance, i.e. wash basin or sink. Valves installed internally should be located in sound-insulated boxing, accessible for maintenance and clearance of blockages, etc. and fitted with a 225 × 75 mm louvred vent. Valves should not be installed in dusty environments. An open vent should always be fitted wherever possible at the head of the drainage system, particularly where a septic tank or sewage treatment system is installed.

### Airtightness and testing

Pipes, fittings and joints should be capable of withstanding an air test of positive pressure of at least 38 mm water gauge for at least 3 minutes. Every trap should maintain a water seal of at least 25 mm.

Smoke testing should be used to identify defects where the water test has failed. Note: Smoke testing is not recommended for uPVC pipes.

### Pumping installations

Where gravity drainage is impractical, or protection is required against flooding due to surcharging in downstream sewers, pumped drainage solutions may be required – subject to building control approval.

Proprietary packaged pumping systems are to consist of a watertight GRP/polyethylene chamber, lockable pedestrian/vehicle covers, pumps, high-level alarm, preset automatic level control, float switch, non-return valve, discharge pipe and connections, etc. Domestic sewage pump sets located within buildings should conform to BS EN 12050, designed in accordance with BS EN 12056-4 and installed in accordance with the manufacturer's details. Domestic sewage pump sets located outside buildings should be designed in accordance with BS EN 752-6 and installed in accordance with the manufacturer's details. Pumped installations must contain 24 hours' inflow storage. The minimum daily discharge of foul drainage should be taken as 150 litres per person per day for domestic use. Auto-changeover duty/standby duplex (twin pumps) pump stations may be accepted as an alternative to 24 hours' storage, subject to approval by building control.

## H2: Septic tanks, sewage treatment systems and cesspools

### Existing septic tank and effluent drainage

Where additional drainage effluent is to be connected to the existing septic tank/treatment system, it should be checked by a specialist and sizes/condition of tank/system are to be confirmed as suitable for treatment of additional effluent.

### Non-mains foul-drainage waste water treatment systems

Non-mains drainage systems are to be used only where connection to the mains drainage system is not possible. Either a septic tank or sewage treatment system is to be installed to suit specific ground conditions, as agreed with building control. No septic tank/sewage treatment system and associated tertiary (secondary) treatment is permitted by the Environment Agency in prescribed Zone 1 groundwater source protection zones. Where no other option is feasible, the installation of a cesspool is to be agreed with building control and the Environment Agency.

## Septic tanks

Septic tanks are to consist of a watertight chamber (watertight from both sides, to prevent the ingress of water and contain the effluent). The sewage is liquefied by anaerobic bacteria action in the absence of oxygen, assisted by the natural formation of a surface scum or crust. Sludge settlement at the base of the tank must be removed annually (or more frequently if required). Discharge from tanks is to be taken to drainage fields, drainage mounds or wetlands/reed beds for secondary treatment, as detailed in the guidance below.

Proprietary, factory-made septic tanks are to be designed and constructed to BS EN 12566 and installed in accordance with the manufacturer's details, or:

Non-proprietary septic tanks constructed in situ are to be designed and constructed to a drainage specialist's design and approved by building control before the works commence on site. Typically the tank consists of two chambers (the first being twice as large as the second), constructed using minimum 150 mm-thick reinforced concrete base C25P mix to BS 5328; 220 mm-thick engineering-quality brickwork walls (or concrete), mortar mix 1:3 cement/sand ratio with waterproof rendering or a suitable proprietary tanking system applied to both sides; and a designed, heavy concrete roof structure. 100 mm-diameter inlet and outlet 'dip pipes' are required and are designed to prevent disturbance of the surface scum; the inlet pipe is laid at a flatter gradient for at least 12 metres before it enters the tank.

Septic tanks are to be fitted with durable lockable lids or covers for emptying and cleaning, and an inspection chamber is to be fitted on the discharge side of the tank for sampling of the effluent.

Septic tanks are to be sited at least 7 m from any habitable part of any building, preferably down slopes, and within 30 m of a suitable vehicle access for emptying and cleaning sludge, which must not be taken through a dwelling or place of work and must not be a hazard to the building's occupants. If the tank invert is more than 3.0 m, the 30 m distance should be reduced.

A septic tank should have a minimum capacity of 2,700 litres for up to four users, increased by 180 litres for each additional user. (Recommended minimum size of a septic tank is 3,800 litres, to accommodate discharges from washing machines, dishwashers, etc.). A notice plate must be fixed within the building to include the following information: address of the property; location of the treatment system; description of the septic tank and effluent drainage installed; necessary maintenance to be carried out (including monthly checks of the apparatus and emptying of the tank every 12 months by a licensed contractor); and a statement that the owner is legally responsible to ensure that the system does not cause pollution, health hazard or nuisance.

Consultations are to be carried out with building control and the Environment Agency before any works commence on site. It is the occupier's responsibility to register the effluent discharge as an exempt facility with the Environment Agency for discharges of $2\,m^3$ or less per day to the ground from a septic tank, or to obtain an Environmental Permit from the Environment Agency. Septic tanks must not discharge to a watercourse. For more information contact the Environment Agency at: www.environment-agency.gov.uk.

## Sewage treatment systems

Proprietary sewage treatment systems treat sewage by an accelerated (aerobic) process to higher standards than that of septic tanks, and are to be factory-made, designed and constructed to BS EN 12566 (if fewer than 50 persons, otherwise to BS 6297:2007 Code of Practice for design and installations of small sewage treatment works and cesspools and BBA certification (or other approved accreditation). They are to be installed and maintained in accordance with the manufacturer's details and fitted with an uninterruptible power supply (or 6 hours' power back-up). Note: Only treatment systems suitable for intermittent use should be used for holiday lets or similar uses where the system is unused for periods of time.

A sewage treatment system is to be sited at least 7 m from any habitable part of any building, preferably down slopes, and within 30 m of a suitable vehicle access for emptying and cleaning sludge, which must not be taken through a dwelling or place of work and must not be a hazard to the building's occupants. If the tank invert is more than 3.0 m, the 30 m distance should be reduced.

A sewage treatment system should be designed to British Water design criteria, based on the maximum occupancy of the property and the final effluent-quality requirements of the Environment Agency.

Discharges from sewage treatment systems can be taken to a water course or alternatively a designed drainage field, drainage mound, wetlands or reed beds as detailed below.

A notice plate must be fixed within the building and include the following information: address of the property; location of the treatment system; description of the sewage treatment system and effluent drainage installed; necessary maintenance to be carried out in accordance with the manufacturer's details; and a statement that the owner is legally responsible to ensure that the system does not cause pollution, health hazard or nuisance.

Consultations should be carried out with building control and The Environment Agency before any works commence on site. It is the occupier's responsibility to register the effluent discharge as an exempt facility with the Environment Agency for discharges of $5\,m^3$ or less per day to a surface watercourse, or $2\,m^3$ or less per day to the ground from a sewage treatment system, or to obtain an Environmental Permit from the Environment Agency. For more information, contact the Environment Agency at: www.environment-agency.gov.uk.

## Disposal of sewage from septic tanks and sewage treatment systems

### Drainage fields

Drainage fields consist of irrigation pipes laid below ground, allowing partially treated effluent to percolate into the ground and further biological treatment to take place in the aerated soil layers. Construction of drainage fields is to be carried out as the tank/system manufacturer's details and BS6297:2007 + A1:2008. See typical guidance section detail and drainage field layout in Figures 2.57 and 2.58 below. The drainage field area is calculated from the percolation test results; the suggested minimum area is $30\,m \times 0.6\,m$ wide, subject to percolation test results and number of users and approved by building control before works commence on site. See below for percolation test procedure.

Drainage fields are to be located 10 m from any watercourse, 50 m from any point of water abstraction, 15 m from any building, 2 m from any boundary and sufficiently far from any other drainage areas so the overall soakage capacity of the ground is not exceeded. Water supply pipes, access roads, drives or paved areas, etc. must not be located within the drainage areas.

See Diagram 1 of ADH2 for typical drainage field construction details.

### Drainage mounds

Drainage mounds consist of drainage fields constructed above the ground, allowing further biological treatment of the partially treated effluent in the aerated soil layers. Drainage mounds are to be used where there is a high water-table level, impervious or semi-waterlogged ground. Drainage mounds and drainage mound areas should be designed by a drainage specialist for particular ground problems and approved by building control before works commence on site. See Diagram 2 of ADH2 for typical drainage mound construction details.

Drainage mounds are to be located 10 m from any watercourse, 50 m from any point of water abstraction, 15 m from any building, 2 m from any boundary and sufficiently far from any other drainage areas so the overall soakage capacity of the ground is not exceeded. Water supply pipes, access roads, drives or paved areas, etc. must not be located within the drainage areas.

See Diagram 2 of ADH2 for typical drainage mound construction details.

### Wetlands/reed beds

Constructed manmade wetlands/reed-bed treatment systems can be used to provide secondary or tertiary treatment of effluent from septic tanks or sewage treatment systems. Reed beds can be constructed as either vertical- or horizontal-flow reed-bed systems (see Diagrams 3 and 4 of ADH2 for full details) for the purification of the partially treated effluent by filtration, biological oxidation, sedimentation and chemical precipitation as the partially treated effluent passes through gravel beds and root systems of wetland plants. Wetlands should not be constructed in shaded, windblown or severe-winter areas. Vertical- or horizontal-flow wetland treatment systems should be designed by a drainage specialist for particular ground problems and approved by building control before works commence on site.

A notice plate must be fixed within the building to include the following information: address of the property; location of the treatment system; description of the sewage treatment system and effluent drainage installed; necessary maintenance to be carried out in accordance with the drainage specialist's details; and a statement that the owner is legally responsible to ensure that the system does not cause pollution, health hazard or nuisance.

## Percolation tests

A percolation test is required to calculate the area of a drainage field for a septic tank or sewage treatment system. A preliminary assessment of the site should be carried out, including consultation with the Environment Agency and building control to determine the suitability of the site.

### Ground conditions

Ground conditions should be assessed to determine the suitability of subsoils. Examples of suitable subsoils with good percolation include sand, gravel, chalk, sandy loam and clay loam. Examples of poor subsoils are sandy clay, silty clay and clay. It is important that percolation characteristics are suitable in both summer and winter conditions and that the subsoil is well drained and not saturated with water. A trial hole should be excavated 1.5 m below the invert of the proposed effluent drainage pipework, to determine the position of the standing ground-water table. The ground-water level in summer and winter should be at least 1.0 m below the invert of the effluent drainage pipework.

### Percolation test method

Percolation tests should not be carried out in abnormal weather conditions such as heavy rain, severe frost or drought.

**Step 1:** Excavate a test hole 300 mm square × 300 mm deep below the proposed invert level of the drainage field trench bottom.
**Step 2:** Fill the test hole with water and allow it to drain away overnight.
**Step 3:** Refill to a depth of 300 mm and note the time taken in seconds to drain away from 75 percent full to 25 per cent full (i.e. 150 mm drop in level, from 225 mm to 75 mm).
**Step 4:** Carry out the procedure a second and a third time (can be on the same day if the hole empties completely and quickly enough).
**Step 5:** Repeat the procedure in two more test holes and calculate the average of the three results as follows:

$$\frac{(\text{test 1} + \text{test 2} + \text{test 3})}{3} = \textbf{average time taken for each test hole}$$

**Step 6:** Find the average of these results as follows:

$$\frac{(\text{hole } 1 + \text{hole } 2 + \text{hole } 3)}{3} = \textbf{average time taken for all test holes}$$

**Step 7:** Calculate the Vp (average time in seconds for the water to drop 1 mm) as follows:

For example: If average time above took 2100 seconds:

(i)  Divide 2100 seconds by 150 mm depth of water
(ii) $\frac{2100}{150} = 14$ Vp* (see note below*)
(iii) Area of trench = number of persons to use property × Vp × 0.25 (0.25 figure is used for septic tanks and can be reduced to 0.20 for treatment systems)
Therefore: 5 persons × 14 × 0.25 = 17.5 m² of effluent drainage field is required.
(iv) To calculate actual length of drainage trench required, divide 17.5 m² by width of the trench required; therefore:

$$\frac{17.5 \text{ m}^2}{0.6 \text{ m wide}} = \underline{\textbf{29.16}} \text{ (suggested minimum area 30 m long × 0.6 m wide)}$$

*Vp should range between 12 and 100 to be successful; otherwise the system should be designed by a drainage specialist.

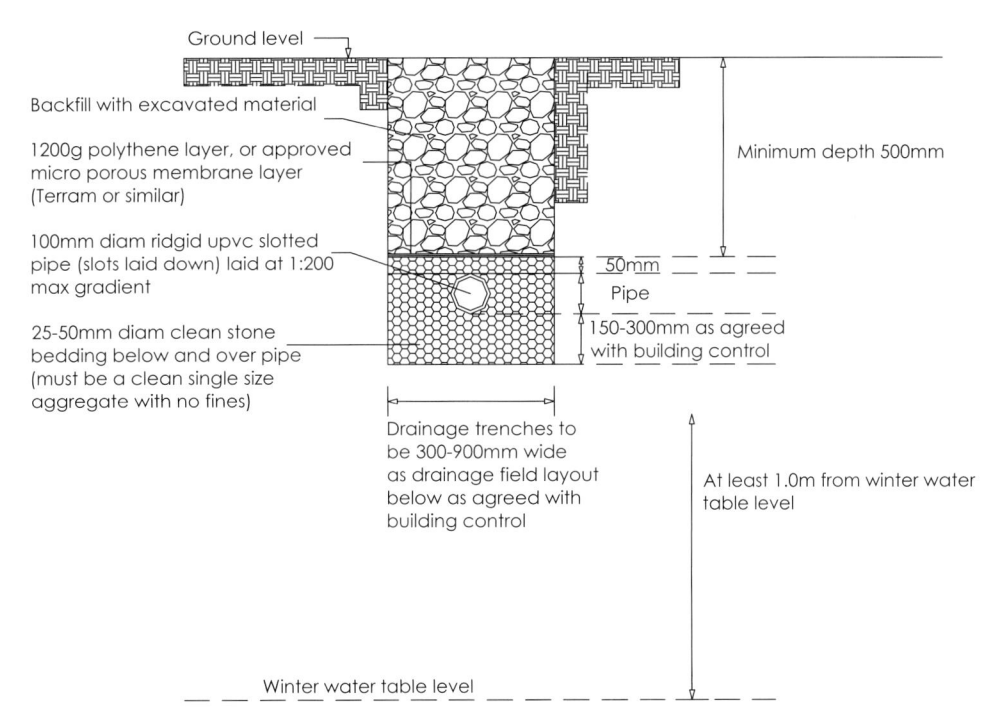

**Figure 2.57:** Typical section through a septic tank/sewage treatment system drainage field *(not to scale)* (See Diagram 1 of ADH2 for full details.)

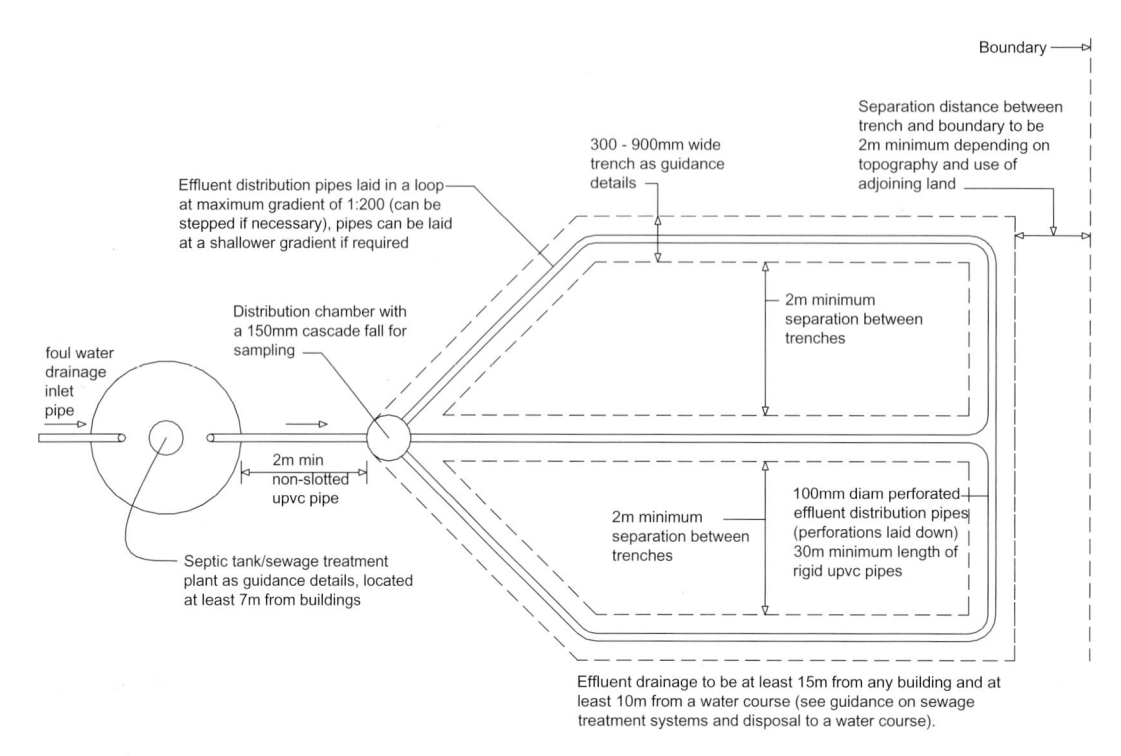

**Figure 2.58:** Typical drainage field plan layout *(not to scale)*
(See Diagram 1 of ADH2 for full details.)

## Cesspools

Cesspools are sealed, watertight tanks used for the containment of domestic sewage and must be emptied regularly by a licensed contractor. Cesspools are used in locations without main drainage in locations acceptable to the Environment Agency, where the discharge of treated effluent is not permissible owing to unsuitable ground conditions, or where infrequent use or seasonal use would prevent the functioning of a septic tank or sewage treatment system.

Proprietary, factory-made cesspools are to be designed and constructed to BS EN 12566-1 and installed in compliance with the manufacturer's details, **or:**

Non-proprietary cesspools can be constructed in situ to a drainage specialist's design and approved by building control before the works commence on site. Cesspools must be watertight to prevent leakage of the contents and ingress of subsoil water, Typically the tank consists of one chamber constructed using minimum 150 mm-thick reinforced concrete base designed by a suitably qualified specialist and suitable for storing aggressive effluents; 215 mm-thick engineering-quality brickwork walls (or dense concrete bricks), bond to be agreed with building control; mortar mix 1:3 cement/ sand ratio with waterproof render or suitable proprietary tanking system applied to both sides; and a designed, heavy concrete roof structure.

Cesspools are to be ventilated and fitted with durable lockable lids or covers for emptying and cleaning, and the inlet side of the tank should be fitted with a lockable access for inspection. No other openings are permitted. A high-level alarm should be fitted for monitoring the cesspool for optimum usage.

Cesspools are to be sited:

- at least 7 m from any habitable part of any building, preferably down slopes and lower than any existing building; and
- within 30 m of a vehicle access suitable for emptying and cleaning the effluent. The contents should not be taken through a dwelling or place of work and must not be a hazard to the building's occupants.

Cesspools should have a minimum capacity of 18,000 litres (18.0 m$^3$) for up to two users and increased by 6800 litres (6.8 m$^3$) for each additional user.

A notice plate must be fixed within the building describing the necessary maintenance, and the following is an example of suitable wording:

- 'The foul drainage system from this property is served by a cesspool.'
- 'The system should be emptied approximately every (insert frequency) by a licensed contractor and inspected fortnightly for overflow.'
- 'The owner is legally responsible to ensure that the system does not cause pollution, a health hazard or a nuisance.'

Consultations are to be carried out with building control and the Environment Agency before any works commence on site. Cesspools normally do not need registration with the Environment Agency, as they are sealed systems with no discharge to the environment. For more information contact the Environment Agency at: www.environment-agency.gov.uk.

## H3: Rainwater drainage and harvesting

### Rainwater gutters and down pipes

Rainwater gutters and down-pipe sizes and number are to be suitable for the roof area to be drained, in compliance with Table 2.48 below, and fixed in compliance with the manufacturer's details. (See H3 of ADH for further information)

**Table 2.48:** Gutter sizes and pipe outlet sizes for drainage of roof areas
(See Table 2 of ADH3 for full details.)

| Maximum effective roof area (m$^2$) | Gutter sizes (mm diameter) | Outlet sizes (mm diameter) |
|---|---|---|
| 18.0 | 75 | 50 |
| 37.0 | 100 | 63 |
| 53.0 | 115 | 63 |
| 65.0 | 125 | 75 |
| 103.0 | 150 | 89 |

**Note:** The sizes above refer to half-round gutters and round rainwater pipes.

### Rainwater/grey water harvesting storage tanks and systems

A rainwater harvesting system is to be designed, installed and commissioned by a specialist to supply rainwater to sanitary appliances. Below-drainage pipework is to be carried out in accordance with the foul-water pipe guidance details above. Overflow from the rainwater storage tank is to discharge to a designed soak-away system constructed at least 5 m from any building.

Grey water (consisting of recycled bath, shower and basin waste water) systems designed for use within the building are to be designed, manufactured, installed and commissioned by a suitably qualified and experienced specialist. Grey water is to be treated prior to use in toilets etc. by an approved method and overflow to discharge to the foul-water drainage system.

Grey water and rainwater tanks and systems should:
- prevent leakage of the contents and ingress of subsoil water, and should be ventilated
- have an anti-backflow device on any overflow connected to a drain or sewer
- have a corrosion-proof, locked access cover for emptying and cleaning
- have supply pipes from the grey water or rainwater collector tanks to the dwelling that are clearly marked as either 'GREY WATER' or 'RAIN WATER'
- follow the guidance in paras 1.69–1.72 of App Doc H2, App Doc G of the Building Regulations, the Water Regulations Advisory Scheme Leaflet No: 09-02-04, and BS 8515:2009
- be proprietary manufactured systems, installed in compliance with manufacturer's details.

### Surface water drainage around the building

Paths and paved areas around the building are to have a non-slip finish and be provided with a surface cross fall of 1:40–1:60 to dispose of rain/surface water, and a reverse gradient of at least 500 mm away from walls of building (unless the paved/path area is a proprietary system designed to be porous and installed in accordance with the manufacturer's details). Surface water is to be disposed of by an adequately sized and roddable drainage system via soak-aways, or by other approved means.

### Rain-/surface-water disposal

Rain/surface water is to be piped away from buildings, as detailed in the guidance above, and discharged into new or existing surface water soak-away, storm-water or combined storm-/foul-water drainage facilities using an existing or new inspection chamber connection, as shown on plans/specification, or as agreed with building control on site. New connections to existing storm-or combined storm-/foul-water systems may require consent from the relevant water authority before works commence on site. Rain/surface water should connect into a combined system only with the consent of the relevant water authority, and only into a foul system under exceptional circumstances and subject to written approval from the water authority.

Rain/surface water disposed of in a separate surface-water sewer or combined sewer should be connected via trapped gullies, with inspection chamber positions as detailed in the guidance for foul-water drainage. Drainage systems to low-lying buildings or basements that carry storm water or other vulnerable drainage systems should be provided with anti-flood protection, such as one-way valves etc., to prevent flooding and sewage from entering the building.

### Existing soak-aways

Where additional rain-/surface-water systems are to be connected to the existing soak-away system, it should be checked by a specialist and the sizes of the soak-away should be confirmed and agreed with building control as adequate for percolation into the surrounding ground.

### New soak-aways

New surface-water soak-away(s) are to be designed, sited and constructed to provide adequate short-term storage for rain/surface water and adequate percolation into the ground. Soak-aways should be sited at least 5 m from any buildings and constructed on land lower than, or sloping away from, the foundations of the buildings.

Soak-aways are to have a minimum capacity of 1–2 m$^3$ (in free-draining granular-type subsoils) per rainwater pipe serving a roof area up to 30 m$^2$, as agreed with building control, and should be constructed of clean stone/rubble with particle sizes ranging in size from 20 to 150 mm, then covered with polythene (or suitable geotechnical membrane) and topsoil, typically as Figure 2.59 below. Soak-aways in clay subsoils or serving roof areas exceeding 30 m$^2$ per rainwater pipe are to be designed in accordance with BRE Digest 365 or by a drainage specialist (i.e. a hydrologist).

## Oil/fuel separators

Under the requirements of the Water Industries Act, it is an offence to discharge fuels into watercourses, coastal water or underground water. Oil separators are required where fuel is stored or in other high-risk areas or car parks, and the Environment Agency has issued guidance on the provision of oil separators. For paved areas around buildings or car parks a bypass separator is required, with a nominal size of 0.0018 times the contributing area and silt storage area (in litres) equal to 100 times the nominal size.

In fuel storage areas and other high-risk areas full-retention separators are required, with a nominal size equal to 0.018 times the contributing area and silt storage area (in litres) equal to 100 times the nominal size. Separators discharging to infiltration devices or surface-water sewers should

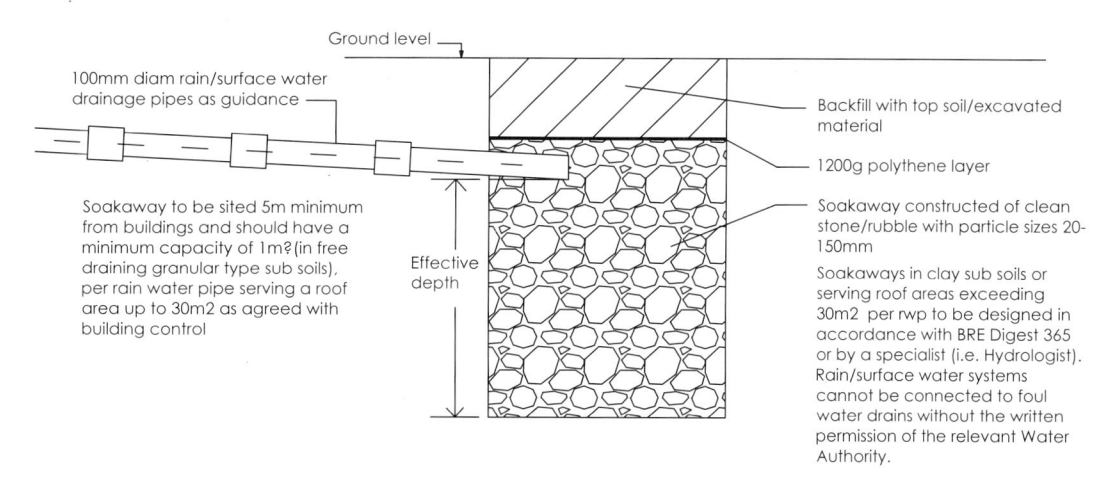

**Figure 2.59:** New soak-away design (section detail – *not to scale*)

be class 1 (and capable of accommodating the whole content volume of one compartment of a delivery tanker).

Proprietary oil separators should be factory-made, waterproof, and designed and constructed to the requirements of the Environment Agency, the licensing authority's requirements (where the Petroleum Act applies), prEN858 and BBA certification (or other approved accreditation). Separators must be installed and maintained in compliance with the manufacturer's details and inlet arrangements should not be directly to the water surface. Adequate ventilation must be provided. The separator must be cleaned out and emptied regularly by a licensed contractor. See Appendix H3-A of ADH3 for further information

## H4: Building over or close to, and connections to, public sewers

## Building over or close to a public sewer

The relevant water authority (WA), being the sewerage undertaker, is responsible for maintaining public sewers, and the owner/developer of a building being constructed, extended or underpinned within 3 m of a public sewer – as indicated on the relevant WA sewer maps – is required to consult with the WA to ensure that:

(i) No damage occurs to the sewer. The extra weight of a building being constructed, extended or underpinned, or a new building above a sewer, could cause the sewer to collapse, resulting in structural damage to the new building, interrupted drainage from other properties and wastewater flooding. In these instances the sewer will need to be repaired quickly and that could involve taking down the building.
(ii) Suitable access is available to carry out any maintenance, repair or replacement works to the public sewer.
(iii) Consent is obtained and an agreement is entered into to build close to or over the public sewer before works commence on site.

### Locating a public sewer

Copies of the sewer record maps are held by the WA and local authority for the location of public sewers, and checks should be carried out at an early stage to ensure that the proposed works do not affect a public sewer.

### *Options*

If you find that your plans could affect a public sewer, you should consult the relevant WA and discuss with them the following options:

- Avoiding the sewer through modifications of plans, so that the building is at least 3 metres away from the sewer. This is often the easiest and cheapest option.
- Diverting the sewer. If the plans cannot be modified, the WA will usually require the sewer to be diverted. In most cases the diversion works are carried out at the property owner's expense, normally by contractors approved by the WA.

**The WA will not normally allow construction directly over a manhole or pressurised pumping main.**

### The build-over process

If the only option is to apply to build over a public sewer, the building owner should make an application to the WA, who may allow a sewer to be built over, subject to the sewer being in satisfactory condition and to obtaining their written agreement before works commence.

### Typical procedure

- A closed-circuit television (CCTV) survey is carried out by the WA before works commence, to ascertain whether any repair work is required.
- Another survey is required when the building is completed, to check that the sewer has not been damaged.
- In certain circumstances, if the building owner does not obtain the WA's agreement, the WA has the right to discontinue the works and take down the building erected over the public sewer.
- Consultations should be carried out early on in the design process to avoid any abortive costs, delays or other problems.
- The WA makes a charge for applications.

## Private Sewer Transfer Regulations

Since the implementation of the Private Sewer Transfer Regulations on 1 October 2011, all lateral drains and sewers, i.e. those serving two or more properties and that connect to the public sewer network, will be adopted by the relevant water authority/sewerage provider and the above requirements for building over/close to and/or making new connections to public sewers will apply. As these lateral drains and sewers may not yet show up on the sewer maps, it is important that consultations with the WA are carried out at an early stage.

## Protection

Protection of the sewer pipes and systems is to be carried out in compliance with the WA's requirements.

## Further information

More information is available from the relevant water authority or www.defra.gov.uk/environment/quality/water/sewage/sewers or www.water.org.uk/home/policy/private-sewer-transfer.

# Connections to public sewers

Owners/developers of a building with new drainage connections or indirect drainage connections being made to a public sewer, as indicated on the relevant water authority's sewer maps, are required to consult with the WA and where necessary obtain consent before works commence on site.

## H5: Separate systems of drainage

The building owner/agent must carry out all necessary consultations with the relevant water authority before works commence on site. Rain-/surface-water systems cannot be connected to foul-water drains without the written permission of the relevant water authority. See H5 of ADH for further information.

## H6: Solid waste storage

This applies only to new dwellings and conversions to create new dwellings. See H6 of ADH and 'Section 3: New dwellings' in this guidance for further information

## PART J: COMBUSTION APPLIANCES AND FUEL STORAGE SYSTEMS

Please refer fully to the relevant Approved Document.

### Space and hot-water/heat-producing appliances in general

All space and hot-water systems must be in accordance with BS 5449, BS 5410 and BS 8303, installed, commissioned, calibrated and certified by a suitably qualified person or installer registered with an appropriate competent persons' scheme, and details supplied to building control and the owner along with the operating manuals etc. before the building is completed/ occupied.

Boilers are to have a SEDBUK efficiency above 90 per cent to comply with Building Regulations as amended in October 2010 for gas/LPG/oil, and must be provided with separate controls for heating and hot water, with a boiler interlock and timer. Separate temperature control of zones within the dwelling should be provided as follows: room thermostat or programmable room thermostats in all zones, and individual radiator control such as thermostatic radiator valves (TRVs) on all radiators other than in reference rooms (with a thermostat) and bathrooms.

**Table 2.49:** Typical minimum design guide temperatures for rooms

| Room | Design room temperature °C |
| --- | --- |
| Living room (including study or similar room) | 21 |
| Dining/breakfast room | 21 |
| Bed-sitting room/open-plan flat | 21 |
| Bedroom | 18 |
| Hall and landing | 18 |
| Kitchen | 18 |
| Bathroom/shower room/en suite | 22 |
| Toilet/cloakroom | 18 |

**Note:** Design room temperatures above are based on an external temperature of −3°C.

Unvented hot-water systems require safety devices, including non-self-setting energy cut-out and temperature release valve and thermostat. Safety valves from vented hot-water systems must discharge safely.

Hot-water vessels are to be insulated with 75 mm minimum thickness of PU foam and both heating and water pipes are to be insulated with proprietary foam covers, equal to their outside diameter, within 1 m of the vessel and in unheated areas.

## Solid fuel appliances up to 50 kW rated output

### Construction of open fire recessed and hearth

Fireplace walls are to consist of non-combustible material of at least 200 mm in thickness to the side and at least 100 mm thick in the back wall recess, lined with suitable fire bricks or proprietary fire back. The constructional hearth is to be at least 125 mm thick (or of 25 mm minimum thickness for decorative, non-combustible, superimposed hearth with changes in levels to mark safe perimeter, fixed over 100 mm minimum concrete floor slab). The hearth is to project at least 150 mm from the side jambs and 500 mm in front of the jambs, as detailed in Figure 2.60 below

### Construction of solid-fuel masonry chimneys

Chimneys are to be constructed as detailed in Figure 2.61 below, in external-quality, frost-resistant materials of 100 mm minimum thickness (increased to 200 mm where they separate another fire compartment or another dwelling), using brick, dense blocks or reconstituted/natural stone to match the existing, with suitable mortar joints for the masonry – as the masonry manufacturer's details – with any combustible material kept at least 200 mm away from the flue and 40 mm away from the walls containing flues, in compliance with Diagram 21 of ADJ. Line the chimney with manufactured flue liners, installed in compliance with the manufacturer's details as follows:

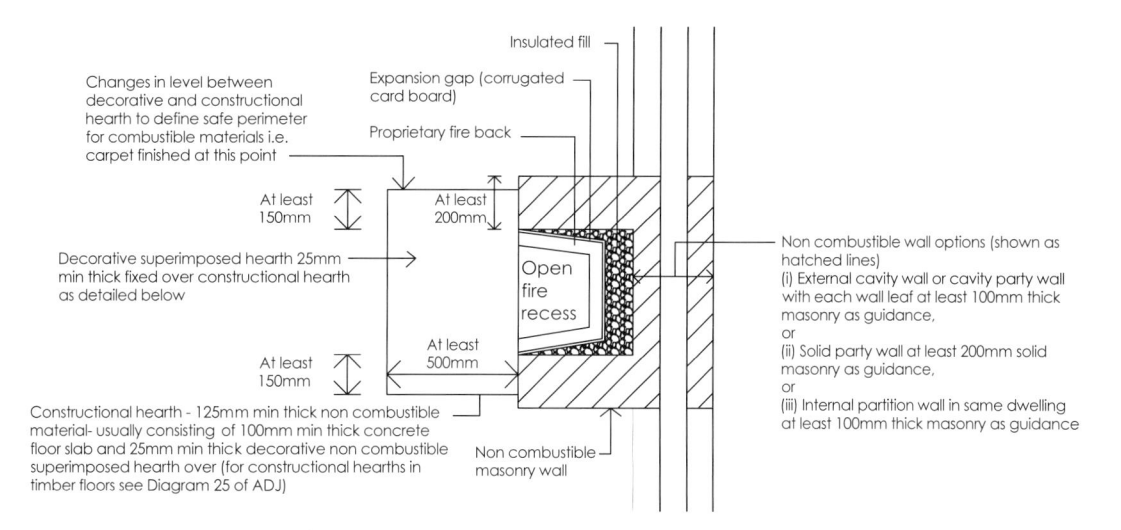

**Figure 2.60:** Non-combustible hearth details for recessed open fire (plan – *not to scale*)
(See Diagrams 28 and 29 of ADJ for full details.)

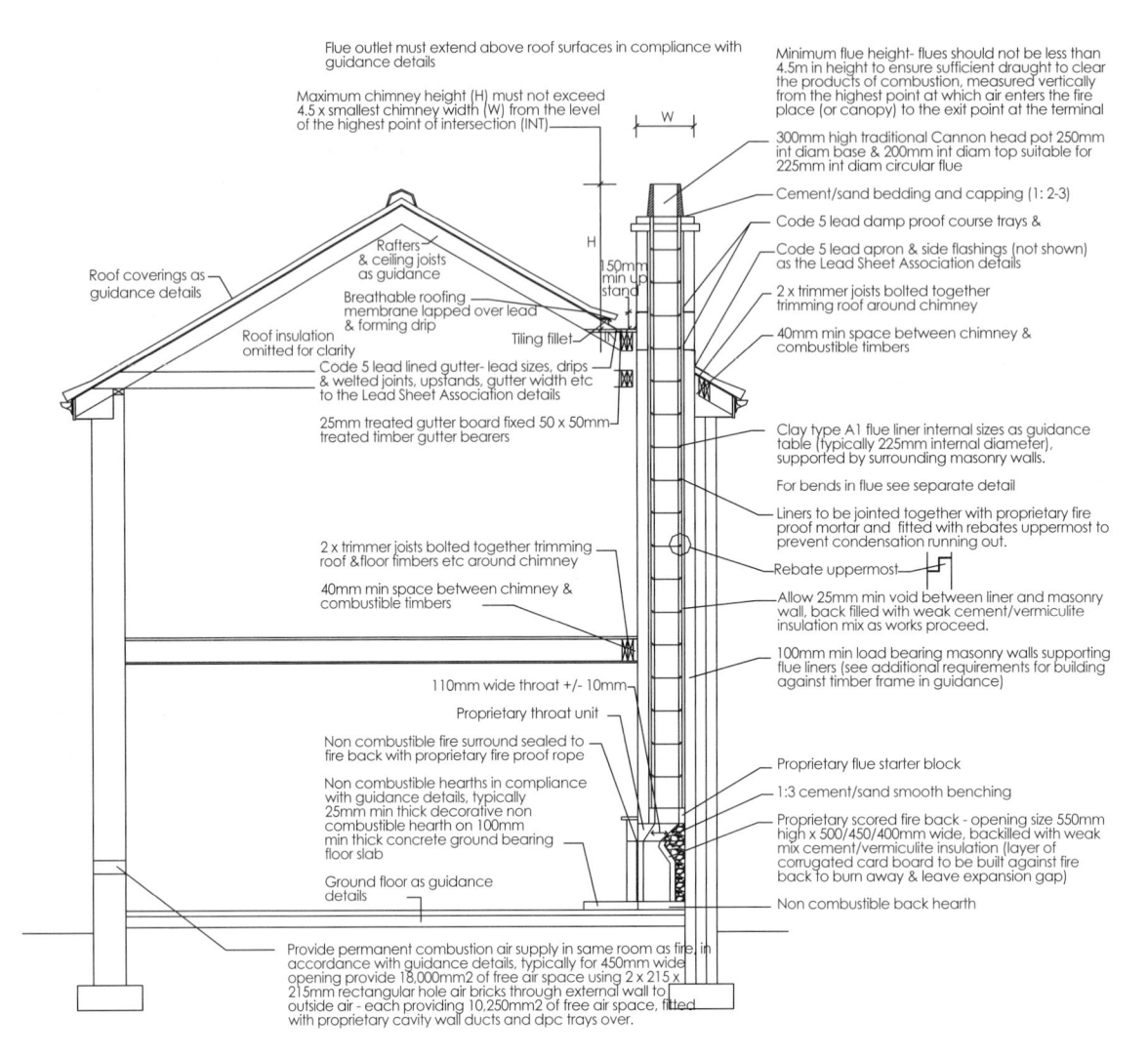

**Figure 2.61:** Solid-fuel masonry chimney construction (section detail – *not to scale*)

(i) Clay flue liners to BS EN 1457:2009: Class A1 N1 or Class A1 N2, to be laid vertically and continuously with socket up (jointed with fireproof mortar) from appliance, with a minimum diameter in compliance with Table 2.50 below.
(ii) Concrete flue liners to BS EN 1857:2003: Type A1, A2, B1 or B2, to be laid vertically and continuously socket up (jointed with fireproof mortar) from appliance, with a minimum diameter in compliance with Table 2.50 below.
(iii) Liners whose performance complies to BS EN 1443:2003: designation: T400 N2 D 3 G, with a minimum diameter in compliance with Table 2.50 below.

### Free-standing solid-fuel stove and hearth

Free-standing solid-fuel stoves are to be installed in accordance with the manufacturer's details, fixed to a non-combustible hearth, sizes at least $840 \times 840$ mm, positioned 150 mm minimum away

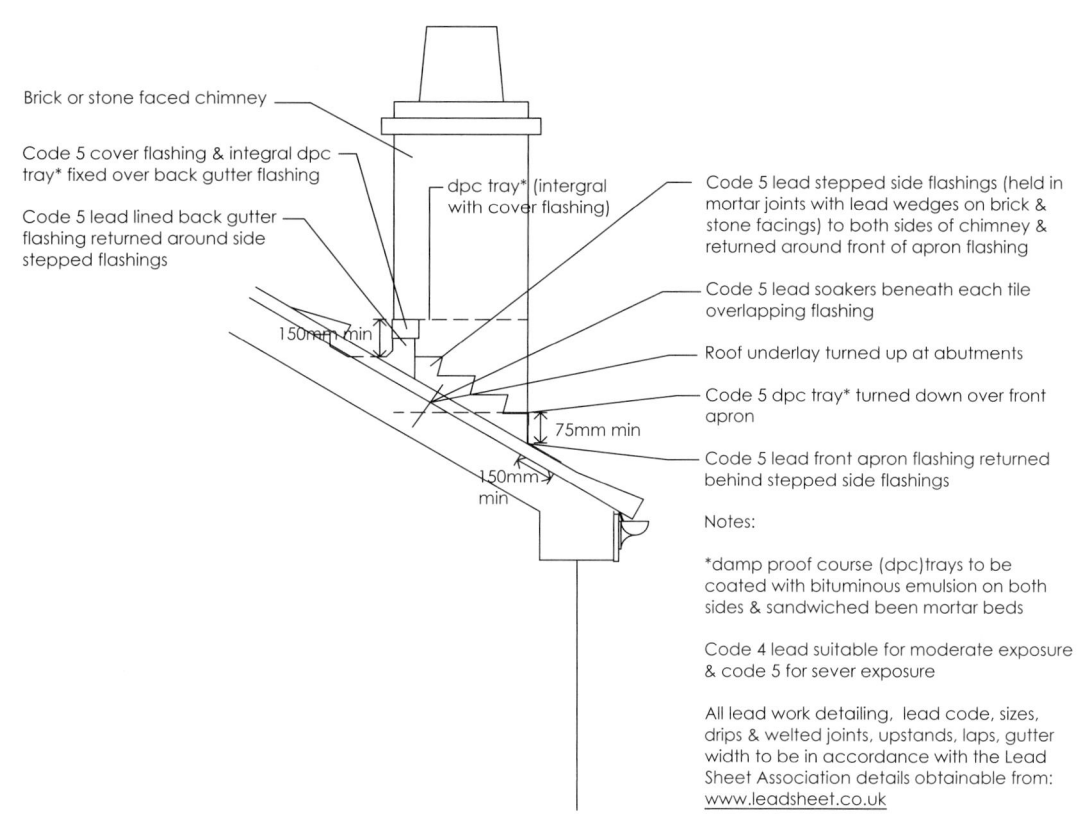

**Figure 2.62:** Lead flashing detail to solid-fuel masonry chimney (elevation detail – *not to scale*)

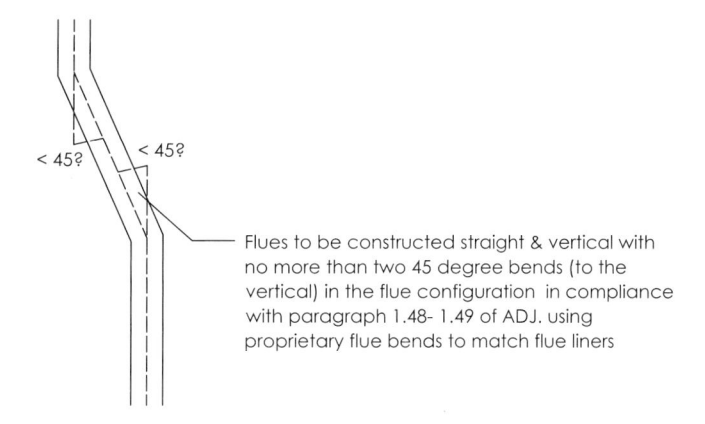

**Figure 2.63:** Solid-fuel chimney construction with bends (section detail – *not to scale*) (See Diagram 15 of ADJ for full details.)

from enclosing non-combustible walls (walls to be at least 100 mm thick). The constructional hearth is to be at least 125 mm thick (or can be a decorative, non-combustible, superimposed hearth of 25 mm minimum thickness, fixed over minimum 100 mm-thick concrete floor slab with changes in levels to mark a safe perimeter). The hearth should project at least 150 mm to the sides and rear of the appliance and 300 mm in front of the operable appliance door, as detailed in Figure 2.65 below.

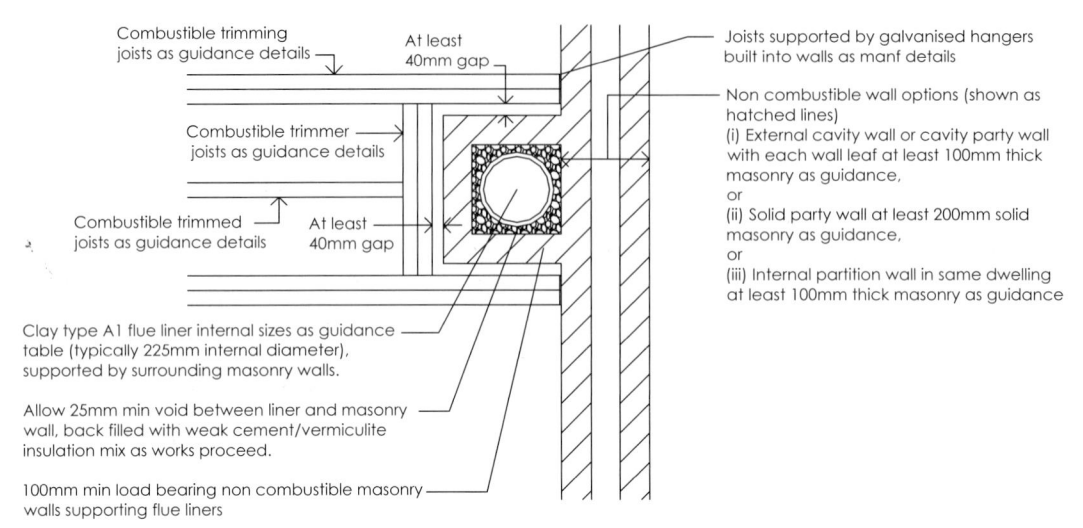

## 2.128 Domestic extensions

**Figure 2.64 labels (upper diagram):**

- Combustible trimming joists as guidance details
- At least 40mm gap
- Joists supported by galvanised hangers built into walls as manf details
- Combustible trimmer joists as guidance details
- Non combustible wall options (shown as hatched lines)
  (i) External cavity wall or cavity party wall with each wall leaf at least 100mm thick masonry as guidance,
  or
  (ii) Solid party wall at least 200mm solid masonry as guidance,
  or
  (iii) Internal partition wall in same dwelling at least 100mm thick masonry as guidance
- Combustible trimmed joists as guidance details
- At least 40mm gap
- Clay type A1 flue liner internal sizes as guidance table (typically 225mm internal diameter), supported by surrounding masonry walls.
- Allow 25mm min void between liner and masonry wall, back filled with weak cement/vermiculite insulation mix as works proceed.
- 100mm min load bearing non combustible masonry walls supporting flue liners

Additional space separation requirements:
(i) Combustible material i.e. floor joists, etc built into or fixed to chimney wall must be at least 200mm from the flue liner
(ii) Timber frame construction built against a chimney wall must be at least 200mm from the flue liner
(ii) Combustible material supported by metal fastening or support bracket built into chimneys must be at least 50mm from the flue liner
Note: Decorative trims i.e. skirting board, picture rail, architrave etc can be fixed directly to 100mm min thick chimney wall with no additional space separation requirements

**Figure 2.64:** Minimum separation distances from combustible material in or near to a solid-fuel chimney (plan detail – *not to scale*)
(See Diagram 21 of ADJ for full details.)

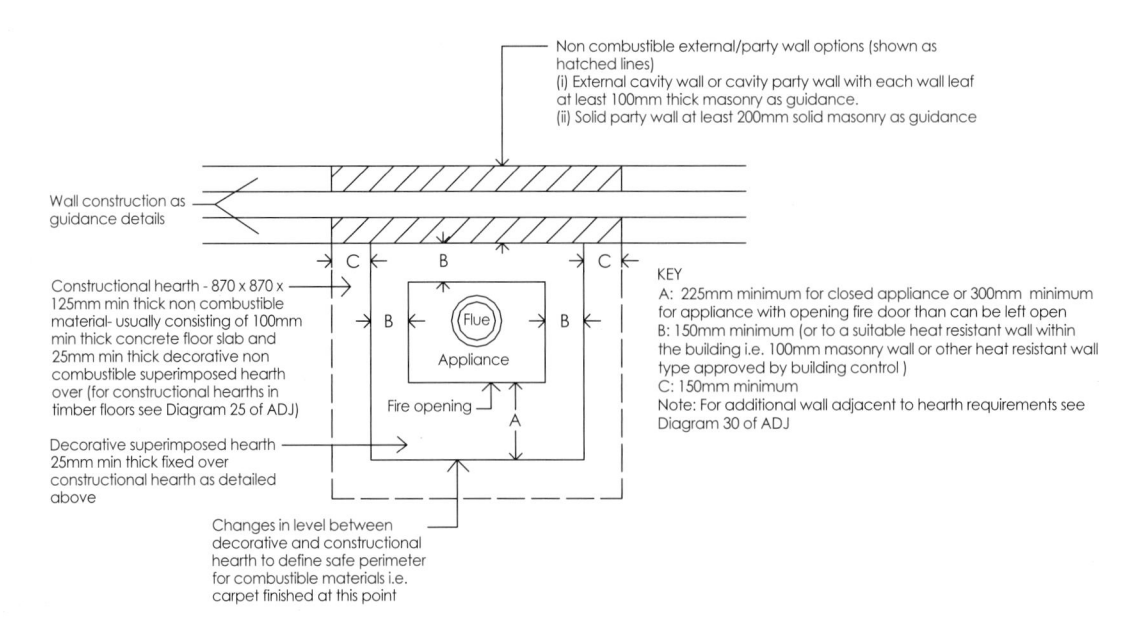

**Figure 2.65 labels:**

- Non combustible external/party wall options (shown as hatched lines)
  (i) External cavity wall or cavity party wall with each wall leaf at least 100mm thick masonry as guidance.
  (ii) Solid party wall at least 200mm solid masonry as guidance
- Wall construction as guidance details
- Constructional hearth - 870 x 870 x 125mm min thick non combustible material- usually consisting of 100mm min thick concrete floor slab and 25mm min thick decorative non combustible superimposed hearth over (for constructional hearths in timber floors see Diagram 25 of ADJ)
- Decorative superimposed hearth 25mm min thick fixed over constructional hearth as detailed above
- Changes in level between decorative and constructional hearth to define safe perimeter for combustible materials i.e. carpet finished at this point
- (Flue) Appliance
- Fire opening

KEY
A: 225mm minimum for closed appliance or 300mm minimum for appliance with opening fire door than can be left open
B: 150mm minimum (or to a suitable heat resistant wall within the building i.e. 100mm masonry wall or other heat resistant wall type approved by building control)
C: 150mm minimum
Note: For additional wall adjacent to hearth requirements see Diagram 30 of ADJ

**Figure 2.65:** Non-combustible hearth detail under free-standing solid-fuel stove (plan detail – *not to scale*)
(See Diagrams 26, 27 and 30 of ADJ for full details.)

### *Flue pipe connections to free-standing stove and chimneys*

Single flue pipes connecting the appliance to a chimney should not extend beyond the room in which the appliance is located, and should not pass through any roof space, partition, internal wall or floor (unless it connects to the chimney at that point). The maximum recommended length is 1–1.5 m, to prevent heat transfer and improve flue efficiency. Minimum flue length is 0.6 m.

Single flue pipes should be guarded if they could be at risk of damage, or if the burn hazard is not immediately apparent to people. A single flue pipe must be located to avoid igniting combustible materials and must be at least three times its internal diameter from any combustible materials ($3 \times 150$ mm = 450 mm); **or:**

The combustible material can be heat-shielded; the flue must be at least 1.5 times its diameter from the heat shield. The heat shield (typically 12 mm-thick proprietary fire-resistant board) must extend at least 1.5 times the flue's internal diameter to each side of the flue, and there must be an air gap of at least 12 mm (formed with strips of fire board) between the shield material and the combustible material; **or:**

The connecting flue pipe is to be factory-made in compliance with T 400 N2 D3 G, according to BS EN 1856-2:2004, and installed to BS EN 15827-1

### Construction of factory-made, insulated, twin-walled metal chimneys

Construction of factory-made metal chimneys is to be carried out in compliance with Paragraphs 1.42–1.46 of ADJ and the appliance manufacturer's details. The separation of combustible materials from a factory-made, twin-walled metal chimney is to be carried out in compliance with Diagram 13 of ADJ. Where a metal chimney passes through a cupboard, storage space or roof space it must be fully separated by at least 50 mm from combustible materials with a non-combustible steel-mesh guard. Factory-made metal chimneys concealed in the building are to be accessible for inspection, in compliance with Paragraph 1.47 and Diagrams 13 and 14 of ADJ. Chimneys passing through combustible floors and roofs should be fitted with proprietary fire-stop shields. For chimneys passing through fire compartment walls or floors – contact building control for further advice.

### Carbon monoxide alarms

A carbon monoxide alarm is required at ceiling level in the same room as the solid-fuel appliance, which must be either battery-operated in compliance with BS EN 5029: 2001, or mains-operated with sensor failure warning device in compliance with BS EN 5029: Type A carbon monoxide alarms are to be positioned on the ceiling at least 300 mm from walls or, if located on the wall, as high up as possible (above any doors or windows), but not within 150 mm of the ceiling, and between 1 m and 3 m horizontally from the appliance

### Air supply (ventilation) to solid-fuel appliances

Permanently open combustion air vents ducted to outside are to be provided in the same room as the solid-fuel appliance, with a total free area in compliance with Table 2.51 below (see Table 1 of ADJ for further information).

**2.130** Domestic extensions

Where a metal chimney passes through a cupboard, storage space or roof space it must be fully separated with at least 50mm* from combustible materials with an approved non combustible steel mesh guard.

Ceiling joists

Where a metal chimney passes through a room it must be boxed in (or guarded), the flue must be separated with at least 50mm* from combustible materials & must be accessible for inspection

Floor joists

Single flue pipe connection appliance to twin walled chimney should not extend beyond the room in which appliance is located. Recommended maximum length 1-1.5m. Minimum length 0.6m

Single flue pipe must be at least 3 times its internal diameter from combustible materials, or 1.5 times its diameter from a suitable heat shield in compliance with guidance details

Permanent combustion air to be in compliance with guidance details

* Actual distance should be calculated in compliance with BS EN 1856 & BS 4543-1

Roof structure

Flue outlet height in compliance with guidance details

Proprietary roof penetration flashing suitable for roof pitch installed in compliance with flue manf details

Proprietary fire stop shield installed through ceiling to provide 50mm* minimum air gap clearance between flue & combustible materials in compliance with flue manf details

Twin walled insulated stainless steel multi-fuel chimney/flue system, installed in compliance with manf details. chimney/flues concealed in the building are to be accessible for inspection in compliance with manf details & paragraph 1.47 & diagrams 13 & 14 of **ADJ**

150mm minimum internal flue sizes as guidance details

Proprietary fire stop shield installed through floor to provide 50mm* minimum air gap clearance between flue & combustible materials in compliance with flue manf details

Flues to be constructed straight & vertical with no more than a 90 degree bend with cleaning access where the flue connects to the appliance & no more than two 45 degree bends (to the vertical) in the flue configuration in compliance with paragraph 1.48- 1.49 of **ADJ**.

Free standing stoves installed to manf details, & positioned on non combustible hearths in compliance with guidance details, typically 25mm min thick decorative non combustible hearth on 100mm min thick concrete ground bearing floor slab

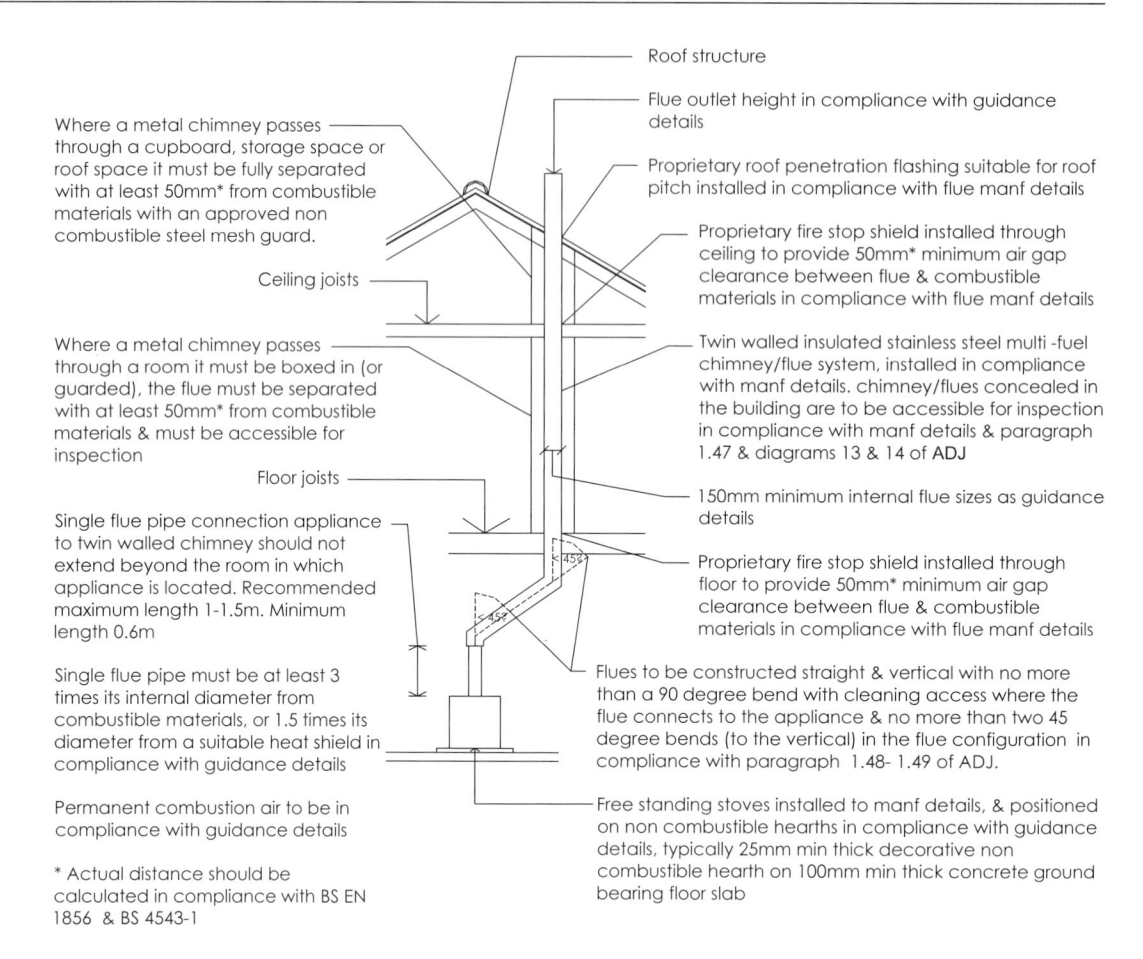

**Figure 2.66:** Free-standing stove and twin-walled insulated stainless steel chimney detail through a building (section detail – *not to scale*)
(See Diagrams 14, 15, 16, 17, 18 and 19 of ADJ for full details.)

### *Construction of factory-made flue block chimneys*

Construction of factory-made flue block chimneys is to be carried out in compliance with Paragraphs 1.29–1.30 of ADJ, and the appliance manufacturer's details.

### *Configuration of flues serving open-flue appliances*

Flues are to be constructed straight and vertical, with no more than a 90-degree bend and with cleaning access where the flue connects to the appliance and no more than two 45-degree bends (to the vertical) in the flue configuration, in compliance with Paragraphs 1.48–1.49 of ADJ.

### *Inspection and cleaning openings in chimneys and flues*

Where a chimney/flue cannot be cleaned through the appliance, an airtight, accessible inspection and cleaning opening should be fitted using proprietary factory-made components compatible with the flue system, fitted and located to allow sweeping of the flue in compliance with the appliance manufacturer's details.

Non combustible wall options as guidance details:
(i) External cavity wall
(ii) Cavity or solid party wall
(iii) Internal partition wall

Timber frame

12.5mm plaster board

Note: Proprietary fire stop shield to be installed through upper timber floor(s) to provide 50mm minimum air gap clearance between flue & combustible materials in compliance with flue manf details (Actual distance should be calculated in compliance with BS EN 1856 & BS 4543-1)

Where a metal chimney passes through a room it must be boxed in (or guarded)

Twin walled insulated stainless steel multi-fuel chimney/flue system, installed in compliance with manf details. chimney/flues concealed in the building are to be accessible for inspection in compliance with manf details & paragraph 1.47 & diagrams 13 & 14 of ADJ

Flue must be separated with at least 50mm from combustible materials (Actual distance should be calculated in compliance with BS EN 1856 & BS 4543-1) & must be accessible for inspection.

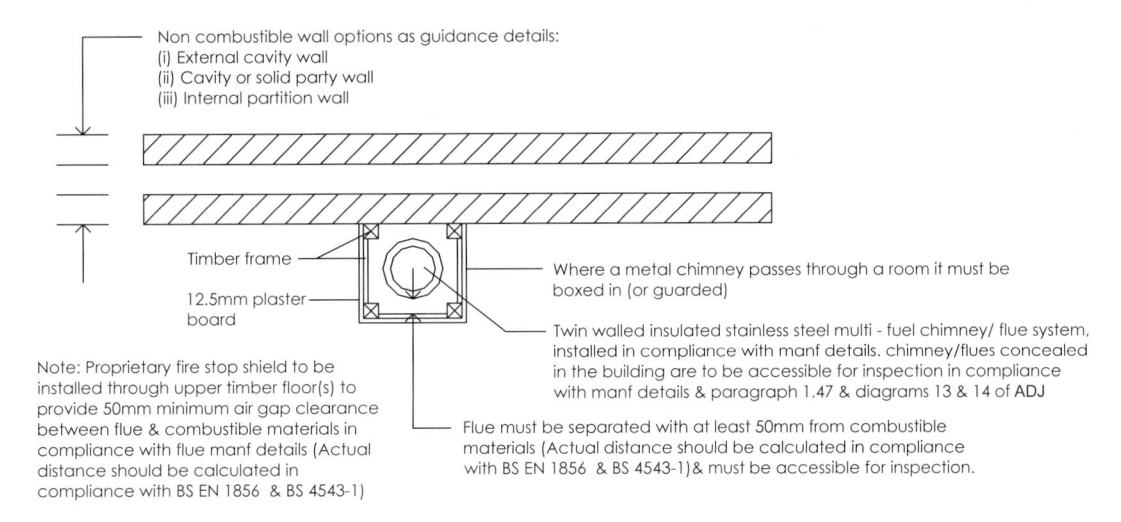

**Figure 2.67:** Separation of twin-walled insulated flue from combustible materials (plan detail – *not to scale*)
(See Diagram 13 of ADJ for full details.)

**Table 2.50:** Sizes of flues in chimneys
(See Table 2 of ADJ for full details.)

| Installation | Minimum internal flue sizes |
| --- | --- |
| Fireplace with opening up to 500 × 500 mm | 200 mm diameter (or rectangular/square flue having same cross-sectional area and minimum dimension of 175 mm) |
| Fireplace with opening more than 500 × 500 mm or exposed on both sides | Area equal to 15% of the total face area of the fireplace opening. (note: total face areas more than 15% or 0.12 m$^2$ are to be designed by a heating specialist). |
| Closed appliances (stove, cooker, room heater and boiler) up to: 30 kW rated output | 150 mm diameter (or rectangular/square flue having same cross-sectional area and minimum dimension of 125 mm) |
| 30–50 kW rated output | 175 mm diameter (or rectangular/square flue having same cross-sectional area and minimum dimension of 150 mm) |
| Closed appliances up to 20 kW rated output that burn smokeless/low-volatile fuel, or comply with the Clean Air Act | See Table 2 of Approved Document J. |
| Pellet burner that compiles with the Clean Air Act | See Table 2 of Approved Document J. |

### *Interaction of mechanical extract vents and open-flue combustion appliances*

Where a kitchen etc. contains an open-flue solid-fuel appliance and a mechanical extract vent, the appliance should be tested and certificated by a suitable qualified and registered HETAS engineer to see that the combustion appliance operates safely, whether or not the fans are running. Alternatively, the ventilation from the passive stack effect of an open-flue appliance may negate the need for a mechanical extract fan to be fitted in the same room, subject to approval by building control.

**Table 2.51:** Air supply (permanent ventilation) to solid-fuel appliances
(See Table 1 of ADJ for full details.)

| Type of appliance | Minimum amount of ventilation |
| --- | --- |
| Open fireplace with no throat (i.e. under a canopy) | 50% of the cross-sectional area of the flue |
| Open fireplace with throat | 50% of the cross-sectional area of the throat opening area |
| or for fire openings sizes: | |
| 500 mm wide | 20,500 mm$^2$ |
| 450 mm wide | 18,500 mm$^2$ |
| 400 mm wide | 16,500 mm$^2$ |
| 350 mm wide | 14,500 mm$^2$ |
| Enclosed stove with flue draught stabiliser*: | |
| (i) In new building/extension (good airtightness) | 850 mm$^2$/kW of appliance's rated output |
| (ii) In existing older building (if airtightness improved – use figure for new extension) | 300 mm$^2$/kW for first 5 kW and 850 mm$^2$/kW of balance of appliance's rated output |
| Enclosed stove with **no** flue draught stabiliser: | |
| (i) In new building/extension (good airtightness) | 550 mm$^2$/kW of appliance's rated output |
| (ii) In existing building (if airtightness improved – use figure for new extension) | 550 mm$^2$/kW for appliance's rated output above 5 kW |

**Note:** *A draught stabiliser is a factory-made counterbalance flap device admitting air to the flue, from the same space as the combustion air, to prevent excessive variations in the draught. It is usual for it to be in the flue pipe or chimney, but it may be located in the appliance. (see Diagram 3 of ADJ).

## Chimney/flue heights

Chimney height is not to exceed 4.5 times its narrowest thickness above the highest point of intersection (density of masonry is to be greater than 1500 kg/m$^3$). The chimney/terminal is to discharge at a minimum height in compliance with Diagram 17 of ADJ, as follows:

- 1.0 m above flat roofs
- 1.0 m above opening windows or roof lights in the roof surface
- 0.6 m above the ridge
- outside a zone measured 2.3 m horizontally from the roof slope
- 0.6 m above an adjoining or adjacent building that is within 2.3 m measured horizontally (whether or not beyond the boundary).

Please refer to Diagram 18 of ADJ for flue positions on easily ignited roofs (i.e. thatch).

### *Repair/relining of existing flues*

Repair/relining of existing flues is to be carried out by a suitably qualified and experienced specialist. Existing flues are to be inspected, tested and certified by a suitably qualified and experienced specialist prior to reuse as being suitable for solid-fuel appliances.

Relining of existing flues is to be carried out in compliance with BS EN 1443:2003: designation: T400 N2 D 3 G, with minimum diameters in compliance with Table 2.50 and using lining systems suitable for solid-fuel appliances as follows:

(i)   factory-made flue lining systems in compliance with BSEN1856-1:2003 or BSEN1856-2:2004
(ii)  cast in-situ flue lining system in compliance with BSEN1857:2003+A1:2008.

### Notice plates for hearths and flues

Notice plates for hearths and flues must be permanently displayed next to the flue (or electricity consumer unit or water stop-tap) detailing the property address; location of installation (room); type of installation the flue is suitable for; size and construction of flue, whether suitable for condensing appliance or not, installation date, and any other information (optional).

## Appliances other than solid fuel

### Gas heating appliances up to 70 kW

Gas-burning appliances up to 70 kW are beyond the scope of this guidance and are to be installed, commissioned and tested in compliance with Section 3 of ADJ, and BS 5440, BS 5546, BS 5864, BS 5871, BS 6172, BS 6173, BS 6798, and the Gas Safety (installation and use) Regulations. All works are to be to be carried out by an installer registered with Gas Safe (www.gassaferegister.co.uk). Copies of the commissioning certificates are to be issued to building control on completion of the works.

### Interaction of mechanical extract vents and open-flue gas combustion appliances

Where a kitchen etc. contains an open-flue gas appliance and a mechanical extract vent, the rate of the extract fan should not exceed 20 l/s (73 $m^3$/hour) and the appliance should be tested and certificated by a suitable qualified and registered gas safe engineer so that the combustion appliance operates safely, whether or not the fans are running.

### Oil heating appliances up to 45 kW

Oil-burning appliances up to 45 kW are beyond the scope of this guidance and are to be installed, commissioned and tested in compliance with Section 4 of ADJ and BS 5410, BS 799. All works should be carried out by an installer registered with OFTEC (www.oftec.org). Copies of the commissioning certificates are to be issued to Building Control on completion of the works.

### Interaction of mechanical extract vents and open-flue oil combustion appliances

Where a kitchen etc. contains an open-flue oil appliance and a mechanical extract vent, the rate of the extract fan should not exceed 40 l/s for an appliance with a pressure jet burner and 20 l/s for an appliance with a vaporising burner, and the appliance should be tested and certified by a suitable qualified and registered OFTEC engineer that the combustion appliance operates safely, whether or not the fans are running.

## Fuel storage tanks

### LPG tanks and cylinders up to 1.1 tonnes

LPG tanks up to 0.25 tonne capacity are to be positioned in the open air at least 2.5 m from buildings or boundaries, and 1.1 tonne tanks must be positioned 3 m from buildings or boundaries. Cylinders are to be positioned in the open air on a minimum 50 mm-thick concrete base, securely chained to the wall and positioned at least 250 mm below and 1 m from any openings horizontally into the building, such as windows, combustion vents or flue terminals, and 2 m from untrapped drains or cellar entrances. See Section Diagrams 43 and 44 and Section 5 of ADJ for full details.

### Oil tanks up to 3500 litres

Oil tanks up to 3500 litres are to be positioned in the open air on a concrete base with a minimum thickness of 50 mm, extending a minimum of 300 mm beyond the tank base, and must be positioned a minimum of 1.8 m from buildings or flues and 760 mm from boundaries. They should also be provided with a proprietary fire-resistant pipe-and-valve system. Where there is a risk of pollution to watercourses or open drains, including inspection chambers with loose covers, the tank should be either internally bunded or be provided with an impervious masonry bund equal to a capacity of 110 per cent of its volume. Where any of the above requirements cannot be met, please contact building control for further guidance. See Section 5 of ADJ for full details.

## Renewable energy/micro regeneration installations

Renewable energy systems must be installed, commissioned, calibrated and certified by a suitably qualified person or specialist installer registered with an appropriate competent persons' scheme (where applicable) and details supplied to building control and the owner, along with the operating manuals etc., for the following installations:

- solar photovoltaic (pv) roof tiles, roof/wall panels for producing electricity
- biomass boiler for space heating and hot-water systems
- wind energy turbines for producing electricity
- hydro-power systems for producing electricity
- solar thermal water heating roof/wall panel systems, fitted with an additional heating source to maintain an adequate water temperature
- ground/air source heat pumps for space heating and hot-water systems
- micro – combined heat and power (CHP) – systems (low-carbon technology that are similar to conventional gas boilers but also produce electricity).

All roof/wall structures must be adequate to support the above installations, in compliance with the manufacturer's details; additional calculations/details may also be required from a suitably qualified person if requested by building control, and must be approved before works commence on site (unless the installer is registered with a Competent Persons' Scheme). Installations must be installed in accordance with the manufacturer's details to prevent ingress of water/moisture into the building.

## Further information

More information on renewable energy/micro regeneration Installations is available from the following sources:

- BS EN 12975-2:2006: Thermal solar systems and components
- ER G59/2: Recommendations for the Connection of Generating Plant to the Distribution Systems of Licensed Distribution Network Operators
- ER G83/1: Recommendations for the Connection of Small-scale Embedded Generators (up to 16 A per phase) in Parallel with Public Low-voltage Distribution Networks
- BRE Digest 489: Wind Loads on Roof-based Photovoltaic Systems
- BRE Digest 495: Mechanical Installation of Roof-mounted Photovoltaic Systems
- The HVCA Guide to Good Practice Installation of Biofuel Heating (TR/38)
- The HVCA Guide to Good Practice Installation of Heat Pumps (TR/30)
- British Wind Energy Association: Small Wind Turbine Performance and Safety Standard
- Photovoltaics in Buildings: Guide to the installation of PV systems. 2nd edition (DTI publication 06/1972)
- CE72: Energy Efficiency Best Practice in Housing – Installing small wind-powered electricity generating systems
- CE131: Energy Efficiency Best Practice in Housing – Solar water-heating systems.

### Provision of information – commissioning certificates (testing)

A copy of the installer's commissioning certificate is to be sent to building control on completion of the work.

## PART K: PROTECTION FROM FALLING, COLLISION AND IMPACT

Please refer fully to the relevant Approved Document.

## Internal stairs, guarding and landings for changes in level of 600 mm or more

Private stairs (used for only one dwelling) are to be constructed in accordance with BS 5395 and BS 585, as detailed in the following guidance details and the Figures below: (Spiral and helical stairs are to be designed to BS 5395: Part 2.)

### Stair pitch

Stair pitch must not exceed 42°.

### Headroom

Stairs are to have minimum headroom of 2 m above the pitch line of the stairs

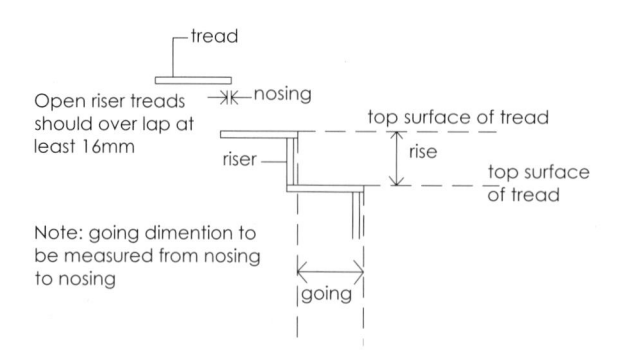

**Figure 2.68:** Measuring rise and goings *(not to scale)*
(See Diagram 1 of ADK for full details.)

### *Rise and going*

Rise and going are to be level and equal to all steps, and to fall within the following separate classes:

- Any rise between 155 mm and 220 mm used with any going between 245 mm and 260 mm, **or**
- Any rise between 165 mm and 200 mm used with any going between 223 mm and 300 mm.

(The sum of twice the rise plus the going must be between 550 and 700 mm.)

## Landings

Landings are to be provided at the top and bottom of the stairs, equal in length to the width of the stairs and clear of any door opening onto them. If a door opens across the bottom of a landing (or cupboard doors open in a similar way at the top and bottom of a flight), a clear 400 mm space must be maintained between the riser and edge of the open door across the width of the flight, in compliance with Diagrams 6 and 7 of ADK.

### *Stair width*

There is no minimum stair width for new extensions or replacement stairs in existing dwellings, but stairs should be safe and practicable. Treads should be slip-resistant where open to the weather or in wet areas.

### *Handrails*

Handrails must be provided on one side of the stairs if they are less than 1 m wide, and there should be one on each side if stairs are wider. Handrails are to provide a firm handhold with a minimum clearance of 25–50 mm between the handrail and wall, to prevent trapping of hands; should be securely fixed at a height of 900–1000 mm above floor/nosing levels; and must be continuous throughout their length.

## Guarding

Stair flights, landings, ramps and edges of internal floors are to be guarded at a minimum height of 900 mm, measured from the floor/pitch line of the stairs (across the nosings) to the top of the handrail. Guards should be continuous throughout their length, fitted with non-climbable vertical balustrading, with no gaps exceeding 100 mm (so that a 100 mm-diameter sphere cannot pass through) and all constructed to resist a horizontal force of 0.36 kN/m. Any open treads, gaps, etc. should not exceed 100 mm. See Diagram 11 of ADK for full details and BS6180 for protective barrier details.

## *Length of flights*

Stairs having more than 36 risers in consecutive flights should have a landing between flights that should be equal in length to the width of the stairs and should make a change of direction of at least 30°, in compliance with Diagram 5 of ADK.

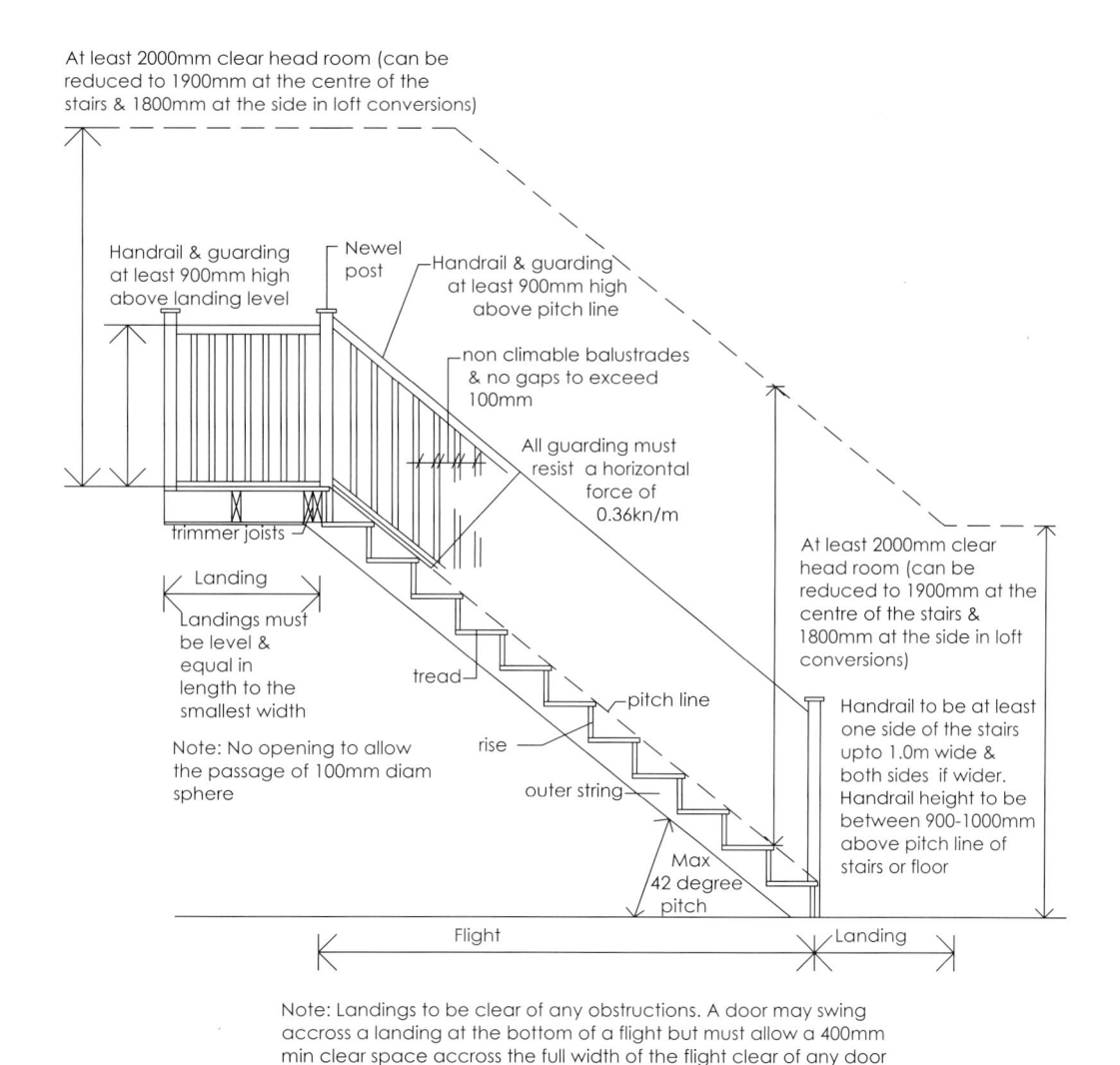

**Figure 2.69:** Typical internal staircase and guarding construction details *(not to scale)* (See Diagram 2 of ADK for full details.)

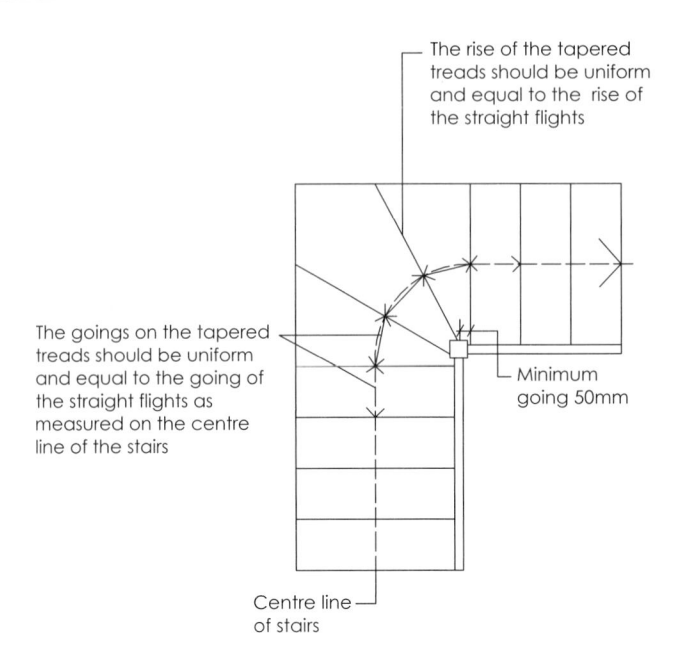

**Figure 2.70:** Typical internal tapered-tread staircase *(not to scale)*
The rise of tapered treads should be uniform and equal to the rise of the straight flight. The going on the tapered treads should be uniform and equal to the going of the straight flights as measured on the centre line of the stairs. (See Diagram 8 of ADK for full details.)


### *Typical internal staircase construction details*

Typical staircase construction details: side strings ex. 230 × 35 mm; capping ex. 32 × 63 mm; treads 25 mm thick; risers in 12.5 mm-thick plywood; newel posts ex. 75 × 75 mm; handrails ex. 75 × 63 mm; balustrades ex. 32 × 32 mm at 125 mm centres fixed into proprietary timber head and base-rebated capping.


## External stairs, guarding and landings for changes in level of 600 mm or more

### External stairs and landings

As internal stair guidance details above (guarding to external stairs and landings etc. to be in accordance with the details below).

### External guarding

Stair flights, landings, ramps and edges of external floors are to be guarded at a minimum height of 1100 mm, measured from the floor/pitch line of the stairs (across the nosings) to the top of the handrail and should be continuous throughout their length, fitted with non-climbable vertical balustrading, with no gaps exceeding 100 mm (so that a 100 mm-diameter sphere cannot pass through), and all constructed to resist a horizontal force of 0.74 kN/m. Any open treads, gaps, etc. should not exceed 100 mm. See Diagram 11 of ADK for full details and BS6180 for protective barrier details.

## Guarding to upper-storey window openings/other openings within 800 mm of floor level

Opening windows located above the ground-floor storey with openings within 800 mm of floor level must be provided with non-climbable containment/guarding or proprietary catches, which should be removable (but childproof), to means-of-escape windows in the event of a fire. Any gaps etc. to containment/guarding should not exceed 100 mm.

## Loft conversion stairs

### *Reduced headroom to stairs in loft conversions*

Where there is not enough space to achieve 2.0 m clear headroom it can be reduced to 1.9 m at the centre of the stairs and 1.8 m at the side in loft conversions, as detailed in Figure 2.71.1 below.

### Alternating tread stairs for loft conversions

Alternating tread stairs are suitable only for loft conversions and should be installed only in one or more straight flights – and then only where there is not enough space to accommodate stairs in

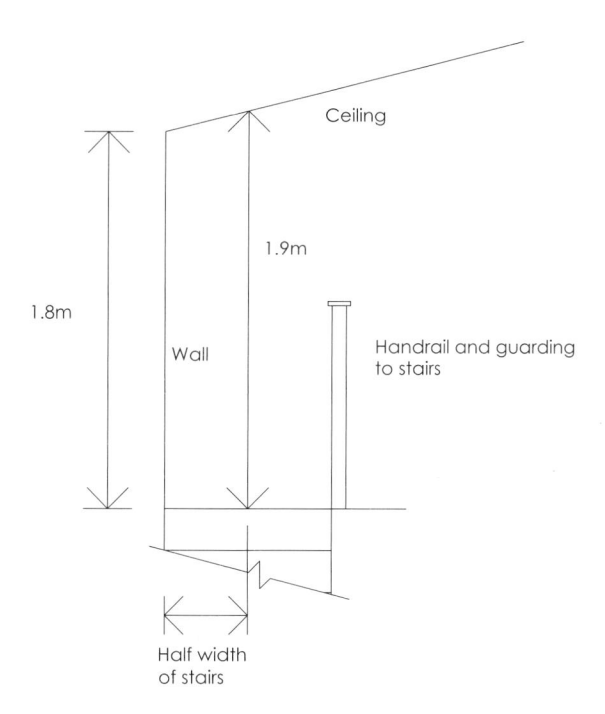

**Figure 2.71.1:** Reduced headroom to stairs in loft conversions *(not to scale)* (See Diagram 3 of ADK for full details.)

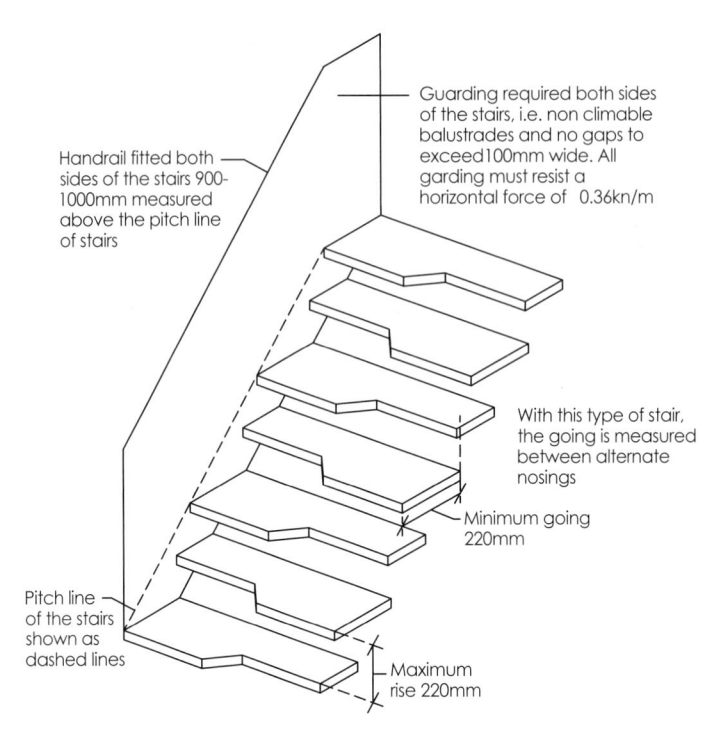

**Figure 2.71.2:** Alternating tread stairs for loft conversions *(not to scale)*
(See Diagram 9 of ADK for full details.)

accordance with Figures 2.69 and 2.70 above. The stairs should be used only to access one habitable room together with a bath/shower room or WC, providing it is not the only WC in the dwelling. The user relies on familiarity and regular use for reasonable safety. The alternating tread stairs should be constructed as follows and in accordance with Figure 2.71.2: (See diagram 9 of ADK for full details.)

- Steps should be uniform, with parallel nosings.
- Treads should be slip-resistant.
- Tread sizes over the wider part of the step should be in accordance with the dimensions in the guidance above, with a maximum rise of 220 mm and a minimum going of 220 mm.
- Handrails are o be fitted to both sides and guarded in accordance with the above guidance details for internal stairs.

### Fixed ladders for loft conversions

Fixed ladders should only be used in certain circumstances, in accordance with Paragraph 1.25 of ADK and subject to building control approval.

## Ramps

See Section 2 of ADK2 for full details. See ramp requirements for new dwellings in Part M of this guidance and in ADM.

# PART L: CONSERVATION OF FUEL AND POWER IN EXISTING DWELLINGS

Please refer fully to the relevant Approved Document.

### Listed buildings, conservation areas and ancient monuments

If the proposed energy efficiency requirements will unacceptably alter the character or appearance of a historic/listed building/ancient monument or building within a conservation area, then the energy efficiency standards may be exempt or improved to what is reasonably practical or acceptable without increasing the risk of deterioration of the building fabric or fittings, in consultation with the local planning authority's conservation officer and in compliance with Paragraphs 3.6–3.14 of AD L1B.

## Areas of external windows, roof windows and doors

Area of external windows, roof windows and doors should not exceed the sum of:

(i) 25 per cent of the floor area of the extension, plus
(ii) the total area of any windows or doors that, as a result of the extension works, no longer exist or are no longer exposed.

*Notes:*

(1) Area of glazing less than 20 per cent of the total floor area may result in poor levels of daylight in the extension and dwelling.
(2) Areas of glazing greater than 25 per cent may be acceptable in certain circumstances, i.e. to make the extension consistent with the external appearance of the host building. In such cases the U-value of the window should be improved in accordance with Para 4.1b of ADL1B, or other compensation measure as Paras 4.4–4.7 of ADL1B.
(3) Where necessary, SAP calculations can be submitted to building control to confirm how compensating measures could provide flexibility where the area of external windows, roof windows and doors exceeds 25 per cent of the floor area of the extension.

## New thermal elements

### External glazing

External glazing insulation details are to comply with U-values for external windows, doors and roof lights in compliance with Paragraphs 4.19–4.23 and Table 1 of ADL1B and the Tables below: (Note: All external doors, windows, roof lights, etc. are to be factory draft stripped.)

#### *Closing around window and door openings*

Checked rebates should be constructed to window/door reveals, or alternatively proprietary finned insulated closers should be used. Checked rebates are where the outer skin masonry/skin projects

**Table 2.52:** U-value requirements for external windows and doors including roof windows. (See Table 1 of ADL1B for full details.)

| Fitting | Insulation standard U-value not worse than: |
|---|---|
| Windows, roof window or roof light | 1.6 (or Window Energy Rating (WER) as Band C of Para 4.22 of ADL1B) |
| Doors with more than 50% glazing | 1.8 |
| Other doors | 1.8 |
| Replacement windows/doors | As above or 1.2 centre pane – if external appearance of the facade or character of the building is to be maintained |

**Table 2.53:** U-values for double glazing

| **Pilkington Glass** | **Outer pane (4 mm)** | **Cavity/spacer/gas** | **Inner pane (4 mm)** | **U-value** |
|---|---|---|---|---|
| IGU | Optifloat | 16 mm air-filled | K-Glass | 1.7 |
| EnergiKare Classic | Optiwhite | 16 mm argon-filled with aluminium spacer bar | K-Glass | 1.5 |
| EnergiKare Classic | Optiwhite | | K-Glass OW | 1.5 |
| EnergiKare Classic | Optiwhite | 16 mm argon-filled with warm edge spacer bar | K-Glass | 1.5 |
| EnergiKare Classic | Optiwhite | | K-Glass OW | 1.5 |

Source: a representative selection of values taken from *Technical Note 10, U-Values of Elements (Approved Document L1B 2010)*, produced by Hertfordshire Technical Forum for Building Control. Reproduced by permission of Hertfordshire Technical Forum for Building Control.

**Table 2.54:** U-values for triple glazing – Pilkington EnergiKare glazing system

| **Outer pane** | **Cavity** | **Middle pane** | **Cavity** | **Inner pane** | **U-value** |
|---|---|---|---|---|---|
| Optiwhite | 12 mm argon | K Glass T | 12 mm argon | K-Glass | 1.0 |
| Optiwhite | 16 mm argon | K Glass T | 16 mm argon | K-Glass | 0.8 |
| Optiwhite | 12 mm argon | K Glass OWT | 12 mm argon | K-Glass OW | 1.0 |
| Optiwhite | 16 mm argon | K Glass OWT | 16 mm argon | K-Glass OW | 0.8 |
| Optiwhite | 12 mm krypton | K Glass OWT | 12 mm krypton | K-Glass OW | 0.7 |

Source: a representative selection of values taken from *Technical Note 10, U-Values of Elements (Approved Document L1B 2010)*, produced by Hertfordshire Technical Forum for Building Control. Reproduced by permission of Hertfordshire Technical Forum for Building Control.

across the inner skin by at least 25 mm, the cavity is closed by an insulated closer and the window or door is fully sealed externally with mastic or similar

### *Sealing and draught-proofing measures*

All external door and window frames, service penetrations to walls, floors and ceilings, etc. should be sealed both internally and externally with proprietary sealing products such as proprietary waterproof mastic, expanding foam or mineral wool or tape to ensure airtightness.

## Energy-efficient lighting

### Fixed internal lighting

Fixed internal energy-efficient lighting in new extensions must not be less than 75 per cent of all the fixed low-energy light fittings (fixed lights or lighting units) in the main dwelling spaces (excluding cupboards and storage areas), fitted with lamps that must have a luminous efficiency greater than 45 lumens per circuit-watt and a total output greater than 400 lamp-lumens. Light fittings are to be either dedicated fittings that take only low-energy lamps, or standard fittings that take low-energy lamps. (Note: Light fittings with supplied power of less than 5 circuit-watts are excluded from the overall count of the total number of light fittings.)

### Fixed external lighting

Fixed external energy-efficient lighting in new extensions must consist of either:

(i) lamp capacity not greater than 100 lamp-watts per light fitting, fitted with automatic switch-off between dawn and dusk and when lit area becomes unoccupied; **or**
(ii) lamp efficacy greater than 45 lumens per circuit-watt; and fitted with automatic switch-off between dawn and dusk and fitted with manual controls.

## Insulation of pipework to prevent freezing

All hot and cold water service pipework, tanks and cisterns should be located within the warm envelope of the building to prevent freezing.

Where hot and cold water service pipework, tanks and cisterns are located in unheated spaces they should be insulated to prevent freezing, in compliance with BS 6700 and BS 8558, and typically as follows:

(i) All tanks and cisterns should be thermally insulated with proprietary insulated systems to prevent freezing, in compliance with the manufacturer's systems (insulation is normally omitted from below tank, where it benefits from heat in the heated area below).
(ii) Pipework should be insulated with proprietary insulated sleeves of phenolic/polyisocyanurate/polyurethane foam (or other approved material) having a minimum wall thickness of 30 mm for 15 mm-diameter pipes and 12 mm for 22 mm-diameter pipes, and fixed in accordance with the manufacturer's details.

## External walls, roofs, floors and swimming-pool basin

External walls, roofs, floors and swimming-pool basin are to comply with new thermal element requirements, in compliance with Table 2.55 below (see Paragraphs 5.1–5.6 and Table 2 of ADL1B for full details).

### 2.144 Domestic extensions

**Table 2.55:** U-values for external walls, roofs, floors and swimming-pool basin (See Table 2 of ADL1B for full details.)

| Element[1] (see construction details in Part A) | Insulation standard U-value: W/m².K |
|---|---|
| Walls (exposed and semi-exposed) | 0.28[2] |
| Pitched roof and dormer windows with insulation at ceiling level | 0.16 |
| Pitched roof and dormer windows with insulation at rafter level | 0.18 |
| Flat roof or roof with integral insulation | 0.18 |
| Floors[3] | 0.22[4] |
| Swimming-pool basin (walls and floor) | 0.25 |

**Notes:**
1. 'Roof' includes the roof parts of dormer windows, and 'wall' includes the wall parts (cheeks) of dormer windows.
2. Area-weighted average values.
3. A lesser provision may be appropriate where meeting such a standard would reduce the floor area by 5% in the room bounded by the wall.
4. A lesser provision may be appropriate where meeting such a standard would cause significant problems in relation to adjoining floor levels. The U-value of the floor of the extension can be calculated using the exposed perimeter and the floor area of the whole enlarged dwelling.

## Renovation/upgrading of existing thermal elements

Where the existing walls, roof or floor is to be retained and become part of the thermal envelope, or renovated, or subject to a material change of use, and is insulated below the threshold values in column (a) of Table 2.56 below, then the thermal elements should be thermally renovated/upgraded to the U-values in column (b) in the Table. (Note: Renovation of existing thermal elements applies only where the area to be renovated is more than 50 per cent of the surface area of the individual element and 25 per cent of the total building envelope; and also renovation or upgrading of the existing thermal elements applies only where it is technically and functionally feasible, with a simple payback of 15 years-plus.). See Section 5 of ADL1B for full details.

## Consequential improvements (applies to existing buildings with a total useful floor area exceeding 1000 m²)

Consequential improvements (additional works) are required to make an existing building more energy-efficient when it has a total useful floor area exceeding 1000 m² and is subject to an extension or provision of fixed building service, in compliance with Paragraphs 6.1–6.5 of ADL1B and Section 6 of ADL2B.

### Commissioning of fixed building services

A copy of the commissioning certificate for fixed building services is to be sent to building control within five days of completion of the commissioning work being carried out (or within 30 days for works commissioned by a person registered with a Competent Persons' Scheme).

**Table 2.56:** Renovation/upgrading of existing thermal elements
(See Table 3 of ADL1B for full details.)

| Element[1] | (a) Threshold U-value W/m².K | (b) Upgraded U-value W/m².K |
|---|---|---|
| Cavity walls[2] (where suitable for filling with insulation) | 0.7 | 0.55 |
| Solid walls (external or internal insulation)[3] | 0.7 | 0.30 |
| Floors [4,5] | 0.7 | 0.25 |
| Pitched roof- insulation at ceiling level | 0.35 | 0.16 |
| Pitched roof- insulation between rafters[6] | 0.35 | 0.18 |
| Flat roof or roof with integral insulation[7] | 0.35 | 0.18 |

**Notes:**
1. 'Roof' includes the roof parts of dormer windows, and 'wall' includes the wall parts (cheeks) of dormer windows.
2. This only applies if the cavity wall is suitable for the installation of cavity wall fill as ADD; otherwise, insulation should be fixed internally or externally.
3. A lesser provision may be appropriate where meeting such a standard would reduce the floor area by 5% in the room bounded by the wall.
4. The U-value of the floor of the extension can be calculated using the exposed perimeter and the floor area of the whole enlarged dwelling.
5. A lesser provision may be appropriate where meeting such a standard would cause significant problems in relation to adjoining floor levels.
6. A lesser provision may be appropriate where meeting such a standard would create limitations on headroom. In such cases the depth of insulation and required air gap should be at least to the depth of the rafter, using insulation to achieve the best practical U-value
7. A lesser provision may be appropriate if there are problems associated with the load-bearing capacity of the frame or up-stand height.

### *Providing information – building log book*

A log book containing the following information is to be provided in the dwelling on completion:

- Operating and maintenance instructions for fixed building services
- Instructions on how to make adjustments to timing and temperature control settings, etc.
- Instructions on routine maintenance requirements for fixed building services, in compliance with the manufacturer's details.

## PART M: ACCESS TO AND USE OF BUILDINGS FOR DISABLED

Please refer fully to the relevant Approved Document.

ADM of the Building Regulations does not apply to extensions to existing buildings, unless it is an extension of a dwelling where ADM of the Building Regulations would have applied and the proposed extension will make things worse – for example, removal of an access ramp or a downstairs WC, unless it is to be reinstated as part of the proposed works, in compliance with ADM. Please contact building control for their specific requirements.

## PART N: SAFETY GLAZING, OPENING AND CLEANING

Please refer fully to the relevant Approved Document.

## Safety glass and glazing

Doors and adjacent sidelights/windows in critical locations within 1500 mm of floor/ground level and 300 mm of doors and windows within 800 mm of floor/ground are to be safety glazed to: BS EN 12150, BS EN 14179 and BS EN 14449 (which supersedes BS 6206), in compliance with Figure 2.72 below:

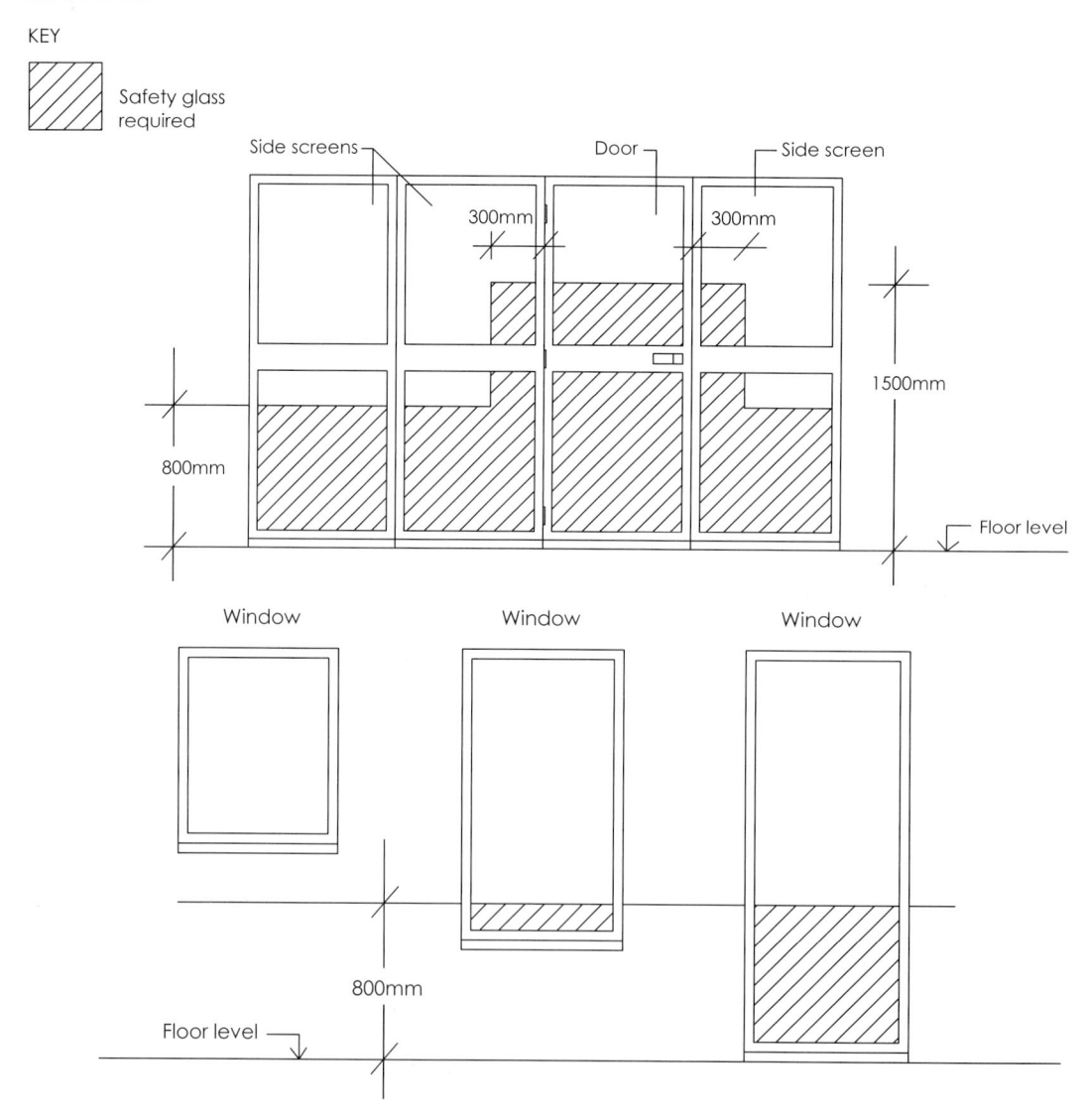

**Figure 2.72:** Safety glass requirements in doors/side screens and windows (See Diagram 1 of ADN for full details.)

*Notes:*

(1) Where safety glazing is required in part of an opening, as indicated by hatched lines in the above guidance diagram, that complete pane must be in safety glass.
(2) Glass thickness must be suitable for dimension limits and opening sizes, in accordance with the glass manufacturer's details. See Diagram 2 of ADN for full details.

### Glazing in small panes

Small panes of glass should not exceed $0.5\,m^2$ in area and should have one dimension smaller than 250 mm, measured between glazing beads. Glass should be annealed and not less than 6 mm thick. See Diagram 3 of ADN for full details.

### New system of marking

Safety glazing must comply with the new system of marking, which requires visible, clear and indelible markings on each piece of safety glazing within critical locations, in compliance with: BS EN 12150, BS EN 14179 and BS EN 14449.

## PART P: ELECTRICAL SAFETY

Please refer fully to the relevant Approved Document.

## Electrical installations

Fixed electrical wiring installed in dwellings in England and Wales must comply with Part P of the Building Regulations. Work performed on new or existing electrical circuits or systems must be designed, installed, inspected, tested and certified by a competent person, in accordance with the current version of the IEE Regulations as documented in BS 7671.

A competent electrician or a member of a Competent Persons' Scheme must test and certify all such works. The electrician must provide signed copies of an electrical installation certificate conforming to BS 7671 for the owner of the property and a copy must be forwarded to the building control surveyor for approval at completion, so the building control completion certificate can be issued. Guidance on electrical installations that need not be notified to building control is provided in Table 2.57.1 below, and guidance on special locations and installations is provided in Table 2.57.2 below. Additional guidance and specific examples are given in the additional notes below the Tables.

Consumer units must be fixed above any flood level and must be generally accessible for use by responsible persons in the household, and they should not be installed where young children might interfere with them.

**Table 2.57.1:** Guidance on electrical installations that need *not* be notified to building control
(as Table 1 of ADP)

| |
|---|
| **Work consisting of:**
• Replacing of any fixed electrical equipment (for example, socket outlets, control switches and light roses) that does not include any new fixed cabling
• Replacing the cabling for a single circuit only, where damaged, for example, by fire, rodent or impact [a]
• Refixing or replacing the enclosures of existing installation components [b]
• Providing mechanical protection to existing fixed installations [c]
• Installing or upgrading main or supplementary equipotential bonding. [d] |
| **Work that is not in a kitchen or special location and does not involve a special installation [e] and consists of:**
• Adding lighting points (light fittings and switches) to an existing circuit [f]
• Adding socket outlets and fused spurs to an existing ring or radial circuit. [f] |
| **Work not in a special location, on:**
• Telephone or extra-low-voltage wiring and equipment for communications, information technology, signalling, control and similar purposes
• Prefabricated equipment sets and associated flexible leads with integral plug-and-socket connections. |

**Notes:**
(a) On condition that the replacement cable has the same current-carrying capacity and follows the same route
(b) If the circuit's protective measures are unaffected
(c) If the circuit's protective measures and current-carrying capacity of conductors are unaffected by increased thermal insulation
(d) Such work will need to comply with other applicable legislation, such as Gas Safety (Installation and Use) Regulations.
(e) Special locations and installations are listed in Table 2.57.2 below (and Table 2 of ADP).
(f) Only if the existing circuit protective device is suitable and provides protection for the modified circuit, and other relevant safety provisions are satisfactory.

**Table 2.57.2:** Special locations and installations [a]
(As Table 2 of ADP)

| |
|---|
| **Special locations:**
• Locations containing a bath tub or shower basin
• Swimming pools or paddling pools
• Hot-air saunas. |
| **Special installations:**
• Electric floor or ceiling heating systems
• Garden lighting or power installations
• Solar photovoltaic (PV) power supply systems
• Small-scale generators such as micro CHP units
• Extra-low-voltage lighting installations, other than pre-assembled, CE-marked lighting sets. |

**Note:**
[a] See IEE Guidance Note 7, which gives more guidance on achieving safe installations where risks to people are greater.

## Additional notes (as ADP)

Tables 2.57.1 and 2.57.2 above (Tables 1 and 2 of ADP) give the general rules for determining whether or not electrical installation work is notifiable. The rules are based on the risk of fire and injury and what is practical. The following notes provide additional guidance and specific examples:

(a) Notifiable jobs include new circuits back to the consumer unit, and extensions to circuits in kitchens and special locations (bathrooms, etc.) and associated with special installations (garden lighting and power installations, etc.).

(b) Replacement, repair and maintenance jobs are generally not notifiable, even if carried out in a kitchen or special location or associated with a special installation.
(c) Consumer unit replacements, however, are notifiable.
(d) In large bathrooms the location containing a bath or shower is defined by the walls of the bathroom.
(e) Conservatories and attached garages are not special locations. Work in them is therefore not notifiable unless it involves the installation of a new circuit or the extension of a circuit in a kitchen (or special location or associated with a special installation.
(f) Detached garages and sheds are not special locations. Work within them is notifiable only if it involves new outdoor wiring.
(g) Outdoor lighting and power installations are special installations. Any new work in, for example, the garden, or that involves crossing the garden, is notifiable.
(h) The installation of fixed equipment is within the scope of Part P, even where the final connection is by a 13A plug and socket. However, work is notifiable only if it involves fixed wiring and the installation of a new circuit or the extension of a circuit in a kitchen or special location or associated with a special installation.
(i) The installation of equipment attached to the outside wall of a house (for example, security lighting, air conditioning and radon fans) is not notifiable providing that there are no exposed outdoor connections and the work does not involve the installation of a new circuit or the extension of a circuit in a kitchen or special location or associated with a special installation.
(j) The installation of a socket outlet on an external wall is notifiable, since the socket outlet is an outdoor connection that could be connected to cables that cross the garden and requires RCD protection.
(k) The installation of prefabricated 'modular' systems (for example, kitchen lighting systems and armoured garden cabling) linked by plug-and-socket connections is not notifiable, providing that the products are CE-marked and that any final connections in kitchens and special locations are made to existing connection units or points (possibly a 13A socket outlet).
(l) Work to connect an electric gate or garage door to an existing isolator is not notifiable, but installation of the circuit up to the isolator is notifiable.
(m) The fitting and replacement of cookers and electric showers is not notifiable unless a new circuit is needed.
(n) New central heating control wiring installations are notifiable, even when work in kitchens and bathrooms is avoided.

## External works – paths, private drives, patios and gardens

The guidance below for external surface finishes does not form part of the Building Regulations and is for domestic guidance use only; associated commercial uses should be designed by a suitably qualified specialist.

### Concrete areas and paths etc.

- Use 100 mm-thick concrete, shuttered with temporary or permanent edge restraint or kerbs. Mix type PAV 1, maximum bay size 6 m, with bitumen-impregnated fibre board isolated joints to BS 8110/5328, laid over:

- Minimum 100–150 mm-thick course of Type 1 sub-base, sand-blinded and mechanically compacted to refusal in 150 mm-thick layers, with a geotechnical membrane underneath, laid over firm subsoils.

### *Tarmac areas*

- 20 mm-thick mechanically rolled wearing course of 100–150 pen-grade bituminous coated macadam using 0–6 mm aggregate sizes (to BS 4987), with permanent edge restraint or kerbs, laid over:
- 60 mm-thick mechanically rolled base course of 100–150 pen-grade bituminous coated macadam using 0–20 mm aggregate sizes (to BS 4987), laid over:
- minimum 100–150 mm-thick course of Type 1 sub-base, sand-blinded and mechanically compacted to refusal in 150 mm-thick layers with a geotechnical membrane underneath, laid over firm subsoils.

### *Block pavers*

- 60 mm precast self-draining concrete block paving to client's choice, laid in compliance with manufacturer's details, to BS 6717 with permanent edge restraint or kerbs, laid over:
- minimum 100–150 mm-thick course of Type 1 sub-base, sand-blinded and mechanically compacted to refusal in 150 mm-thick layers with a geotechnical membrane underneath, laid over firm subsoils.

### *Precast concrete or natural stone slabs*

- 50 mm precast concrete/natural stone slabs. laid in compliance with manufacturer's details to BS 7263:1 (typically fully bedded and pointed in 25 mm-thick sand/cement mortar 1:4 mix, or other approved material in accordance with manufacturer's details)
- minimum 100–150 mm-thick course of Type 1 sub-base, sand-blinded and mechanically compacted to refusal in 150 mm-thick layers with a geotechnical membrane underneath, laid over firm subsoils.

### *Gravel*

- 100 mm gravel, laid in compliance with manufacturer's details to BS 7263:1
- minimum 100–150 mm-thick course of Type 1 sub-base, sand-blinded and mechanically compacted to refusal in 150 mm-thick layers with a geotechnical membrane underneath, laid over firm subsoils.

### Drainage of paved areas

This is to be carried out in accordance with BS 6367:1983 A1 84, ADH and Part H of this guidance. Paths and paved areas are to have a non-slip finish with a fall of 1:80 and a reverse gradient of at least 500 mm away from walls of building, unless using proprietary porous self-draining systems. Surface water is to be disposed of by an adequately sized and roddable drainage system via soakaways, or other approved means. Paved areas are normally set 5 mm above drainage channels or gullies, etc. The local authority's Planning Department may have additional requirements for the drainage of paved areas and should be consulted before works commence.

# Section 3  New dwellings

| | |
|---|---|
| Parts A and L: Starting point | 3.4 |
| Conservation of fuel and power in new dwellings | 3.4 |
| Criterion 1 – Achieving the Target Emission Rate (TER) | 3.4 |
|    The Target Emission Rate (TER) | 3.4 |
|    The Dwelling Design Emission Rate (DER) | 3.5 |
|       Submission of information to building control | 3.5 |
| Criterion 2 – Limits on design flexibility | 3.5 |
|    Worst acceptable standards | 3.5 |
| Criterion 3 – Limiting the effects of solar heat gain | 3.5 |
| Criterion 4 – Calculation of the Dwelling Design Emission Rate (DER) | 3.6 |
|    Base standards for calculation of DER for new dwellings | 3.6 |
|       U-values for party walls | 3.7 |
|       External glazing | 3.7 |
|       Closing around window and door openings | 3.8 |
|       Sealing and draught-proofing measures | 3.8 |
|       Limiting the effects of solar heat gains | 3.8 |
|       Limiting thermal bridging around external openings | 3.8 |
|       Space heating and hot water | 3.8 |
|    Ventilation, natural or mechanical | 3.8 |

*Guide to Building Control: For Domestic Buildings*, First Edition. Anthony Gwynne.
© 2013 John Wiley & Sons, Ltd. Published 2013 by John Wiley & Sons, Ltd.

| | |
|---|---:|
| Energy-efficient lighting | 3.9 |
|     Fixed internal lighting | 3.9 |
|     Fixed external lighting | 3.9 |
|   Air permeability and pressure testing | 3.9 |
|     Limiting air leakage | 3.9 |
|     Pressure testing of new dwellings | 3.9 |
|     Alternative to pressure testing | 3.10 |
|     Photovoltaic panels | 3.10 |
| Criterion 5 – Provision for energy-efficient operation of the dwelling | 3.10 |
|   Providing information | 3.10 |
|     Model designs | 3.10 |
|     Further reading | 3.10 |
| Insulation guidance details for floors, walls and roofs | 3.10 |
|   1. Ground Floors | 3.11 |
|   2. External walls | 3.11 |
|   3. Intermediate upper floor(s) | 3.11 |
|   4. Pitched roofs | 3.11 |
|   5. Flat-roof construction | 3.11 |
|     Roof insulation and ventilation gaps | 3.12 |
| Guidance on the Code for Sustainable Homes for new dwellings | 3.12 |
| Guidance on PassivHaus | 3.18 |
| Part B: Fire safety and means of escape | 3.19 |
|   Domestic sprinklers in new Welsh Homes from 2013 | 3.19 |
| Part C: Site preparation and resistance to contaminants and moisture | 3.20 |
| Part D: Cavity wall filling with insulation | 3.20 |
| Part E: Resistance to the passage of sound | 3.20 |
|   Party walls and floors separating new dwelling/flats | 3.20 |
|   Party walls and floors separating conversion of buildings into new dwellings/flats | 3.20 |
|     Exemptions/relaxations for conversion of historic buildings into dwellings/flats | 3.20 |
|   Pre-completion sound testing of party walls and floors | 3.20 |
|     Testing rate requirements | 3.21 |
|     Remedial works and retesting | 3.21 |
|   New internal walls and floors | 3.21 |
| Part F: Ventilation to new dwellings | 3.21 |
| Ventilation systems | 3.21 |
|   System 1: Purge (natural) ventilation with background ventilation and intermittent extract fans | 3.22 |
| Purge (natural) ventilation to habitable rooms: system 1 – new dwellings | 3.22 |
|   Mechanical extract ventilation and fresh air inlets for rooms without purge ventilation | 3.22 |
|   Purge ventilation to wet rooms | 3.22 |
|   General requirements for purge (natural) ventilation | 3.22 |
| Background ventilation: system 1 – new dwellings | 3.23 |
| Intermittent mechanical extract ventilation: system 1 – new dwellings | 3.24 |
|   Mechanical extract ventilation rates | 3.24 |
|   Additional requirements for mechanical extract ventilation | 3.24 |

| | |
|---|---|
| Part G: Sanitation, hot-water safety and water efficiency | 3.24 |
|     Safety valves, prevention of scalding and energy cut-outs | 3.24 |
|     Discharge pipes from safety devices | 3.24 |
|     Water efficiency and calculations | 3.25 |
|         Commissioning certificates | 3.25 |
| Part H: Drainage and waste disposal | 3.25 |
| Part J: Combustion appliances and fuel storage systems | 3.26 |
| Part K: Protection from falling, collision and impact | 3.26 |
| Part L: Conservation of fuel and power | 3.26 |
| Part M: Access to and use of buildings for disabled | 3.26 |
|     External ramped approach | 3.26 |
|     External stepped approach | 3.26 |
|         External door opening widths | 3.27 |
|         Internal passageways/corridors widths | 3.27 |
|         Accessible level door thresholds into the building | 3.27 |
|         Internal stairs | 3.27 |
|         Accessible switches, sockets, controls, etc. | 3.28 |
|         Provision of a ground-floor WC | 3.29 |
| Guidance on Lifetime Homes Standard for new dwellings | 3.31 |
|     Lifetime Homes in the context of government policy and regulation | 3.31 |
| Part N: Safety glazing, opening and cleaning | 3.32 |
| Part P: Electrical safety | 3.32 |

**3.4** New dwellings

## PARTS A AND L: STARTING POINT

New dwellings are normally complex projects and unless you are experienced in construction you will need to get some professional advice/help from a suitably qualified and experienced property professional and an energy consultant.

New dwellings have a significant number of design considerations for the conservation of fuel and power that need to be addressed at the design stage.

These guidance details for new dwellings commence with the requirements of Approved Document L1A (Conservation of fuel and power for new dwellings) and are intended to be used as a starting point to provide specification details that can be incorporated into the detailed design for the project. (For conversions of other buildings into dwellings, follow the guidance in Section 5.0.)

Depending on the building type and specification used, additional renewable energy technologies may be required to achieve compliance – for example, photovoltaic panels. The details given in this guidance are suggested details only and there is no obligation to adopt any of the specification details stated. You should consult an accredited Energy Assessor for advice and always check with building control that your proposals comply with the requirements of the building regulations.

## Conservation of fuel and power in new dwellings

This section applies to new dwellings, and since 31 December 2011 the Welsh Assembly Government has required that all new residential properties in Wales meet an 8 per cent improvement over the 2010 Code level 3 for sustainable homes (ENE.1). Guidance on the Code for Sustainable Homes is contained at the end of this section.

**Approved Document L1A requires five criteria for establishing compliance, as follows:**

**Criterion 1 – Achieving the Target Emission Rate (TER)**
**Criterion 2 – Limits on design flexibility**
**Criterion 3 – Limiting the effects of solar heat gain**
**Criterion 4 – Achieving the Dwelling Design Emission Rate (DER)**
**Criterion 5 – Provision for energy-efficient operation of the dwelling**

The above five criteria are considered in more detail below.

## Criterion 1 – Achieving the Target Emission Rate (TER)

The Dwelling Emission Rate (DER) must not be greater than the corresponding Target Emission Rate (TER), in accordance with Paragraphs 4.7 to 4.17 of ADL1A, as follows:

### The Target Emission Rate (TER)

This is a benchmark figure and is calculated by an accredited Energy Assessor in accordance with the government's Standard Assessment Procedure (SAP 2009) for a particular dwelling, which is measured in kg of $CO_2$ per $m^2$ per year. The calculation is based on a notional dwelling of the same size and shape as the proposed dwelling and has a minimum overall improvement of 25 per cent relative to 2006 standards.

### The Dwelling Design Emission Rate (DER)

The DER is a measure of the carbon dioxide emissions arising from the use of energy in the dwelling and is calculated by an accredited Energy Assessor in accordance with the government's Standard Assessment Procedure (SAP 2009), which is measured in kg of $CO_2$ per $m^2$ per year and takes into account the energy used for space heating, hot water, fixed lighting, fans and pumps for a particular dwelling. There are also minimum area-weighted U-value requirements for walls, floors, roofs and openings, as well as a minimum number of low-energy light fittings, minimum boiler efficiencies, minimum air leakage rates and minimum controls for the heating system. Guidance on achieving the Dwelling Design Emission Rate (DER) is given in Criterion 4 below.

### *Submission of information to building control*

(i) Before works commence: the person carrying out the work must give the design SAP calculations containing TER and DER to building control (normally at the time the building regulations application is made).
(ii) After the work has been completed: the person carrying out the work must notify building control whether the building has been constructed in accordance with the design SAP calculations, if not, an as-built SAP calculation should be given to building control containing the amended details.
(iii) Within 5 days after the work has been completed: the person carrying out the work shall give an Energy Performance Certificate (EPC) for the building to the building owner and give notice to building control that the EPC has been issued including the reference number under which the EPC has been registered in accordance with the Building Regulations 2010 (as amended). The EPC must contain information in accordance with the Building Regulations 2010 (as amended).

Note: From April 2008 all new dwellings (and conversions into dwellings) must have an Energy Performance Certificate (EPC) to fulfil Building Regulations requirements, and these can only be produced by an accredited Energy Assessor.

## Criterion 2 – Limits on design flexibility

### *Worst acceptable standards*

The thermal performance of building elements and the building services' efficiencies must not fall below the minimum values as contained in Paragraphs 4.18 to 4.24 of ADL1A. The worst acceptable fabric standards are to be in compliance with Table 2 of ADL1A and Table 3.1 below.

## Criterion 3 – Limiting the effects of solar heat gain

Limiting the effects of solar heat gain should be carried out in accordance with Paragraphs 4.25 to 4.27 of ADL1A, and excessive solar gains should be checked in accordance with Appendix P of SAP 2009. Adequate levels of daylight are recommended to be maintained in accordance with BS 8206-2:2008 Code of Practice for Day Lighting.

## Criterion 4 – Calculation of the Dwelling Design Emission Rate (DER)

The Dwelling Emission Rate (DER) is explained in Criterion 1 above and the building performance standards consistent with the Design Emission Rate should be carried out in accordance with Paragraphs 5.1 to 5.31 of ADL1A, and guidance details as follows.

### Base standards for calculation of DER for new dwellings

To assist in the complex design specification choices required for the DER calculation, Table 3.2 below contains recommended starting base U-values and standards for new dwellings.

**Table 3.1:** Worst acceptable fabric standards
(See Table 2 of ADL1A for full details.)

| Element | Worst acceptable fabric standard U-value | |
|---|---|---|
| | Area weighted average (W/m².K) | Individual element (W/m².K) |
| Roof | U-value 0.20 | 0.35 |
| Wall | U-value 0.30 | 0.70 |
| Floor | U-value 0.25 | 0.70 |
| Party wall | U-value 0.20 | – |
| Windows, roof lights, doors and curtain walling | U-value 2.00 | – |
| Air permeability | 10.00 m³/h.m² at 50 pa | – |

Source: amended in consultation with Kingspan Insulation Ltd. Reproduced by permission of Kingspan Insulation Ltd.

**Table 3.2:** Recommended starting base U-values/standards for new dwellings

| Element[1] | Starting base U-values and standard |
|---|---|
| All roof types | U-value 0.13 to 0.16 W/m².K |
| External walls | U-value 0.18 to 0.24 W/m².K |
| Sheltered wall | U-value 0.19 W/m².K |
| Party walls | U-value 0.00 W/m².K (see additional values in Table 2.60 below) |
| Floors | U-value 0.13 to 0.16 W/m².K |
| External glazing | U-value 1.2 to 1.6 W/m².K (see range of values in Table 2.61 below) |
| Thermal bridging | *ACD Psi or better (y-value no longer used other than a default of 0.15) |
| Air permeability | 5.00 to 6.00 m³/h.m² at 50 pa |
| Thermal mass parameters | 250 kJ/ m².K for masonry and 100 kJ/ m².K for timber frame (or structurally insulated panels) |
| Space heating and hot-water system | Natural gas with A-rated boiler (90% efficient) |

Note: Insulation manufacturer's details for these values are contained in the Tables below for new dwellings.
*ACD = Accredited construction details.
[1] Use BRE SAP Conventions for guidance www.bre.co.uk/sapconventions.

***U-values for party walls*** (also see recommended base specification above)

Party walls' U-values are to be in compliance with Table 3 of AD L1A and Table 3.3 below:

***External glazing*** (also see recommended base specification above)

External glazing insulation details are to comply with fabric standards for external windows, doors and roof lights, in compliance with Paragraphs 4.20–4.22 of ADL1A and the Tables below. (Note: All external doors, windows, roof lights, etc. are to be factory draft stripped.)

**Table 3.3:** U-values for party walls
(See Table 3 of ADL1A for full details.)

| Element | U-value |
| --- | --- |
| Solid walls or structurally insulated panels | $0.0\,W/m^2.K$ |
| Unfilled cavity with effective edge sealing | $0.2\,W/m^2.K$ |
| Filled cavity with effective edge sealing | $0.0\,W/m^2.K$ |

**Note:** Unfilled cavity with no effective edge sealing (U-value of $0.5\,W/m^2.K$) will not achieve the minimum required worst-case U-value of $0.2\,W/m^2.K$ and will cause the dwelling to fail.

**Table 3.4:** U-values for double glazing

| Pilkington Glass | Outer pane | Cavity/spacer/gas | Inner pane | U-value |
| --- | --- | --- | --- | --- |
| IGU | Optifloat | 16 mm air-filled | K-Glass | 1.7 |
| EnergiKare | Optiwhite | 16 mm argon-filled with aluminium spacer bar | K-Glass | 1.5 |
| Classic | | | K-Glass OW | |
| EnergiKare | Optiwhite | 16 mm argon-filled with warm edge spacer bar | K-Glass | 1.5 |
| Classic | | | K-Glass OW | |

**Note:** Preferable to use U-window values instead of centre pane U-values for SAP calculations, as centre pane U-values may fail limiting requirements when including the frame.

**Table 3.5:** U-values for triple glazing (Pilkington EnergiKare glazing system)

| | Cavity | Middle pane | Cavity | Inner pane | U-value |
| --- | --- | --- | --- | --- | --- |
| Optiwhite | 12 mm argon | K Glass T | 12 mm argon | K-Glass | 1.0 |
| Optiwhite | 16 mm argon | K Glass T | 16 mm argon | K-Glass | 0.8 |
| Optiwhite | 12 mm argon | K Glass OWT | 12 mm argon | K-Glass OW | 1.0 |
| Optiwhite | 16 mm argon | K Glass OWT | 16 mm argon | K-Glass OW | 0.8 |
| Optiwhite | 12 mm krypton | K Glass OWT | 12 mm krypton | K-Glass OW | 0.7 |

**Note:** Preferable to use U-window values instead of centre pane U-values for SAP calculations, as centre pane U-values may fail limiting requirements when including the frame.

### Closing around window and door openings

Checked rebates should be constructed to window/door reveals, or proprietary finned insulated closers should be used. Checked rebates are where the outer skin/masonry skin projects across the inner skin by at least 25 mm, the cavity is closed by an insulated closer and the window or door is fully sealed with mastic or similar externally.

### Sealing and draught-proofing measures

All external door and window frames, service penetrations to walls, floors and ceilings, etc. should be sealed both internally and externally with proprietary sealing products such as proprietary waterproof mastic/expanding foam to ensure airtightness.

### Limiting the effects of solar heat gains

Details of measures taken to limit solar gains are to be provided in energy assessment, in compliance with Paragraphs 4.25–4.27 of ADL1.

### Limiting thermal bridging around external openings (also see recommended base specification above)

Thermal bridging around external openings are to be provided in compliance with Paragraphs 5.9–5.13 of ADL1A as follows:

(i) Adopt a quality-assured, accredited construction detail approach approved by the Secretary of State; the calculated linear thermal transmittance can be used directly in the DER calculation, in compliance with Paragraph 5.12 a.
(ii) Use non-assessed details, in compliance with BRE Report BR 497 'Conventions for Calculating Linear Thermal Transmittance and Temperature Factors 2007' (ISB N 978 1 86081 986 5); the increased calculated values can be used in the DER calculations, in compliance with Paragraph 5.12 b. (Note: Items (i) and (ii) – thermal bridging must include detail length and Psi value for all junctions (values to be taken from ACD (accredited construction details) or EST (Energy Saving Trust details) or accredited details provided by trade or product manufacturer).
(iii) Use unaccredited details with no specific quantification of the thermal bridge values. In such cases a conservative default y-value of 0.15 must be used in the DER calculation, in compliance with Paragraph 5.12 c. (This is the only permissible y-value.)

### Space heating and hot water (also see recommended base specification above)

See Part J of this guidance: Combustion appliances and fuel storage systems

## Ventilation, natural or mechanical

See Part F of this guidance: Ventilation to new dwellings

## Energy-efficient lighting

Fixed internal and external energy-efficient lighting systems are to be provided in compliance with Paragraph 4.13 of ADL1A, as follows:

### *Fixed internal lighting*

Fixed internal energy-efficient lighting in new dwellings must not form less than 75 per cent of all the fixed low-energy light fittings (fixed lights or lighting units) in the main dwelling spaces (excluding cupboards and storage areas); the fitted lamps must have a luminous efficiency greater than 45 lumens per circuit-watt and a total output greater than 400 lamp lumens. Light fittings are to be either dedicated fittings that take only low-energy lamps, or standard fittings that take low-energy lamps. (Note: Light fittings with supplied power of less than 5 circuit-watts are excluded from the overall count of the total number of light fittings.)

### *Fixed external lighting*

Fixed external energy-efficient lighting in new dwellings will consisting of either:

(i) Lamps with capacity not greater than 100 lamp-watts per light fitting and fitted with automatic movement-detecting control devices (PIR) and automatic daylight cut-off sensors; **or**
(ii) Lamps with efficacy greater than 45 lumens per circuit-watt; and fitted with automatic daylight cut-off sensors and manual controls.

## Air permeability and pressure testing

### *Limiting air leakage*

Air leakage is to be limited in the new dwelling fabric as indicated in the guidance details/diagrams and in compliance ADL1A and Accredited Construction Details (see information available from: www.planningportal.gov.uk). All external door and window frames, service penetrations to walls, floors, joists and ceilings, etc. should be sealed both internally and externally with proprietary sealing products such as proprietary waterproof mastic, expanding foam or mineral wool or tape to ensure airtightness.

### *Pressure testing of new dwellings* (also see recommended base specification above)

Where required by the SAP rating, pressure testing is to be carried out by a specialist registered by the British Institute of Non-destructive Testing, in compliance with the TER design limits and with Paragraphs 5.14–5.23 of ADL1A. Test results are o be provided by the specialist and a copy sent to building control within seven days after the test is carried out.

Pressure testing is to be carried out on three units of **each dwelling type**, or 50 per cent of all instances of **that dwelling type**, whichever is the less, in compliance with Paragraph 5.18 of ADL1A. (Note: Each block of flats should be treated as a separate dwelling type, and ground-/first-/top-floor flats are to be treated as separate dwelling types.)

### Alternative to pressure testing

Alternatives to pressure testing on small sites with fewer than two dwellings to be erected are:

(i) Confirm with test results that a dwelling of the same dwelling type has been constructed by the same developer within the last 12 months and has achieved the required design air permeability, in compliance with Paragraphs 5.14–5.18 of ADL1A;.**or**

(ii) Avoid the need for any pressure testing by using a value of $15\,m^3/(h.m^2)$ at 50 Pa for the air permeability when calculating the DER.

### Photovoltaic panels

Area of panels (typically $7-10\,m^2$ per peak KW, but can be more – although the manufacturer's declaration may better this approximate estimate) is to be in accordance with the SAP calculations. See Part J of this guidance: Combustion appliances and fuel storage systems, for details.

## Criterion 5 – Provision for energy-efficient operation of the dwelling

### Providing information

The following information is to be provided in the new dwelling on completion, in compliance with Paragraphs 6.1–6.4 of ADL1A:

- Operating and maintenance instructions for fixed building services
- Instructions on how to make adjustments to timing and temperature-control settings etc.
- Instructions on routine maintenance requirements for fixed building services, in compliance with the manufacturer's details
- Energy-assessment data used to calculate the TER and DER and the Energy Performance Certificate.

### Model designs

Model design packages should be carried out in compliance with Paragraphs 7.1–7.2 of ADL1A:

### Further reading

Part L 2010 – 'Where to start: An introduction for house builders and designers', obtainabble at: www.nhbcfoundation.org/partL.

## Insulation guidance details for floors, walls and roofs

Thermal insulation values for new dwellings are to be chosen in accordance with the guidance above and the Tables below. For new dwellings follow the construction details in Part A of Section 2 of this guidance for Domestic Extensions.

## 1. Ground Floors

Follow the construction details in Part A of Section 2 of this guidance for Domestic Extensions (using the insulation values in Tables 3.6 and 3.7).

## 2. External walls

Follow the construction details in Part A of Section 2 of this guidance for Domestic Extensions (using the insulation values in Tables 3.8, 3.9, 3.10, 3.11 and 3.12).

## 3. Intermediate upper floor(s)

Follow the construction details in Part A of Section 2 of this guidance for Domestic Extensions (using the insulation values in Table 3.13).

## 4. Pitched roofs

Follow the construction details in Part A of Section 2 of this guidance for Domestic Extensions (using the insulation values in Tables 3.14, 3.15 and 3.16).

## 5. Flat-roof construction

Follow the construction details in Part A of Section 2 of this guidance for Domestic Extensions (using the insulation values in Tables 3.17 and 3.18).

**Table 3.6:** Examples of insulation for ground-bearing floor slabs
(Construction as extension guidance details, typically 65 mm screed, 100 mm concrete slab, (500 g separating layer between foil-faced insulations and slab), insulation layer, 150 mm sand-blinded hardcore.)
**Note: Where the perimeter-over-area ratio (P/A) has not been calculated, use the insulation thickness stated in column 1.0* below.**

| Insulation product | K value | Required thickness of insulation (mm) | | | | | | | | | |
|---|---|---|---|---|---|---|---|---|---|---|---|
| | | Calculated Perimeter/Area ratio (P/A) | | | | | | | | | |
| | | 1.0* | 0.9 | 0.8 | 0.7 | 0.6 | 0.5 | 0.4 | 0.3 | 0.2 | 0.1 |
| U-value 0.15 W/m²k | | | | | | | | | | | |
| Celotex FR5000 | 0.021 | 120 | 120 | 120 | 120 | 120 | 100 | 100 | 90 | 70 | 25 |
| Kingspan Kooltherm K3 Floorboard | 0.020 (0.021 25–44 mm) | 110 | 110 | 110 | 110 | 110 | 100 | 100 | 90 | 70 | 25 |
| U-value 0.18 W/m²k | | | | | | | | | | | |
| Celotex FR5000 | 0.021 | 100 | 90 | 90 | 90 | 90 | 80 | 75 | 70 | 50 | 25 |
| Kingspan Kooltherm K3 Floorboard | 0.020 (0.021 25–44 mm) | 90 | 90 | 90 | 90 | 80 | 80 | 70 | 70 | 50 | 25 |

**Note 1:** Figures indicated above should be adjusted to the insulation manufacturer's nearest thickness.
**Note 2:** *Where P/A ratio has not been calculated, use insulation thickness stated in 1.0* above.
**Note 3:** Insulation to be installed in accordance with the manufacturer's details.

**Table 3.7:** Examples of insulation for suspended beam-and-block ground floors
(Construction as extension guidance details, typically 75 mm screed, (500 g separating layer between foil-faced insulations and screed), insulation layer, beam-and-block floor, ventilated void.)
**Note:** Where the perimeter-over-area ratio (P/A) has not been calculated, use the insulation thickness stated in column 1.0* below.

| Insulation product | K value | Required thickness of insulation (mm) Calculated Perimeter/Area ratio (P/A) | | | | | | | | | |
|---|---|---|---|---|---|---|---|---|---|---|---|
| | | 1.0* | 0.9 | 0.8 | 0.7 | 0.6 | 0.5 | 0.4 | 0.3 | 0.2 | 0.1 |
| **U-value 0.15 W/m²k** | | | | | | | | | | | |
| Celotex FR5000 | 0.021 | 120 | 120 | 120 | 120 | 120 | 120 | 120 | 100 | 90 | 60 |
| Kingspan Kooltherm K3 Floorboard | 0.020 (0.021 25–44 mm) | 110 | 110 | 110 | 110 | 110 | 100 | 100 | 90 | 80 | 50 |
| **U-value 0.18 W/m²k** | | | | | | | | | | | |
| Celotex FR5000 | 0.021 | 100 | 100 | 100 | 90 | 90 | 90 | 90 | 80 | 70 | 40 |
| Kingspan Kooltherm K3 Floorboard | 0.020 (0.021 25–44 mm) | 90 | 90 | 90 | 90 | 80 | 80 | 75 | 70 | 60 | 30 |

**Notes:**
1. Figures indicated above should be adjusted to the insulation manufacturer's nearest thickness.
2. Insulation to be installed in accordance with the manufacturer's details.

### *Roof insulation and ventilation gaps*

Insulation is to be fixed in accordance with the manufacturer's details and must be continuous with the wall insulation, but stopped back at eaves or at junctions with rafters to allow for a continuous 50 mm air gap above the insulation to underside of the roofing felt where a non-breathable roofing felt is used, or 25 mm air space to allow for sag in felt if using a breathable roofing membrane, in accordance with the manufacturer's details.

## Guidance on the Code for Sustainable Homes for new dwellings

**Note: This is not a Building Regulations requirement, but may be required by some planning authorities and is a requirement in Wales (see guidance details).**

(Reproduced by permission of MES Energy Services, Newark Beacon, Beacon Hill Office Park, Cafferata Way, Newark, Notts NG24 2TN, www.mesenergyservices.co.uk, info@mesenergyservices.co.uk, telephone: 01636 653055.)

Launched by the government in 2007, the Code for Sustainable Homes ('the Code') replaces EcoHomes as the national standard to be used in the design and construction of certain new homes in England, Wales and Northern Ireland.

The Code for Sustainable Homes is an 'all-round' measure of the sustainability of a new residential development. The Code is not a set of regulations and is effectively voluntary, with its aim being to help promote higher standards of sustainable design. However, many local authority planning departments are now imposing planning conditions that require minimum Code levels to be met; housing funded by the Homes and Communities Agency (HCA) requires minimum Code

**Table 3.8:** Examples of partial cavity fill insulation for external walls (Construction as extension guidance details, typically 100 mm dense brick or rendered block external skin, cavity, insulation layer, 100 mm insulated block work and plasterboard on dabs internally.)

| Clear cavity width required | Insulation type and minimum thickness | Overall cavity width required | Internal block type and thickness |
|---|---|---|---|
| **U-value 0.18 W/m²k** | | | |
| 50 mm | 85 mm Celotex CG5000 (K value 0.021) or other approved insulation with same K value | 135 mm | 100 mm insulation block – K value 0.15 or lower (Celcon Standard) |
| 50 mm | 75 mm Celotex CG5000 (K value 0.021) or other approved insulation with same K value | 125 mm | 100 mm insulation block – K value 0.11 or lower (Celcon Solar or Thermalite Turbo) |
| 50 mm | 75 mm Kingspan Kooltherm K8 (K value 0.020) or other approved insulation with same K value | 125 mm | 100 mm insulation block – K value 0.11 or lower (Celcon Solar or Thermalite Turbo) |
| **U-value 0.24 W/m²k** | | | |
| 50 mm | 50 mm Celotex CG5000 (K value 0.021) or other approved insulation with same K value | 100 mm | 100 mm insulation block – K value 0.15 or lower (Celcon Standard) |
| 50 mm | 50 mm Celotex CG5000 (K value 0.021) or other approved insulation with same K value | 100 mm | 100 mm insulation block – K value 0.11 or lower (Celcon Solar or Thermalite Turbo) |
| 50 mm | 50 mm Kingspan Kooltherm K8 (K value 0.020) or other approved insulation with same K value | 100 mm | 100 mm insulation block – K value 0.11 or lower (Celcon Solar or Thermalite Turbo) |

**Notes:**
1. Clear cavities can be reduced to 25 mm in compliance with certain insulation manufacturer's details – subject to any building warranty provider's approval where applicable.
2. Insulation to be installed in accordance with manufacturer's details, subject to the suitability of the wall construction and UK zones for exposure to wind-driven rain in accordance with Diagram 12 and Table 4 of ADC.

**Table 3.9:** Examples of full cavity fill insulation for external walls (Construction as extension guidance details, typically 100 mm dense brick or rendered block external skin, insulation layer, 100 mm insulated block work and plasterboard on dabs internally.)

| Clear cavity width required | Insulation type and minimum thickness | Overall cavity width required | Internal block type and thickness |
|---|---|---|---|
| **U-value 0.18 W/m²k** | | | |
| n/a | 135 mm (85 + 50 mm) Earthwool Dritherm 32 Ultimate K value 0.032 | 135 mm | 100 mm insulation block – K value 0.11 or lower (Celcon Solar or similar) |
| **U-value 0.24 W/m²k** | | | |
| n/a | 100 mm Earthwool Dritherm 32 Ultimate K value 0.032 | 100 mm | 100 mm insulation block – K value 0.11 or lower (Celcon Solar or similar) |

**Note:** Insulation is to be installed in accordance with the manufacturer's details, subject to the suitability of the wall construction and UK zones for exposure to wind-driven rain in accordance with Diagram 12 and Table 4 of ADC.

**Table 3.10:** Examples of insulation for solid walls between heated and unheated areas (Construction as extension guidance details, typically plasterboard on dabs, 100 mm block work (K value 0.15), insulation layer, timber battens at 400 mm centres, vapour check and integral/12.5 mm plasterboard and skim finish over studs internally.)

| Insulation product | Minimum thickness (mm) |
|---|---|
| **U-value 0.18 W/m²k** | |
| Celotex FR5000 (K value 0.021) and Celotex PL4000 (K value 0.022) | 80 mm insulation fixed between studs and 45 mm insulation with integral plasterboard fixed over studs with skim finish |
| Kingspan Kooltherm K12 Framing Board (K value 0.020 and Kingspan K18 Insulated Plasterboard (K value 0.020 | 50 mm insulation between studs and 62.5 mm insulation with integral plasterboard fixed over studs with skim finish |
| **U-value 0.24 W/m²k** | |
| Celotex FR5000 (K value 0.021) and Celotex PL4000 (K value 0.022) | 70 mm insulation fixed between studs and 25 mm insulation with integral plasterboard fixed over studs with skim finish |
| Kingspan Kooltherm K12 Framing Board (K value 0.020 and Kingspan K18 Insulated Plasterboard (K value 0.020 | 50 mm insulation between studs and 42.5 mm insulation with integral plasterboard fixed over studs |

**Note:** Insulation is to be installed in accordance with the manufacturer's details.

**Table 3.11:** Examples of insulation for cavity walls with internal timber frame (Construction as extension guidance details, typically 100 mm dense brick or rendered block external skin, 50 mm cavity with breather membrane, structural board, 100/150 × 50 mm studs at 400/600 mm centres, vapour check, integral/12.5 mm plasterboard and skim finish over studs.)

| Clear cavity width required | Timber stud depth (mm) | Insulation type and minimum thickness | Internal insulation/finish over studs |
|---|---|---|---|
| **U-value 0.18 W/m²k** | | | |
| 50 mm | 150 | 100 mm Celotex FR5000 between studs (K value 0.021) | 40 mm Celotex PL4000 insulation with integral plasterboard (K value 0.022 insulation component) |
| 50 mm | 150 | 100 mm Kingspan Kooltherm K12 Framing Board between studs (K value 0.020) | 32.5 mm Kingspan Kooltherm K18 Insulated Plasterboard (K value 0.023) |
| 50 mm | 150 | 140 mm Frametherm 32 Between studs (K value 0.032) | 45.5 mm Knauf Thermal Laminate insulation (K value 0.030) with integral plasterboard |
| 50 mm | 150 | 140 mm Frametherm 32 Between studs (K value 0.032) | 42 mm Kingspan Kooltherm K18 Insulated Plasterboard (K value 0.023) |
| **U-value 0.24 W/m²k** | | | |
| 50 mm | 150 | 60 mm Celotex FR5000 between studs (K value 0.021) | 25 mm Celotex PL4000 insulation with integral plasterboard (K value 0.022) |
| 50 mm | 150 | 90 mm Kingspan Kooltherm K12 Framing Board between studs (K value 0.020) | 12.5 mm plasterboard |
| 50 mm | 150 | 140 mm Frametherm 38 Between studs (K value 0.038) | 17.5 mm Knauf Thermal Laminate insulation (K value 0.030) with integral plasterboard |
| 50 mm | 150 | 90 mm Frametherm 32 Between studs (K value 0.032) | 32 mm Kingspan Kooltherm K18 Insulated Plasterboard (K value 0.023) |

**Note:** Insulation is to be installed in accordance with manufacturer's details, subject to the suitability of the wall construction and UK zones for exposure to wind-driven rain in accordance with Diagram 12 and Table 4 of ADC.

**Table 3.12:** External timber-framed walls with external render/cladding finish (Construction as extension guidance details, typically render/cladding finishes, drained cavity with breather membrane, structural board, 100/150 × 50 mm studs at 400/600 mm centres, vapour check, intergral/12.5 mm plasterboard and skim finish over studs internally.)

| External finish | Timber stud | Insulation type and minimum thickness | Internal insulation/finish |
|---|---|---|---|
| **U-value 0.18 W/m²k** | | | |
| Render/cladding | 100 | 90 mm Celotex FR5000 between studs (K value 0.021) | 45 mm Celotex PL4000 insulation with integral plasterboard (K value 0.022 insulation component) |
| Render/cladding | 100 | 80 mm Kingspan Kooltherm K12 Framing Board between studs (K value 0.020) | 52.5 mm Kingspan Kooltherm K18 Insulated Plasterboard (K value 0.023) |
| **U-value 0.24 W/m²k** | | | |
| Render/cladding | 100 | 70 mm Celotex FR5000 between studs (K value 0.021) | 25 mm Celotex PL4000 insulation with integral plasterboard (K value 0.022 insulation component) |
| Render/cladding | 100 | 80 mm Kingspan Kooltherm K12 Framing Board between studs (K value 0.020) | 32.5 mm Kingspan Kooltherm K18 Insulated Plasterboard (K value 0.023) |

**Note:** Insulation is to be installed in accordance with manufacturer's details, subject to the suitability of the wall construction and UK zones for exposure to wind-driven rain in accordance with Diagram 12 and Table 4 of ADC.

**Table 3.13:** Insulation requirements to exposed upper floors (Construction to be as extension guidance details.)

| Insulation product | K value | Required thickness of insulation (mm) |
|---|---|---|
| **U-value 0.15 W/m²k** | | |
| Kingspan Kooltherm K10 Soffit Board | 0.020 | 125 mm fixed between floor joists |
| Celotex FR5000 | 0.021 | 180 mm fixed in two layers between floor joists |
| **U-value 0.18 W/m²k** | | |
| Kingspan Kooltherm K10 Soffit Board | 0.020 | 110 mm fixed between floor joists |
| Celotex FR5000 | 0.021 | 150 mm fixed between floor joists |

**Note:** Insulation is to be installed in accordance with the manufacturer's details.

standards; and projects promoted or supported by the Welsh Assembly Government are required to meet these more robust sustainability standards.

Although the code for sustainable homes is not a Building Regulations requirement for new dwellings in England, it is a planning requirement for new dwellings in Wales. Since 31 December 2011 the Welsh Assembly Government has required that all new residential properties meet an 8 per cent improvement over the 2010 Code level 3 for sustainable homes (ENE.1) in Wales, and further information is available from: www.wales.gov.uk/planning.

Although one of the aims of the Code is to reduce carbon emissions from new homes, it also considers many other influences such as water efficiency, materials, surface-water runoff (flooding and flood prevention), waste, pollution, health and well-being, management and ecology.

## 3.16 New dwellings

**Table 3.14:** Insulation fixed between/under rafters (Construction as extension guidance details, typically tiles, battens, breather membrane, rafters at 600 mm centres, insulation layer fixed between rafters, vapour check and integral/12.5 mm plasterboard fixed under rafters).

| Insulation product | K value | Required thickness of insulation (mm) and position in roof |
|---|---|---|
| **U-value 0.13 W/m²k** | | |
| Celotex FR5000 and Celotex PL4000 | 0.021<br>0.022 | 120 mm FR5000 between rafters and 50 mm PL4000 insulated plasterboard |
| Kingspan Kooltherm K7 Pitched Roof Board and Kingspan Kooltherm K18 Insulated Plasterboard | 0.020<br>0.020 | 120 mm K7 between rafters and 62.5 mm K18 Insulated Plasterboard |
| **U-value 0.16 W/m²k** | | |
| Celotex FR5000 and Celotex PL4000 | 0.021<br>0.022 | 120 mm FR5000 between rafters and 25 mm PL4000 insulated plasterboard |
| Kingspan Kooltherm K7 Pitched Roof Board and Kingspan Kooltherm K18 Insulated Plasterboard | 0.020<br>0.020 | 120 mm K7 between rafters and 32.5 mm K18 Insulated Plasterboard |

**Note:** Insulation is to be installed in accordance with the manufacturer's details.

**Table 3.15:** Insulation laid horizontally between and over ceiling joists (Construction as extension guidance details, typically tiles, battens, breather membrane, rafters and ceiling joists at 600 mm centres, ventilated roof void, insulation layer fixed between/over ceiling joists and vapour-checked plasterboard fixed under rafters.)

| Insulation product | K value | Required thickness of insulation (mm) and position in roof |
|---|---|---|
| **U-value 0.13 W/m²k** | | |
| Rockwool Roll or Earthwool | 0.044 | 150 mm between joists and 200 mm laid over joists |
| **U-value 0.16 W/m²k** | | |
| Rockwool Roll or Earthwool | 0.044 | 100 mm between joists and 170 mm laid over joists |

**Note:** Insulation is to be installed in accordance with the manufacturer's details.

**Table 3.16:** Insulation fixed between/over rafters (Warm roof – construction as extension guidance details, typically tiles, battens/counter battens, breather membrane, insulation layer fixed over rafters, rafters at 600 mm centres, insulation layer fixed between rafters, cavity and vapour-checked plasterboard fixed under rafters.)

| Insulation product | K value | Required thickness of insulation (mm) and position in roof |
|---|---|---|
| **U-value 0.13 W/m²k** | | |
| Celotex FR5000 | 0.021 | 80 mm fixed over rafters and 80 mm fixed between rafters, with 12.5 mm plasterboard fixed to underside of rafters |
| Kingspan Kooltherm K7 Pitched Roof Board | 0.020 | 80 mm fixed over rafters and 80 mm fixed between rafters, with 12.5 mm plasterboard fixed to underside of rafters |
| **U-value 0.16 W/m²k** | | |
| Celotex FR5000 | 0.021 | 70 mm fixed over rafters and 70 mm fixed between rafters, with 12.5 mm plasterboard fixed to underside of rafters |
| Kingspan Kooltherm K7 Pitched Roof Board | 0.020 | 70 mm fixed over rafters and 60 mm fixed between rafters, with 12.5 mm plasterboard fixed to underside of rafters |

**Note:** Insulation is to be installed in accordance with the manufacturer's details.

**Table 3.17:** Insulation fixed between/under flat-roof joists (Vented 'cold roof' – construction as extension guidance details, typically waterproofing layer, timber deck, 50 mm ventilated cavity, firring strips, flat roof joists at 600 mm centres, insulation layer fixed between joists, vapour check and insulated plasterboard fixed under joists.)

| Insulation product | K value | Required thickness of insulation (mm) and position in roof |
|---|---|---|
| **U-value 0.13 W/m²k** | | |
| Celotex FR5000 and Celotex PL4000 | 0.021 0.022 | 150 mm FR5000 between joists and 50 mm PL4000 Insulated Plasterboard |
| Kingspan Kooltherm K7 Pitched Roof Board and Kingspan Kooltherm K18 Insulated Plasterboard | 0.020 0.021 | 150 mm K7 between rafters and 52.5 mm K18 Insulated Plasterboard |
| **U-value 0.16 W/m²k** | | |
| Celotex FR5000 and Celotex PL4000 | 0.021 0.022 | 150 mm FR5000 between joists and 25 mm PL4000 Insulated Plasterboard |
| Kingspan Kooltherm K7 Pitched Roof Board and Kingspan Kooltherm K18 Insulated Plasterboard | 0.020 0.023 | 140 mm K7 between rafters and 32.5 mm K18 Insulated Plasterboard |

**Notes:**
1. Insulation is to be installed in accordance with the manufacturer's details.
2. The joist depth must be sufficient to maintain a 50 mm air gap above the insulation, and cross-ventilation is to be provided on opposing sides by a proprietary ventilation strip equivalent to a 25 mm continuous gap at eaves level, with insect grille, for ventilation of the roof space.

**Table 3.18:** Insulation fixed above flat-roof joists (Non-vented 'warm roof' – construction as extension guidance details, typically waterproofing layer, insulation layer, vapour-control layer, timber deck, cavity between joists, flat-roof joists at 600 mm centres, and vapour-checked plasterboard fixed under joists.)

| Insulation product | K value | Required thickness of insulation (mm) and position in roof |
|---|---|---|
| **U-value 0.13 W/m²k** | | |
| Celotex EL3000 (or TC3000) | 0.025 0.027 | 130 mm + 50 mm insulation in two layers (total 180 mm), boards mechanically fixed to the top of the flat-roof joists using thermally broken fixings – roof coverings can be hot-bonded or torched on as appropriate. 12.5 mm plasterboard fixed to underside of joists |
| Kingspan Thermaroof TR27 LPC/FM | 0.024–0.026 | 130 mm + 40 mm bonded together in two layers (total 170 mm thick) |
| **U-value 0.16 W/m²k** | | |
| Celotex EL3000 (or TC3000) | 0.025 | 140 mm insulation boards mechanically fixed to the top of the flat-roof joists using thermally broken fixings – roof coverings can be hot-bonded or torched on as appropriate. 12.5 mm plaster board fixed to underside of joists |
| Kingspan Thermaroof TR27 LPC/FM | 0.024–0.026 | 100 mm + 40 mm bonded together in two layers (total 140 mm thick) |

**Note:** Insulation is to be installed in accordance with the manufacturer's details.

### 3.18 New dwellings

The Code grades dwellings from Level 1 to Level 6, with the top Level 6 being regarded as a new home meeting the very highest levels of sustainability. Meeting a Code requirement can be challenging, particularly at the top levels. Credits are scored for fulfilling certain requirements and these are accumulated to achieve the required Code level. Most of the issues are voluntary, so that a developer can pick and choose where they want to gain the necessary number of points; however, some key issues are mandatory in areas such as energy, water efficiency, surface-water runoff and materials, etc.

Formal Code assessment can be carried out only by qualified and accredited Code Assessors.

Projects are assessed in two stages; the Design Stage assessment is normally carried out early in the design process and is based on detailed documentary evidence and commitments, which lead to an 'interim certificate' being issued. This effectively confirms that, providing the dwelling is built to the design stage plans and specification, it will meet a specific Code level.

On completion of the property a final assessment is carried out (including a site visit by the Code Assessor to confirm the final specification) and the final Code certificate is released.

For more information go to:

www.breeam.org/page.jsp?id=86

www.communities.gov.uk/publications/planningandbuilding/codeguide

www.planningportal.gov.uk/buildingregulations/greenerbuildings/sustainablehomes

## Guidance on PassivHaus

**Note: This is not a Building Regulations requirement**

(Reproduced by permission of MES Energy Services, Newark Beacon, Beacon Hill Office Park, Cafferata Way, Newark, Notts NG24 2TN, www.mesenergyservices.co.uk, info@mesenergyservices.co.uk, telephone: 01636 653055.)

A passive house is defined as a building that has an extremely small heating-energy demand, even in the Central European climate, and therefore needs no (or very little) active heating or cooling. Such buildings can be kept warm 'passively', solely by using the existing internal heat sources and the solar energy entering through the windows, as well as by the minimal heating of incoming fresh air.

Developed in Germany, the PassivHaus Standard was developed in the mid-1990s from a desire to bridge the gap between the 'designed' performance and 'in-use' performance of low-energy buildings. Over 500 certified PassivHaus buildings have been monitored in use, some for as long as 15 years, and the findings have been used to refine the design-modelling software. The result is that the 'in use' energy consumption of a certified PassivHaus is extremely closely correlated with the design calculations – and, therefore, is actually delivering the performance expected.

In order for a building to be certified to the PassivHaus standard it needs to meet a set of limiting design criteria, listed below, as demonstrated by modelling the proposed building in the PassivHaus Planning Package software.

- a maximum space heating and cooling demand of less than $15 \, kWh/m^2/year$
- a maximum heating and cooling load of $10 \, W/m^2$
- a maximum total primary energy demand of $120 \, kWh/m^2/year$
- an air-change rate of no more than 0.6 air changes per hour @ 50 Pa.

In order to achieve these values a building requires very high levels of fabric insulation, no thermal bridges, super-insulated windows, very low air permeability, a highly efficient MVHR system, efficient lighting systems and low-energy-consuming appliances/equipment. Applicable to both domestic and non-domestic buildings, the result is energy consumption less than a quarter of that of an average new build.

The application of all the above not only creates a building with low energy consumption, but also one with an excellent internal environment. High levels of insulation and minimal thermal bridging ensure that surface temperatures within the building are warm all year round. Combined with very low air leakage and the MVHR system, uncomfortable draughts are eliminated. Finally, the MVHR system helps to ensure a continuous supply of fresh air and regulates humidity; reducing the incidence of dust mites and preventing mould growth.

For more information go to: www.passivhaustrust.org.uk/guidance.php

## PART B: FIRE SAFETY AND MEANS OF ESCAPE

Follow the guidance details in Section 2 of this guidance for Domestic Extensions with the following additional requirements:

The building shall be designed and constructed to provide facilities to assist fire-fighters in the protection of life, and provision made for access to and into the building in accordance with Table 3.19 below.

Turning facilities should be provided for fire-fighting vehicles in any dead-end access route that is more than 20 m long. This can be by a hammerhead or turning circle, designed on the basis of Diagram 24 of ADB: Volume 1 Dwelling Houses.

### Domestic sprinklers in new Welsh Homes from 2013

The Welsh Government has announced plans that all new and converted residential properties in Wales will need to be fitted with sprinkler systems. Regulations for this will be introduced in September 2013.

For more information go to: http://wales.gov.uk/newsroom/planning/2012/120530lifesaving sprinklers/?lang=en

**Table 3.19:** Typical fire and rescue service vehicle access route specification for new dwellings (including conversions into dwellings) (See Section 11 and Table 8 of ADB: Volume 1 for full details.)

| Appliance type | Minimum width of road between kerbs (m) | Minimum width of gateways (m) | Minimum turning circle between kerbs (m) | Minimum turning circle between walls (m) | Minimum clearance height (m) | Minimum carrying capacity of road (tonnes) |
|---|---|---|---|---|---|---|
| Pump (up to 11 m) | 3.7 | 3.1 | 16.8 | 19.2 | 3.7 | 12.5 |
| High reach (above 1 m) | 3.7 | 3.1 | 26.0 | 29.0 | 4.0 | 17.0 |

**Notes:**
1. As fire appliance sizes and weights are not standardised, the above details may need to change in consultation with the fire and rescue authority.
2. Roads for high-reach pumps should be designed to support a minimum load of 12.5 tonnes in consultation with the fire and rescue authority, but any bridge must be designed to support 17.0 tonnes.

## PART C: SITE PREPARATION AND RESISTANCE TO CONTAMINANTS AND MOISTURE

Follow the guidance details in Section 2 of this guidance for Domestic Extensions

## PART D: CAVITY WALL FILLING WITH INSULATION

Follow the guidance details in Section 2 of this guidance for Domestic Extensions

## PART E: RESISTANCE TO THE PASSAGE OF SOUND

Follow the guidance details in Section 2 of this guidance for Domestic Extensions, with the following additional requirements:

### Party walls and floors separating new dwelling/flats

Sound insulation details for new party walls and floors are to be carried out in accordance with the relevant details contained within this guidance and Approved Document E, as follows:

- 45 dB minimum values for airborne sound insulation to walls, floors and stairs; and
- 62 dB maximum values for impact sound insulation to floors and stairs.

### Party walls and floors separating conversion of buildings into new dwellings/flats

Sound insulation details for new party walls and floors are to be carried out in accordance with the relevant details contained within this guidance and Approved Document E, as follows:

- 43 dB minimum values for airborne sound insulation to walls, floors and stairs; and
- 64 dB maximum values for Impact sound insulation to floors and stairs.

### *Exemptions/relaxations for conversion of historic buildings into dwellings/flats*

If the proposed sound insulation requirements will unacceptably alter the character or appearance of a historic/listed building, then the sound insulation standards may be exempt or improved to what is reasonably practical, or to an acceptable standard that would not prejudice the character or increase the risk of deterioration of the building fabric or fittings, in consultation with the local planning authority's conservation officer. A notice showing the actual sound test results is to be fixed in a conspicuous place inside the building, in compliance with Paragraph 0.7 of ADE

### Pre-completion sound testing of party walls and floors

Robust details are to be adopted or pre-completion sound testing is to be carried out, to demonstrate compliance with Section 1 of ADE1 as follows:

(i) *Robust details*
Robust Construction Purchase Certificates are to be provided by the building owner on commencement and compliance certificates provided on completion, which must be sent to building control to demonstrate compliance with ADE1. Robust Construction. Details can be obtained through Robust Details Ltd. www.robustdetails.com.

(ii) *Pre completion sound testing by specialist*
Pre-completion sound testing is to be carried out by a suitably qualified person or specialist with appropriate third-party accreditation (UKAS or ANC registration), to demonstrate compliance with ADE1, and a copy of the test results is to be sent to building control.

### Testing rate requirements

At least one set of recorded tests is required for every 10 new dwellings on the same site, and additional testing may be required where requested by building control, in compliance with Paras 1.29–1.31 of ADE. Note: sound testing of internal partition walls between rooms is not required.

### Remedial works and retesting

Remedial works and retesting will be required where the test has failed, in compliance with Section 1 of ADE.

### New Internal walls and floors

Sound insulation details between internal walls and floors separating bedrooms, or a room containing a WC and other rooms, are to be carried out in accordance with the relevant details contained within this guidance and ADE. Note: sound testing of internal partition walls between rooms is not required.

## PART F: VENTILATION TO NEW DWELLINGS

## Ventilation systems

**System 1:** Purge (natural) ventilation, background ventilation and intermittent extract fans, in compliance with Table 5.2a of ADF1 – as detailed below.

**System 2:** Passive stack ventilation, to be in compliance with Table 5.2b of ADF1 (not covered in guidance).

**System 3:** Continuous mechanical extract, to be in compliance with Table 5.2c of ADF1 (not covered in guidance).

**System 4:** Continuous mechanical supply and mechanical extract with heat recovery; to be in compliance with Table 5.2d of ADF1 (not covered in guidance).

(Note: ventilations systems 2, 3 and 4 should be designed/installed/commissioned/certificated by a ventilation specialist and a copy of the completion certificate(s) sent to building control.)

System 1: Purge (natural) ventilation with background ventilation and intermittent extract fans

## Purge (natural) ventilation to habitable rooms: system 1 – new dwellings

Purge (natural) ventilation is to be provided to all habitable rooms equal to 1/20 (5 per cent) of floor area where the external windows/doors open more than 30 degrees, increased to 1/10 (10 per cent) of the floor area where the window opens between 15 and 30 degrees. Window openings which open less than 15 degrees are not suitable for purge ventilation and alternative ventilation details are required, as detailed below and in compliance with Section 5 and Appendix B of Approved Document F1. Purge (natural) ventilation openings to habitable rooms are to be typically 1.75 m above floor level.

### Mechanical extract ventilation and fresh air inlets for rooms without purge ventilation

Mechanical extract ventilation and fresh-air inlet are required for habitable rooms without purge (natural) ventilation and must be designed by a ventilation specialist. They should give a minimum of four air changes per hour and be manually controlled, in compliance with Section 5 of Approved Document F1. Note: openable windows/doors may be required for means of escape as detailed in general requirements below and Part B of this guidance.

### Purge ventilation to wet rooms

(Wet rooms are kitchens/utility/bath/WC/en suite, etc.) Opening windows are required for purge (natural) ventilation to wet rooms that have external walls (no minimum sizes), in compliance with Section 5 of ADF1.

### General requirements for purge (natural) ventilation

Purge (natural) ventilation openings to habitable rooms are to be typically 1.75 m above floor level and all internal doors are to have a 10 mm gap under the door for air-supply transfer.

The area of external windows, roof windows and doors should not exceed 25 per cent of the internal floor, in compliance with ADL1A. See fire safety and means-of-escape details in this guidance for permitted unprotected external openings in relation to relevant boundaries.

Means-of-escape windows are to be fitted with proprietary hinges to open to the minimum required clear width of 450 mm. Escape windows must have minimum clear opening casement dimensions of $0.33 \text{ m}^2$ and 450 mm (typically 450 mm wide $\times$ 750 mm high), located within 800–1100 mm above floor level to all bedrooms and habitable rooms at first-floor level, and inner habitable rooms on the ground floor. Windows above the ground-floor storey and within 800 mm of floor level are to be provided with containment/guarding/proprietary catches, which should be removable (but childproof) in the event of a fire. Where escape windows cannot be achieved, direct access to protected stairs (or a protected route to inner rooms) is acceptable, in compliance with ADB1 Para 2.6 (a) or (b).

## Background ventilation: system 1 – new dwellings

Background ventilation is to be provided to all rooms with external walls – either through walls or in windows via hit-and-miss vents (or two-stage window catches, subject to building control approval), equivalent to 5000 mm$^2$ minimum to each habitable room and 2500 mm$^2$ minimum to each wet room (kitchens/utility/bath/WC/en suite, etc.).

The total area of background ventilation must be in compliance with Table 5.2a of ADF1 as follows:

(i)  For dwellings with any design air permeability, use Table 3.20.1.
(ii) For dwellings with design air permeability worse than 5 m$^3$/(h.m$^2$), use Table 3.20.2.

Background ventilation is to be located typically 1.75 m above floor level.
*Note:* For buildings with no external walls, use appropriate ventilation System 2, 3 or 4, in compliance with Paragraphs 5.8–5.10 of ADF1.

**Table 3.20.1:** Total background ventilation (mm$^2$) required for new dwellings with any design air permeability (See Table 5.2a (A) of ADF for full details.)

| Total floor area (m$^2$) | Number of bedrooms | | | | |
|---|---|---|---|---|---|
| | 1 | 2 | 3 | 4 | 5 |
| Less than 50 | 35 000 | 40 000 | 50 000 | 60 000 | 65 000 |
| 51–60 | 35 000 | 40 000 | 50 000 | 60 000 | 65 000 |
| 61–70 | 45 000 | 45 000 | 50 000 | 60 000 | 65 000 |
| 71–80 | 50 000 | 50 000 | 50 000 | 60 000 | 65 000 |
| 81–90 | 55 000 | 60 000 | 60 000 | 60 000 | 65 000 |
| 91–100 | 65 000 | 65 000 | 65 000 | 65 000 | 65 000 |
| 100+ | Add 7000 mm$^2$ for every additional 10 m$^2$ floor area | | | | |

**Table 3.20.2:** Total background ventilation (mm$^2$) required for new dwellings with design air permeability leakier than 5 m$^3$/(h.m$^2$) (See Table 5.2a (B) of ADF for full details.)

| Total floor area (m$^2$) | Number of bedrooms | | | | |
|---|---|---|---|---|---|
| | 1 | 2 | 3 | 4 | 5 |
| Less than 50 | 25 000 | 35 000 | 45 000 | 45 000 | 55 000 |
| 51–60 | 25 000 | 30 000 | 40 000 | 45 000 | 55 000 |
| 61–70 | 30 000 | 30 000 | 30 000 | 45 000 | 55 000 |
| 71–80 | 35 000 | 35 000 | 35 000 | 45 000 | 55 000 |
| 81–90 | 40 000 | 40 000 | 40 000 | 45 000 | 55 000 |
| 91–100 | 45 000 | 45 000 | 45 000 | 45 000 | 55 000 |
| 100+ | Add 5000 mm$^2$ for every additional 10 m$^2$ floor area | | | | |

## Intermittent mechanical extract ventilation: system 1 – new dwellings

### *Mechanical extract ventilation rates*

Mechanical ventilation is to be provided to the rooms listed below, directly ducted to the outside air and equivalent to the following rates.

| | |
|---|---|
| Kitchen | 30 litres per second over hob or 60 litres elsewhere |
| Utility room | 30 litres per second |
| Bathroom | 15 litres per second (including shower rooms and en-suites) |
| Toilet | 6 litres per second WC (with or without a window) |

Mechanical ventilation to rooms without openable windows is to be linked to light operation and have 15 minutes' overrun and a 10 mm gap under the door for air supply.

Mechanical extract fans should not be installed in rooms containing open-flue appliances unless the interaction of mechanical ventilation and open-flue heating appliances is checked and certified by an approved method and by a suitably qualified person, as contained in ADJ.

### *Additional requirements for mechanical extract ventilation*

Mechanical ventilation is to be ducted in proprietary insulated ducts to the outside through walls to a proprietary external vent, or through the roof space to proprietary matching tile/soffit vents.

Mechanical extractor fans should be tested and certified for compliance, and mechanical ventilation with heat-recovery systems (MVHR) should be installed and commissioned by a suitably qualified specialist and copies of the completion certificates sent to building control.

# PART G: SANITATION, HOT-WATER SAFETY AND WATER EFFICIENCY

Follow the guidance details in Section 2 of this guidance for Domestic Extensions, with the following additional requirements:

## Safety valves, prevention of scalding and energy cut-outs

Safety devices and prevention of scalding are to be carried out in compliance with G3 of ADG. Baths should be fitted with an in-line blending valve fixed at 48 degrees Centigrade or below.

## Discharge pipes from safety devices

Discharge pipes from safety devices should be 600 mm maximum length, constructed of metal (or other material suitable for proposed temperatures to BS 7291-1:2006) and connected to a tundish fitted with a suitable air gap, in compliance with the current 'Water Supply (Water Fittings) Regulations. Any discharge into the tundish must be visible (and where the dwelling is occupied by visibly or physically impaired persons, the device must be electronically operated and able to warn of discharge). Discharge pipes from the tundish should be at least 300 mm in length and fixed vertically below it, before connection to any bend or elbow and at a continuous fall of 1:200 thereafter until the point of termination.

Pipes from the tundish should be at least one pipe size larger than the outlet of the safety device up to 9 m in length (2 × larger 9–18 m and 3 × larger 18–27 m) and constructed of metal (or other material suitable for proposed temperatures to BS 7291-1:2006).

The point of termination from discharge pipes can be either:

(i) to a trapped gully – below grating, but above the water seal;
(ii) downward discharges at low level – up to 100 mm above external surfaces (car parks, hardstandings, grassed areas, etc.) and fitted with proprietary wire guard to prevent contact;
(iii) discharges at high level into metal hopper and metal downpipes at least 3 m from plastic guttering collecting the discharge.

*Note:* Visibility of the discharge must be maintained at all times, and discharges of hot water and steam should not come into contact with materials that could be damaged by such discharges.

### Water efficiency and calculations

The estimated water consumption for new dwellings and flats (including changes of use) is to be calculated using a 'Water Efficiency Calculator', which can be calculated by a specialist (or can be downloaded from the Web). Calculations for the estimated consumption of wholesome water are to be submitted at the application stage and approved by building control before works commence on site.

The estimated consumption of wholesome water should not exceed 125 litres per person per day, including a fixed factor of water for outdoor use of 5 litres per person per day.

Typical guidance for flow rates (actual rates are to be stated in accordance with manufacturer's published figures) is:

- 4/2.6 litre dual-flushing toilets
- All taps fitted with flow regulators to 4 litres per minute
- Shower with flow rates of 6–9 litres per minute
- Standard bath (140 litres capacity to overflow)
- Standard washing machine (use manufacturer's published figures)
- Standard dishwasher (use manufacturer's published figures).

#### *Commissioning certificates*

Commissioning certificates for fixed building services are required on completion, with a copy sent to building control

## PART H: DRAINAGE AND WASTE DISPOSAL

Follow the guidance details in Section 2 of this guidance for Domestic Extensions, with the following additional requirements:

Dwellings and conversions are to be provided with an area for the storage of domestic refuse as follows:

- minimum size of 1.2 m$^2$ for the storage of a domestic refuse container (or 0.25 m$^3$ per dwelling for communal waste containers), and

- situated within 30 metres of a suitable collection point without having to take it through the dwelling (unless it is a garage, porch or other similar uninhabitable part of the dwelling, with a maximum number of three steps and gradient not exceeding 1:12).
- All waste and recycling storage spaces must be agreed with the local authority's Waste Collection Department.

## PART J: COMBUSTION APPLIANCES AND FUEL STORAGE SYSTEMS

Follow the guidance details in Section 2 of this guidance for Domestic Extensions.

## PART K: PROTECTION FROM FALLING, COLLISION AND IMPACT

Follow the guidance details in Section 2 of this guidance for Domestic Extensions.

## PART L: CONSERVATION OF FUEL AND POWER

This Part is located at the beginning of this Section 3 for new dwellings.

## PART M: ACCESS TO AND USE OF BUILDINGS FOR DISABLED

Provision is to be made for access to and into the building and for the use of the building and its facilities and sanitary conveniences by disabled persons, in accordance with ADM1 and 4.

### External ramped approach

A level/ramped approach with a firm, hard, non-slip surface at least 900 mm wide is to be provided from a vehicular parking area or the street/road, not steeper than 1:15, with 1.2 m landings every 10 m; or 1:12 with landings every 5 m – both with top and bottom landings at least 1.2 m clear of a door swing up to the principal entrance door suitable for disabled wheelchair users.

### External stepped approach

Sites with a slope of more than 1:15 may be provided with a stepped approach suitable for ambulant disabled as follows:

- An unobstructed width of at least 900 mm with a firm, hard, non-slip surface to all steps and landings.
- Landings at least 900 mm long are to be provided at the top and bottom of the steps.
- The rise of a flight must not exceed 1.8 m between landings.
- Steps are to have a suitable tread, nosing profile and uniform rise between 75 and 150 mm and a uniform going of 280 mm (tapered treads should be measured at a point 270 mm from the inside of the tread).

- Where there are three or more risers, a continuous handrail is to be provided on one side of the flight, with a grippable profile fixed 850–1000 mm above the step pitch line and extending 300 mm beyond the top and bottom nosings.

Even when a stepped approach is used, an accessible or level threshold as described below should be provided. If a step is unavoidable, the rise should not exceed 150 mm.

### External door opening widths

The principal external entrance door should have a minimum clear opening between the door leaf and doorstops of 775 mm.

### Internal passageways/corridors widths

Internal passageways/corridors should have minimum widths as set out in Table 3.21 and Figure 3.1 below:

### Accessible level door thresholds into the building (see Figure 3.2)

Level landings should be provided at the same level as the entrance door thresholds, with a fall of 1:40–1:60 away from the door, separated from the building by proprietary tanking and a proprietary drained channel with accessible drainage grill or 25 mm maximum drainage slot at least 150 mm deep, linked into the storm-water drainage system.

The door threshold should have a maximum 15° slope into the drainage channel/slot and be provided with a proprietary raised threshold storm-proof weather seal, which should not exceed 15 mm high (any projection more than 5 mm should have chamfered or rounded edges), so as to allow safe, unobstructed wheelchair access to and into the building.

### Internal stairs

Minimum stair width for new dwellings is 900 mm in the entrance storey suitable for disabled persons. Handrails must be provided on both sides of the stairs where there are more than three risers, fixed securely at a height 900–1000 mm above floor/landing/nosing levels and continuous

**Table 3.21:** Minimum widths of corridors and passageways for a range of door widths (See Table 4 and Diagram 28 of ADM1 for full details.)

| Doorway clear opening width (mm) | Corridor/passageway width[1] |
|---|---|
| 750 mm | 900 mm (when approached head on) |
| 750 mm | 1200 mm (when not approached head on) |
| 775 mm | 1050 mm (when not approached head on) |
| 800 mm | 900 mm (when not approached head on) |

**Note:** [1]A corridor/passageway width of 750 mm minimum is acceptable where there is a permanent obstruction not exceeding 2.0 m in length, e.g. a radiator or similar, providing the obstruction is not situated opposite a door that would prevent a wheelchair user from turning.

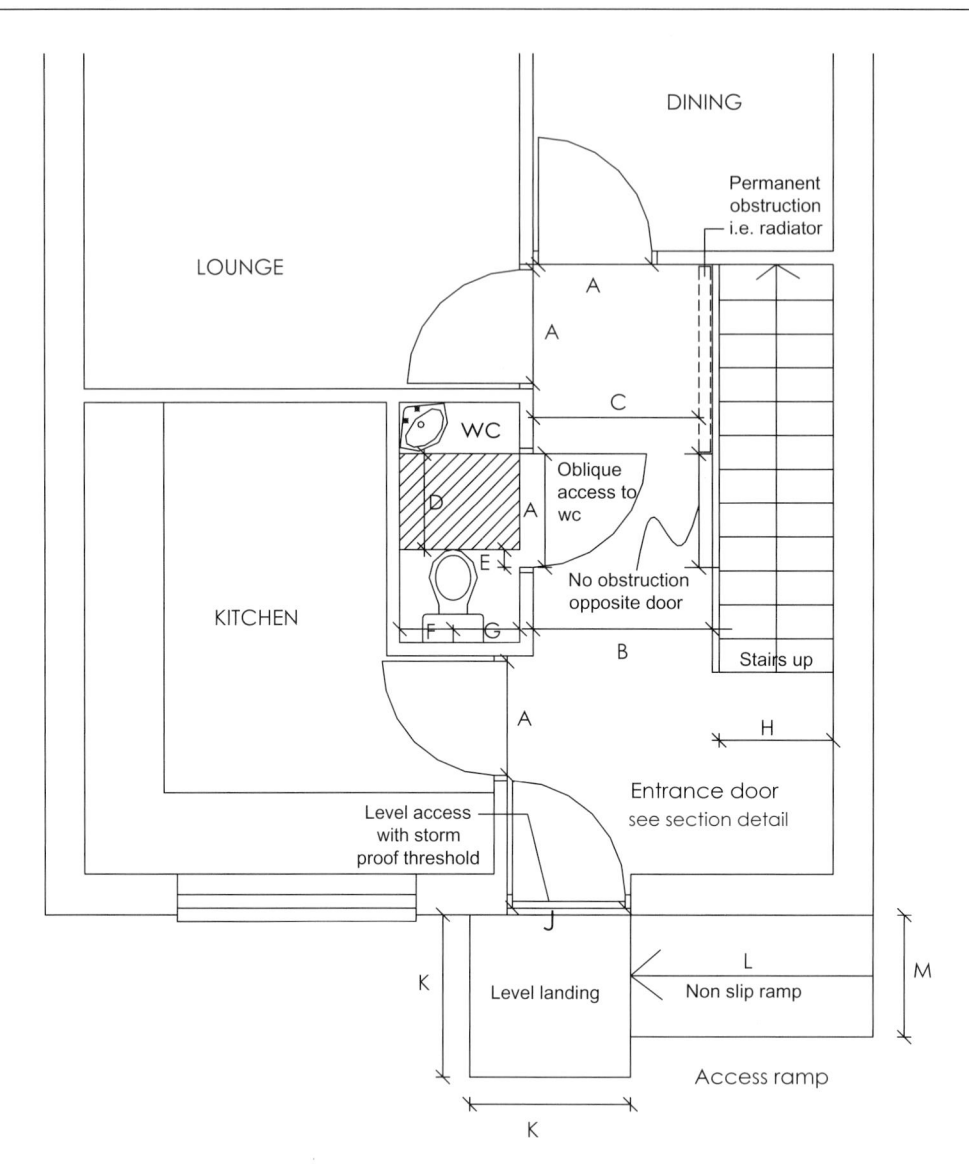

**Figure 3.1:** Typical plan layout of disabled wheelchair access to and into the dwelling. See Table 3.22 below for sizes of openings and ramp gradients *(not to scale)* (See Diagrams 28, 31 and 32 of ADM1 and 4 for full details.)

throughout their length and suitable for disabled persons. Stairs, landings and guarding are to be in compliance with Part K of this guidance

### *Accessible switches, sockets, controls, etc.* (see Figure 3.3)

All switches and sockets, including the consumer unit, ventilation and service controls, doorbells, entry phones, telephone points and tv/computer sockets, etc., should be fixed between 450 and 1200 mm above floor level. Accessible consumer units should be fitted with a childproof cover or installed in a lockable cupboard.

Table 3.22: Minimum clear openings/sizes and maximum ramp gradients for Figure 3.1 and to clarify Table 3.21 above

| Item on layout | Sizes (mm) | |
|---|---|---|
| A and B | **Door minimum clear widths (A)**<br>750–<br>750–<br>775–<br>800– | **Corridor/passageway minimum width (B)**<br>900 (when approached head on)<br>1200 (when not approached head on)<br>1050 (when not approached head on)<br>900 (when not approached head on) |
| C | A corridor width of 750 mm minimum is acceptable where there is a permanent obstruction not exceeding 2.0 m in length, e.g. a radiator or similar, providing the obstruction is not situated opposite a door that would prevent a wheelchair user from turning. | |
| D | 750 min | |
| E | 250 min to edge of door opening | |
| F | 450 min (500 preferred) | |
| G | 400 min | |
| H | 900 min (clear width) | |
| J | 775 min | |
| K | 1200 min | |
| L | Non-slip ramp with maximum gradients 1:15, with 1.2 m landings every 10 m, or 1:12 with landings every 5 m – both with top and bottom landings at least 1.2 m clear of a door swing up to the principal entrance door. | |
| M | 900 mm min | |

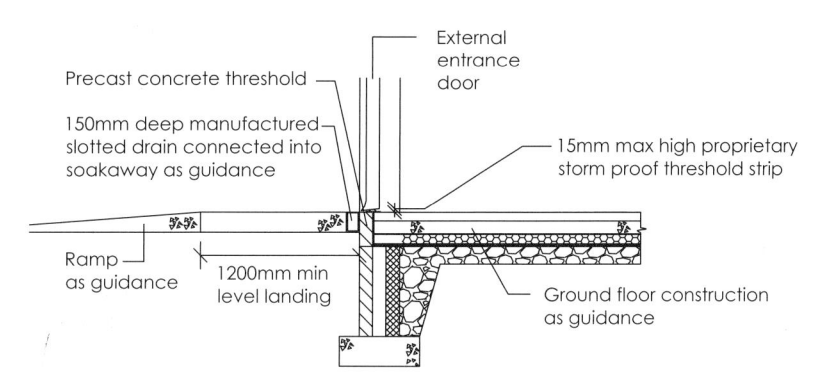

Figure 3.2: Typical section detail of level threshold into dwelling suitable for disabled wheelchair users *(not to scale)*

### *Provision of a ground-floor WC*

A WC must be provided on the principal entrance storey, with an outward-opening door conforming to the above widths. The WC enclosure/position should have a clear space of at least 450 mm each side of the centre of the WC, and a clear space of at least 750 mm in front of the WC pan to allow a wheelchair to approach within 400 mm of the WC from the front (or within 250 mm of the front of the WC pan from the side). The washbasin is to be positioned so as not to impede access or the spaces outlined above.

**3.30** New dwellings

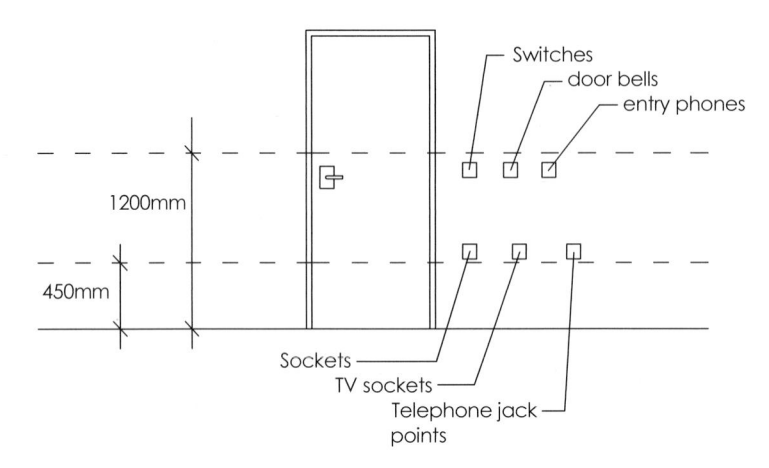

**Figure 3.3:** Heights of switches and sockets, etc. (See Diagram 29 of ADM1 for full details.)

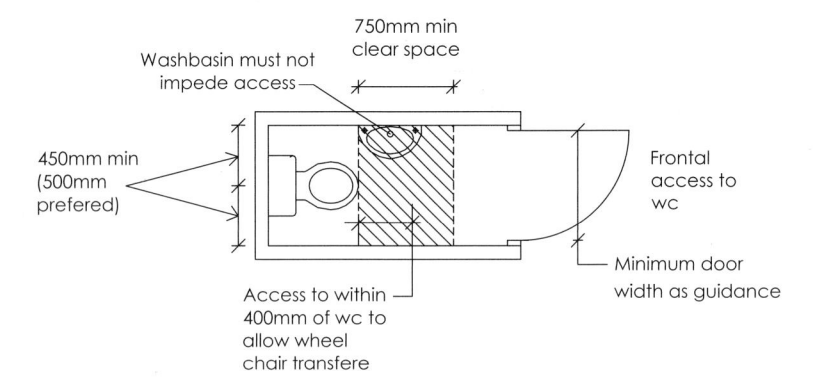

**Figure 3.4:** Typical plan layout of frontal-access WC for disabled wheelchair users *(not to scale)* (See Diagram 31 of ADM1 for full details.)

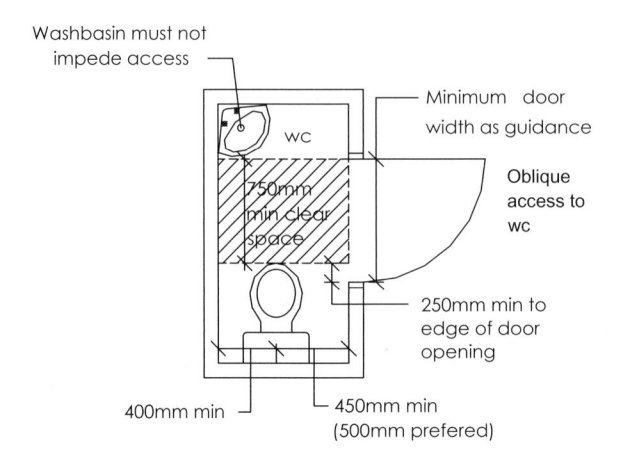

**Figure 3.5:** Typical plan layout of oblique-access WC for disabled wheelchair users *(not to scale)* (See Diagram 32 of ADM1 for full details.)

# Guidance on Lifetime Homes Standard for new dwellings

**Note: This is not a Building Regulations requirement.**

The Lifetime Homes Standard was first established in the mid-1990s, and was subsequently reviewed in 2010. It seeks to enable 'general needs' housing to provide, either from the outset or through simple and cost-effective adaptation, design solutions that meet existing and changing needs of the widest range of households. This can give many households more choice over where they live, and the range of visitors they can accommodate.

The design of a Lifetime Home removes barriers to accessibility. Flexibility and adaptability within the design and structure also enable a Lifetime Home to meet a diverse range of household needs over time by simple and cost-effective adaptation. Dwellings will therefore have potential to provide for the widest cross-section of individuals within the general population and will offer greater 'visitability', so that an individual is less likely to be prevented from visiting a household because of the dwelling's design.

Lifetime Homes need not be complicated, or expensive, for house-builders. The 16 design criteria that make up the Standard have been carefully considered so that they can be incorporated into a dwelling's design and construction from the outset, with only a marginal cost effect. Once occupied, the adaptability of the dwelling can be cost-effective for a household if needs change and adaptations are required.

The Standard is based on the five overarching principles of inclusivity, accessibility, adaptability, sustainability and good value, which inform the detailed design and specification requirements detailed within the 16 Lifetime Homes Design Criteria. These criteria include both internal and external features of individual dwellings, and also internal and external communal features of blocks of dwellings. Detailed information on the design and specification requirements of each of the 16 Design Criteria can be found on the Lifetime Homes website at: http://www.lifetimehomes.org.uk/pages/revised-design-criteria.html.

While a Lifetime Home may offer enough flexibility for some households with wheelchair users, the Lifetime Homes Standard is quite different from the standards for wheelchair housing and wheelchair-adaptable housing, which have much more detailed and demanding spatial and specification requirements.

## Lifetime Homes in the context of government policy and regulation

The Lifetime Homes Standard was discussed extensively during a review of Part M of the Building Regulations for England and Wales during the late 1990s. At that time, the Approved Document M was being reviewed and extended to include provisions for new dwellings. The resultant 1999 edition of Part M (and subsequent editions) includes requirements equivalent to some, but not all, of the Lifetime Homes design and specification standards. Those included in Part M tend, generally speaking, to relate to basic accessibility and 'visitability' into the entrance level of the dwelling, and access to service controls. In some instances the Part M requirements are similar, but do not fully cover the Lifetime Homes design and specification requirements. The Part M requirements exclude the Lifetime Homes provisions for future adaptability. A summary comparison between the Lifetime Homes Criteria and Part M requirements can be found on the Lifetime Homes website at: http://www.lifetimehomes.org.uk/pages/lifetime-homes-and-part-m.html.

Some planning and funding authorities require new dwellings to exceed the Regulatory Part M requirements and achieve the Lifetime Homes Standard. For example, in 2004, the Greater London

Authority adopted the Lifetime Homes Standard in the Supplementary Planning Guidance of the London Plan. This states that all residential units in new housing developments should be Lifetime Homes, including houses and flats of varying sizes, in both the public and private sectors. The Standard has subsequently also been incorporated into the London Mayor's interim edition of the London Housing Design Guide, 2010.

In Wales and Northern Ireland, the Welsh Assembly and the Northern Ireland Housing Executive, respectively, already require the Lifetime Homes Standard in developments they fund. The Standard is also recognised within the Department for Communities and Local Government's Code for Sustainable Homes. This Code currently gives credits within its 'Health and well-being' section to developments that achieve the Lifetime Homes Standard.

More information on Lifetime Homes is available on the Lifetime Homes website www.lifetimehomes.org.uk. (Reproduced by permission of Lifetime Homes.)

## PART N: SAFETY GLAZING, OPENING AND CLEANING

Follow the guidance details in Section 2 of this guidance for Domestic Extensions.

## PART P: ELECTRICAL SAFETY

Follow the guidance details in Section 2 of this guidance for Domestic Extensions.

# Section 4   Domestic loft conversions

| | |
|---|---|
| Converting an existing loft space | 4.3 |
|     Additional technical and practical guidance | 4.3 |
| Assessing the feasibility of your loft for conversion | 4.3 |
|     1. Roof structure and shape | 4.3 |
|     2. Roof coverings and roofing felt | 4.3 |
|     3. Ceilings | 4.4 |
|     4. Internal space available | 4.4 |
|     5. Headroom available | 4.4 |
|     6. Means of escape | 4.4 |
|         Existing single-storey buildings | 4.4 |
|         Existing two-storey buildings | 4.4 |
|     7. Ventilation | 4.4 |

*Guide to Building Control: For Domestic Buildings*, First Edition. Anthony Gwynne.
© 2013 John Wiley & Sons, Ltd. Published 2013 by John Wiley & Sons, Ltd.

| | |
|---|---:|
| Part A: Structure | 4.5 |
| A1: Inspection of the existing roof and building structure | 4.5 |
| A2: Alteration, modification and strengthening of the existing roof structure | 4.5 |
| A3: Roof conversion details | 4.5 |
| Upgrading existing external walls | 4.5 |
|     Option 1: Upgrading cavity walls suitable for cavity wall filling | 4.5 |
|     Option 2: Upgrading cavity walls not suitable for cavity wall filling | 4.5 |
|     Option 3: Upgrading solid masonry walls | 4.7 |
| Internal load-bearing timber stud walls | 4.7 |
| Part B: Fire safety and means of escape | 4.11 |
| Single-storey dwellings with loft conversion | 4.11 |
| Two-storey dwellings with loft conversion (or new third storey) | 4.13 |
|     Option 1: Protected stairway | 4.13 |
|     Option 2: Protected stairway with alternative exits at ground-floor level | 4.14 |
|     Option 3: Fire-separated top storey with alternative external/internal fire exit | 4.15 |
|     Option 4: Residential sprinkler systems for means of escape | 4.16 |
|       Important note | 4.16 |
| Part C: Site preparation and resistance to contaminants and moisture | 4.17 |
| Part D: Cavity wall filling with insulation | 4.17 |
| Part E: Resistance to the passage of sound | 4.17 |
| Part F: Ventilation | 4.17 |
| Part G: Sanitation, hot-water safety and water efficiency | 4.17 |
| Part H: Drainage and waste disposal | 4.17 |
| Part J: Combustion appliances and fuel storage systems | 4.17 |
| Part K: Protection from falling, collision and impact | 4.17 |
| Part L: Conservation of fuel and power in conversions | 4.17 |
| Part M: Access to and use of buildings for disabled | 4.18 |
| Part N: Safety glazing, opening and cleaning | 4.18 |
| Part P: Electrical safety | 4.18 |

## Converting an existing loft space

Converting an existing loft space can be an easy and cost-effective way of increasing living accommodation in most houses. This guide to loft conversions will provide useful guidance on how some of the more common technical design and construction requirements of the Building Regulations can be achieved, where the loft space of an existing one- or two-storey dwelling is being converted into habitable accommodation to form an additional storey to the dwelling. Where the house has three or more storeys before the loft is converted, please contact building control for further guidance.

### Additional technical and practical guidance

This is additional guidance for loft conversions, which should be read in conjunction with the construction details in Section 2 of this guidance for Domestic Extensions.

## Assessing the feasibility of your loft for conversion

Before commencing a loft conversion, it is important to assess the feasibility of the project. This will involve inspection of the existing loft and dwelling to assess the following:

### 1. Roof structure and shape

The overall form, construction and profile of the roof will have a major bearing on whether the roof is suitable for conversion into a usable space. Traditional cut-timber pitched roofs with gable end walls (cavity walls or solid walls at least 250 mm thick) and horizontal ridges are generally easier to convert than hipped roofs or roofs with intersecting pitches and valleys, which may require more complicated structural designs.

Trussed rafter roofs constructed using a series of complex trusses should only be altered, modified and converted in compliance with details and calculations carried out by a suitable qualified and experienced property professional. **No trussed rafter should ever be cut or modified in any way until a new supporting structure is in place that has been designed by a suitably qualified and experienced specialist designer.** Further information for loft conversions with trussed rafter roofs can be obtained from TRADA at: www.trada.org.uk.

All existing timbers should be in a sound condition; any defective timber is to be replaced with new in compliance with guidance span tables or details and calculations carried out by a suitable qualified and experience property professional. Existing timbers are to be inspected, and where necessary treated against insect and fungal attack, by a suitably qualified and experienced specialist.

### 2. Roof coverings and roofing felt

These should be in a good, sound and weather-tight condition; any defective coverings should be replaced with new or existing sound coverings, and defective roofing felt is to be replaced with new, fixed according to the manufacturer's details.

### 3. Ceilings

The underside of the new storey floor should achieve 30 minutes' fire resistance. Normally, existing 12.5 mm plasterboard and skim or sound lath and plaster in older houses will achieve this – otherwise additional upgrading will be required using 12.5 mm plasterboard and skim or a proprietary intumescent product to achieve 30 minutes fire resistance, applied and certified in accordance with manufacturer's details.

### 4. Internal space available

The roof space should not have any chimneys or services passing centrally through the loft space that cannot be easily moved, altered or modified. Any structural alterations/modifications required should be carried out in compliance with details and calculations carried out by a suitably qualified and experienced property professional.

### 5. Headroom available

Headroom available is measured vertically from the top of the new floor (which typically can be 200 mm or more above the existing ceiling joists) to the underside of the new horizontal/sloping ceilings (which can typically down-stand 50–75 mm from the existing roof structure). Minimum headroom of 2.0 m is normally required to the stairs; where there is not enough space to achieve that height it can be reduced to 1.9 m at the centre of the stairs and 1.8 m at the side in loft conversions, in accordance with the details in Section 2 of this guidance for Domestic Extensions.

Although there is no minimum room height requirement for habitable rooms, a ceiling height of 2.2 to 2.3 m is preferred in the centre of the roof for headroom, reducing to 0.8 to 1.2 m at the side walls to sloping ceilings, to allow for the placing of furniture. Normally, if the existing roof pitch is less than 30 degrees and roof span is less than 6 metres, a loft conversion may be impractical and you should consult a loft specialist for further advice.

### 6. Means of escape

#### *Existing single-storey buildings*

Existing single-storey buildings with conversion of the loft space forming a second storey will require means-of-escape windows or access to protected stairs.

#### *Existing two-storey buildings*

Existing two-storey buildings with conversion of the loft space forming a third storey will require access to protected stairs that connect to a hall and final exit at ground-floor level, or else give access to at least two escape routes to final exits at ground level, which will need to be separated by fire-resisting construction and fire doors. Alternatively, the new top storey can be separated by fire-resisting construction and provided with an alternative escape route (subject to planning permission), or a domestic sprinkler system can designed by a fire engineer.

Means-of-escape and fire-safety requirements and diagrams are detailed below in this guidance.

### 7. Ventilation

Adequate provision for ventilation will be required, in accordance with the details in Section 2 of this guidance for Domestic Extensions.

## PART A: STRUCTURE

### A1: Inspection of the existing roof and building structure

Existing foundations, lintels and wall structure that will be built off or used to support the new storey loads may need to be exposed at the discretion of the building control surveyor and structural engineer, to ensure that they are adequate and suitable – this may include opening up or excavating to expose foundations/walls/floors to check their condition. If they do not appear to be adequate to support the proposed works, details/justification of the proposed remedial works/alterations, including necessary engineering calculations and details, will need to be submitted for approval before works commence on site.

### A2: Alteration, modification and strengthening of the existing roof structure

Inspection of the existing roof coverings and structural timbers should be carried out, overhauling and replacing defective and missing tiles, treated timber battens, roofing felt, lead valleys, flashings, facia boards, soffit/ barge boards and rainwater goods, etc. as necessary to match existing ones. Re-point/rebuild defective masonry walls/chimneys etc. as necessary. Repair/ replace defective roof timbers as necessary. Existing timbers are to be inspected, repaired, replaced and treated as necessary by a specialist with a warrant-backed guarantee against insect and fungal attack.

Alteration/modification/strengthening of the existing roof structure/structural members/walls etc. should only be carried out/repaired/ replaced/supported or removed in strict compliance with details and calculations received from a suitably qualified person. These details must be approved by building control before works commence on site.

### A3: Roof conversion details

### Upgrading existing external walls

Where existing external walls are to be used as part of the loft conversion and have a threshold U-value worse than 0.7, the U-values should be upgraded as follows.

#### Option 1: Upgrading cavity walls suitable for cavity wall filling

A U-value of 0.55 is required for upgrading existing cavity walls using a British Board of Agreement (BBA or other third-party accredited) cavity wall insulation system, assessed and installed by a specialist and suitable for the proposed situation.

#### Option 2: Upgrading cavity walls not suitable for cavity wall filling

A U-value of 0.3 is required for existing cavity walls not suitable for cavity wall filling, typically upgraded internally using 50/75 × 50 mm treated soft wood studs, fixed vertically to existing walls at 400 mm centres (damp-proof course fixed between studs and wall). Thermal insulation is to be fixed tightly between/over studs as detailed in Table 4.1 below, with integral/12.5 mm vapour-checked plasterboard fixed to studs with 3–5 mm-thick plaster skim finish.

## 4.6 Domestic loft conversions

**PITCHED ROOF**

Internal sound insulated stud partition as guidance details

Load bearing stud with thermal insulation as detailed below

**Shower room**

12.5mm vapour checked plaster board & skim

50/75 x 50mm timber studs at 400mm ctrs fixed to wall

Insulation fixed between/over studs as guidance table

EXISTING SEPARATING WALL
Upgrade sound insulation to separating walls between dwellings as necessary as guidance details

**Upgrading existing external walls:**
Where existing external walls are to be used as part of the loft conversion and have a threshold U-value worse than 0.7, the U-values should be upgraded as the following options:

**Option 1:** Upgrading cavity walls suitable for cavity wall filling to achieve a U-value of 0.55 as detailed in guidance diagram 79 below

**Option 2:** Upgrading cavity walls not suitable for cavity wall filling to achieve a U-value of 0.30 as detailed in guidance diagram 79 below

**Option 3:** Upgrading solid masonry walls to achieve a U-value of 0.30 as detailed in guidance diagram 79 below

Proposed dormer roof omitted for clarity of layout

**Bedroom**

FD 20 fire door**

Smoke alarm

30 minutes fire resisting** (& sound insulating) stud partition to protect stairs as guidance details

Stairs & handrail as guidance details

**ADJOINING UNHEATED ROOF SPACE**

**Protected stairs is required for three storey buildings (i.e. existing two storey building with loft conversion forming a third storey) and can be used where means of escape windows cannot be provided in two storey dwellings as guidance details

12.5mm vapour checked plaster board & skim

12mm structural plywood skin & fixings to s/engineers details & calculations

100mm timber studs at 400mm ctrs

Thermal insulation fixed between/over studs as guidance details/table

Load bearing studs supported by structural beams to s/engineers details & calculations

**PITCHED ROOF**

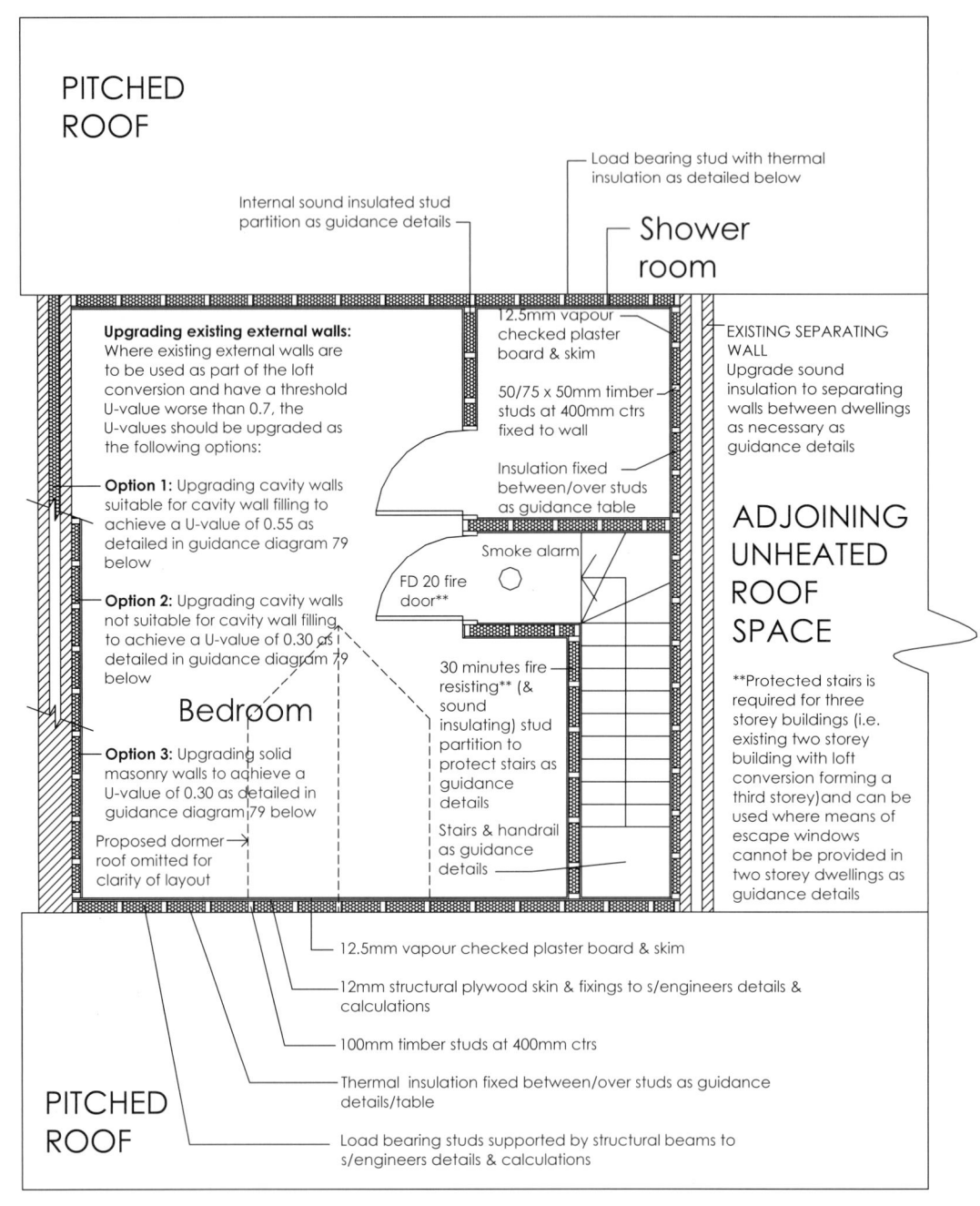

*Solid external stone/brick walls should not be less than 328mm thick with rendered finish or block work less than 250mm thick with rendered finish. Alternatively, walls to be tanked internally with a British Board of Agreement (BBA or other third party accredited) water proof tanking system, applied to prepared walls in accordance with tanking manufacturers details to prevent the ingress of water/ moisture/ condensation into the building.

**Figure 4.1:** Typical plan layout of a loft conversion with intersecting dormer pitched roof *(not to scale)*

## Option 3: Upgrading solid masonry walls

A U-value of 0.3 is required for solid masonry walls typically upgraded internally using 50/75 × 50 mm treated soft wood head and sole plates with vertical studs fixed at 400 mm centres, fixed 25 mm away from the face of the masonry wall. Thermal insulation is to be fixed tightly between/over studs as detailed in Table 4.1 below, with integral/12.5 mm vapour-checked plasterboard fixed to studs with 3–5 mm-thick plaster skim finish. Solid external stone/brick walls should not be less than 328 mm thick, with external rendered finish, or block work less than 250 mm thick with external rendered finish. Insufficiently thick walls are to be tanked internally with a British Board of Agreement (BBA or other third-party accredited) waterproof tanking system, which is to be vapour-permeable when applied above ground, to prevent condensation within the building, and applied to prepared walls in accordance with the tanking manufacturer's details and specification for the particular project, to prevent the ingress of water into the building.

## Internal load-bearing timber stud walls

Load-bearing timber stud walls are to be in compliance with details and calculations by a suitably qualified and experienced person, and must be approved by building control before works commence on site.

PITCHED ROOF

Option 1: Existing cavity walls suitable for cavity wall insulation

Existing cavity walls with a threshold U-value worse than 0.7, upgrade U-value to 0.55 using an injected cavity wall insulation system.

The suitability of the cavity wall for filling must be assessed before the works is carried out by an insulation specialist in accordance with BS 8208:Part 1: 1985 and the insulation system must be British Board of Agreement (BBA or other third party) accredited.

Option 2: Existing cavity walls not suitable for cavity wall filling

Existing cavity walls with a threshold U value worse than 0.7, upgrade U-value to 0.3 internally as follows:

- 12.5mm vapour checked plaster board & skim
- 50/75 x 50mm timber studs at 400mm ctrs fixed to dpc strip & existing wall
- Thermal insulation fixed between/over studs as guidance details/table

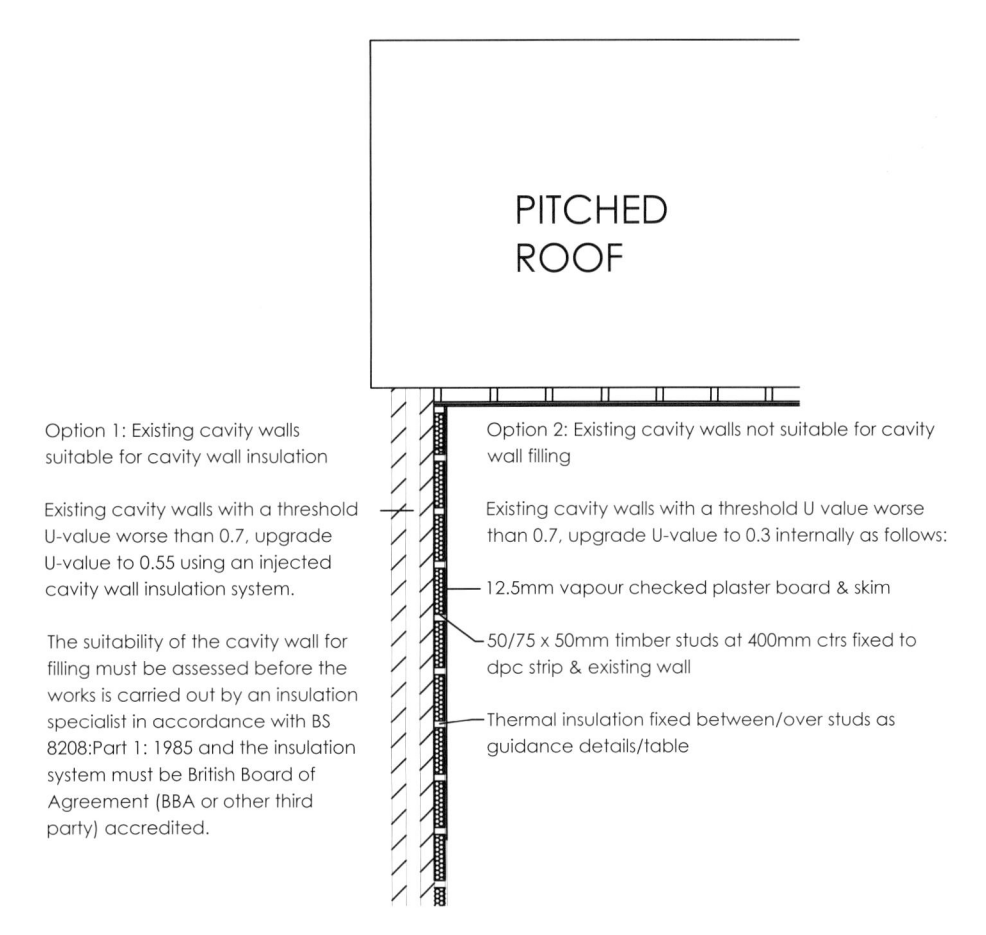

**Figure 4.2.1:** Typical plan layout of upgrading existing cavity walls in loft conversion *(not to scale)*

### 4.8 Domestic loft conversions

**Option 3: Solid walls**

Existing solid walls with a threshold U value worse than 0.7, upgrade U-value to 0.3 internally as follows:

- 12.5mm vapour checked plaster board & skim
- 50/75 x 50mm treated timber studs at 400mm ctrs Thermal insulation fixed between/over studs as guidance details/table
- 25mm air gap
- Repoint/repair existing walls as necessary

Existing solid external walls less than: 328mm thick rendered stone/brick or less than 250mm thick rendered blockwork should be tanked internally with a BBA approved water proof tanking system to prevent ingress of moisture into the building as guidance details

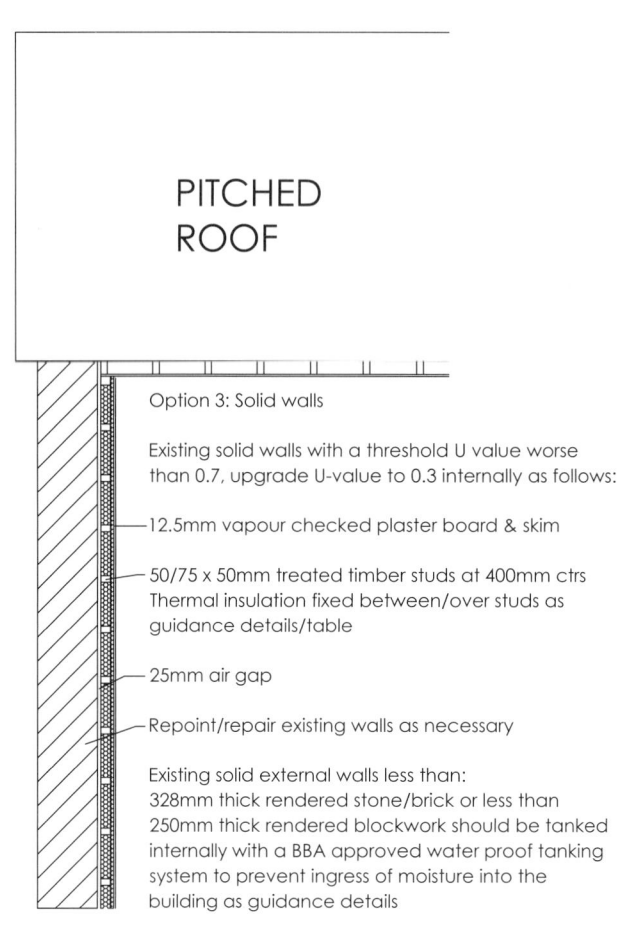

**Figure 4.2.2:** Typical plan layout of upgrading existing solid masonry walls in loft conversion *(not to scale)*

**Table 4.1:** Insulation requirements for upgrading existing external walls

| External wall type | Insulation type | Minimum thickness (mm) |
| --- | --- | --- |
| Cavity walls with threshold U-value worse than 0.7, upgrade U-value to 0.55 | Upgrade using a British Board of Agreement (BBA or other third-party accredited) cavity wall insulation system, assessed and installed by a specialist suitable for the proposed situation, in compliance with Part D of guidance for domestic extensions. | To be confirmed by specialist |
| Solid walls with threshold U-value worse than 0.7, upgrade U-value to 0.30 | Kingspan Kootherm K18 (K value 0.020) | 72.5 mm fixed across studs with integral vapour-checked plasterboard and skim finish |
| Solid walls with threshold U-value worse than 0.7, upgrade U-value to 0.30 | Celotex FR5000 (K value 0.021) and Celotex PL4000 (K value 0.022) | 50 mm between 50 mm deep studs or 40 mm between 75 mm deep studs and 25 mm fixed across studs with integral vapour-checked plasterboard and skim finish |

# Domestic loft conversions

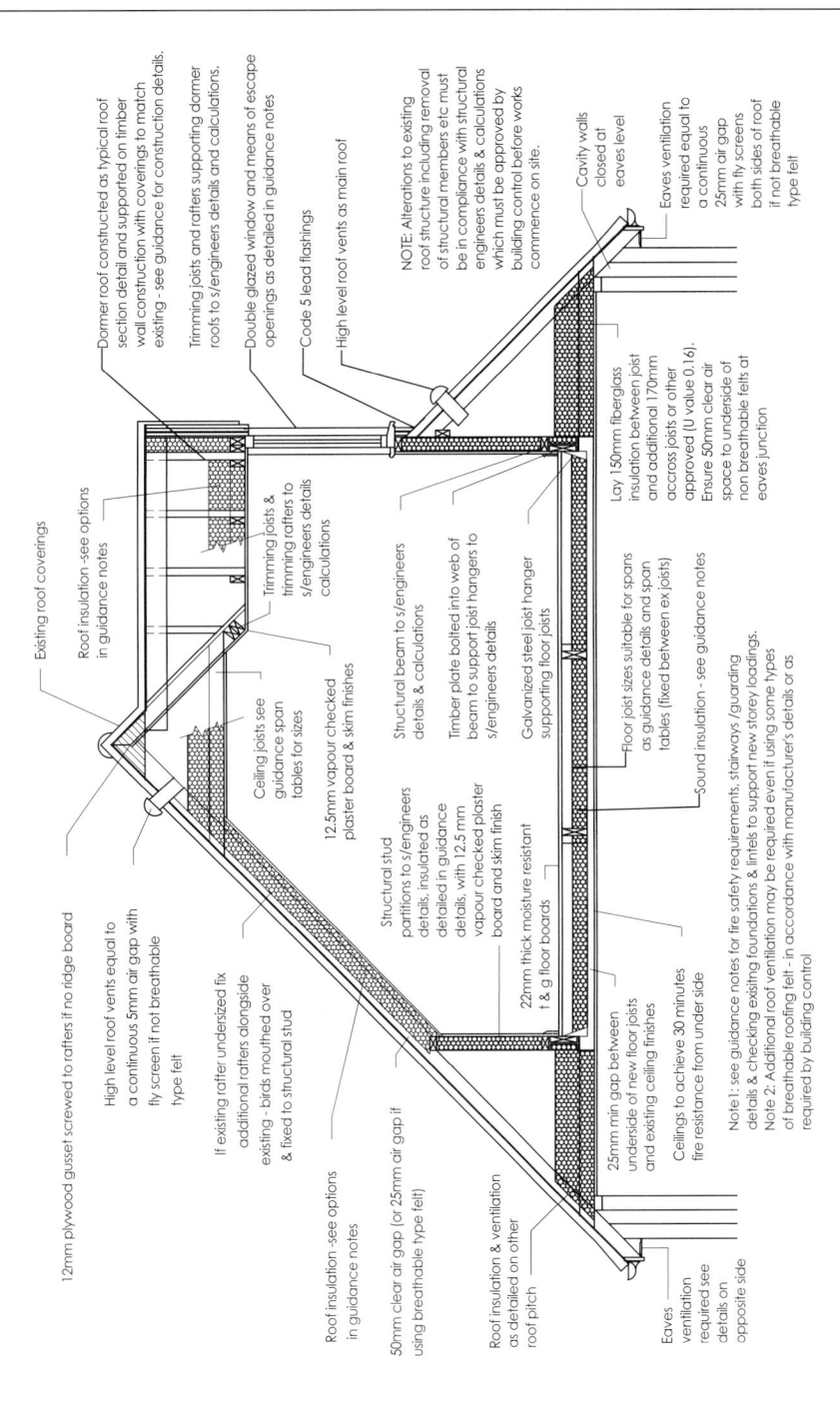

**Figure 4.3:** Typical section through a loft conversion with dormer pitched roof *(not to scale)*

Labels (clockwise from top-left):

- 12mm plywood gusset screwed to rafters if no ridge board
- Existing roof coverings
- Roof insulation - see options in guidance notes
- Dormer roof constructed as typical roof section detail and supported on timber wall construction with coverings to match existing - see guidance for construction details.
- Trimming joists and rafters supporting dormer roofs to s/engineers details and calculations.
- Double glazed window and means of escape openings as detailed in guidance notes
- Code 5 lead flashings
- High level roof vents as main roof
- NOTE: Alterations to existing roof structure including removal of structural members etc must be in compliance with structural engineers details & calculations which must be approved by building control before works commence on site.
- Cavity walls closed at eaves level
- Eaves ventilation required equal to a continuous 25mm air gap with fly screens both sides of roof if not breathable type felt
- Lay 150mm fiberglass insulation between joist and additional 170mm across joists or other approved (U value 0.16). Ensure 50mm clear air space to underside of non breathable felts at eaves junction
- Trimming joists & trimming rafters to s/engineers details calculations
- Structural beam to s/engineers details & calculations
- Timber plate bolted into web of beam to support joist hangers to s/engineers details
- Galvanized steel joist hanger supporting floor joists
- Floor joist sizes suitable for spans as guidance details and span tables (fixed between ex joists)
- Sound insulation - see guidance notes
- Note1: see guidance notes for fire safety requirements, stairways /guarding details & checking existing foundations & lintels to support new storey loadings.
- Note 2: Additional roof ventilation may be required even if using some types of breathable roofing felt - in accordance with manufacturer's details or as required by building control
- Ceilings to achieve 30 minutes fire resistance from under side
- 25mm min gap between underside of new floor joists and existing ceiling finishes
- 22mm thick moisture resistant t & g floor boards
- 12.5mm vapour checked plaster board & skim finishes
- Ceiling joists see guidance span tables for sizes
- Structural stud partitions to s/engineers details, insulated as detailed in guidance details, with 12.5 mm vapour checked plaster board and skim finish
- If existing rafter undersized fix additional rafters alongside existing - birds mouthed over & fixed to structural stud
- High level roof vents equal to a continuous 5mm air gap with fly screen if not breathable type felt
- Roof insulation & ventilation as detailed on other roof pitch.
- 50mm clear air gap (or 25mm air gap if using breathable type felt)
- Roof insulation - see options in guidance notes
- Eaves ventilation required see details on opposite side

## 4.10 Domestic loft conversions

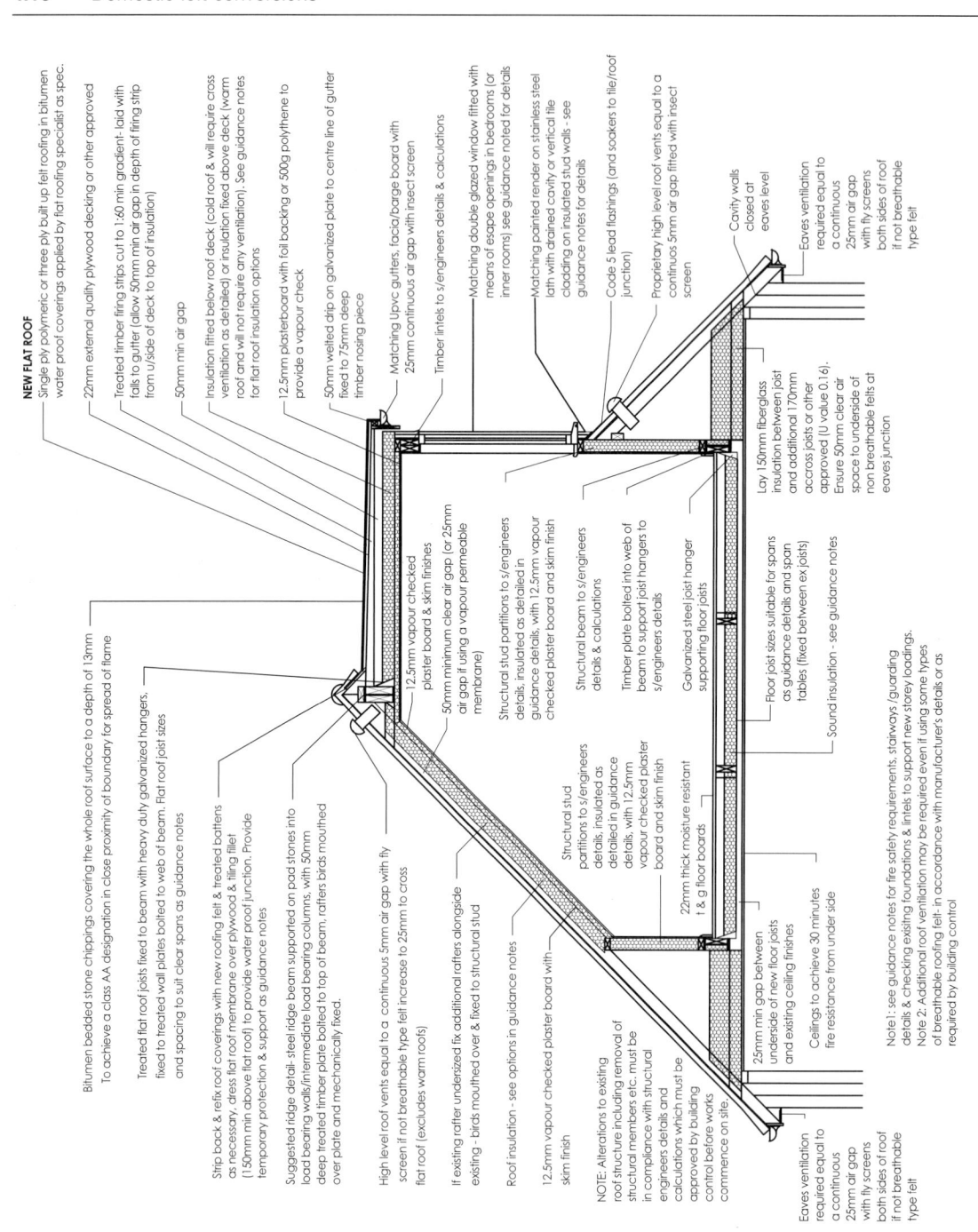

**Figure 4.4:** Typical section through a loft conversion with dormer flat roof (*not to scale*)

Typically, load-bearing stud walls not exceeding 2.5 m high are to be constructed of 100 × 50 mm C24 soft wood studs fixed vertically at 400 mm centres with head and sole plates and intermediate noggins fixed at 600 mm, with thermal insulation friction-fixed airtight between studs as detailed in the guidance table below, with 12 mm-thick structural plywood glued and screwed to the full height and width of the room side face of the stud wall, finished with 12.5 mm plasterboard and skim. Insulated and draught-proofed access hatches are to be formed between studs into the loft space. New rafters to be birds-mouthed over and mechanically fixed to head plate, and mechanically fixed to existing rafters by an approved method.

## PART B: FIRE SAFETY AND MEANS OF ESCAPE

### Single-storey dwellings with loft conversion

The conversion of loft spaces to bungalows or similar single-storey buildings to form a new first floor does not require a protected hall, landing or fire doors, but the supporting upper floor must be provided with 30 minutes' fire-resisting construction from the underside. The stairs may be positioned in a ground-floor room such as the living room, provided there is access to an external door that opens directly to the outside for means of escape. All rooms or bedrooms on the first floor (except bathrooms or toilets) must be directly accessible off the stair landing. Smoke detection must be installed as detailed below, with an additional interlinked heat detector at ceiling level in kitchens that are open to the stairs/circulation areas at ground level. An acceptable typical two-storey building layout is illustrated in Figure 2.43 above.

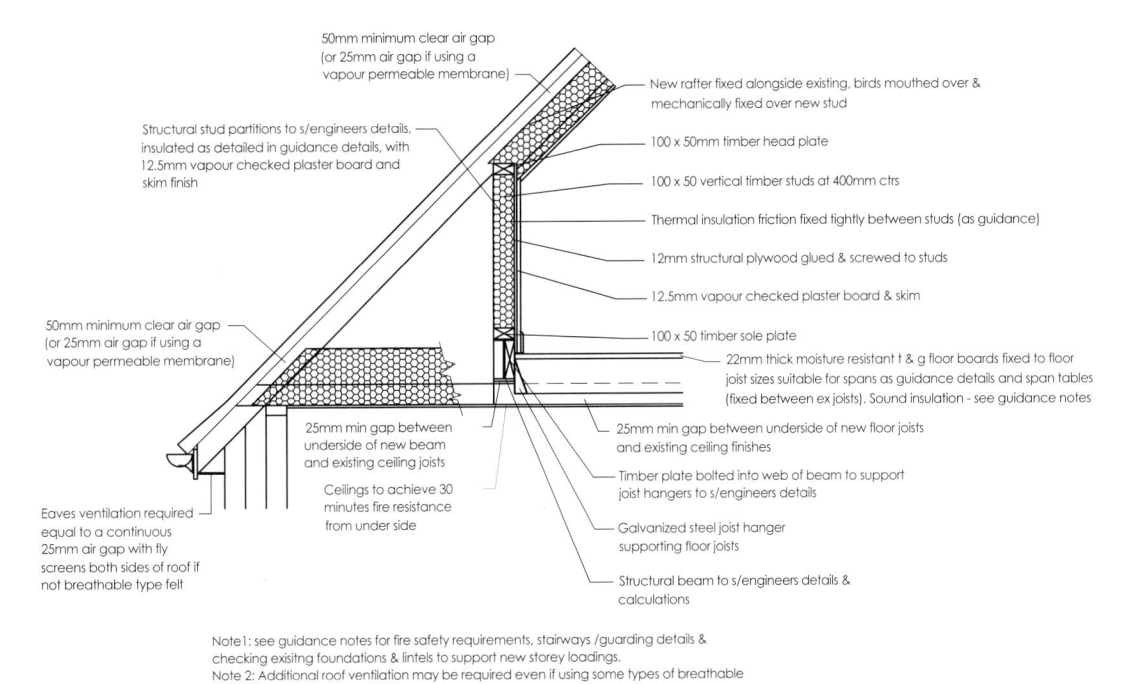

**Figure 4.5:** Typical section through existing roof indicating new internal load-bearing timber stud wall supporting roof loads *(not to scale)*

**Table 4.2:** Insulation requirements to exposed timber-framed walls (U-value no worse than 0.28 W/m²k.)

| Timber stud (mm) | Insulation type and minimum thickness | Internal insulation/finish |
|---|---|---|
| 100 × 50 mm | 50 mm Kingspan Kooltherm K12 Framing Board. K Value 0.20 **or** Kingspan Thermawall TW55 K value 0.022 fixed between studs | 32.5 mm Kingspan Kooltherm K18 Insulated plasterboard. K value 0.023 fixed over studs |
| 100 × 50 mm | 60 mm Celotex FR4000 K value 0.22 fixed between studs | 37.5 mm Celotex PL4000 K value 0.22 fixed with integral plasterboard with lightweight skim fixed over studs |
| 125 × 50 mm | 85 mm Kingspan Kooltherm K12 Framing Board. K value 0.020 fixed between studs | 12.5 mm plasterboard and 3 mm skim finish fixed over studs |
| 125 × 50 mm | 90 mm Celotex FR4000 K value 0.22 fixed between studs | 12.5 mm plasterboard and 3 mm skim finish fixed over studs |

**Note:** Insulation is to be installed in accordance with the manufacturer's details.

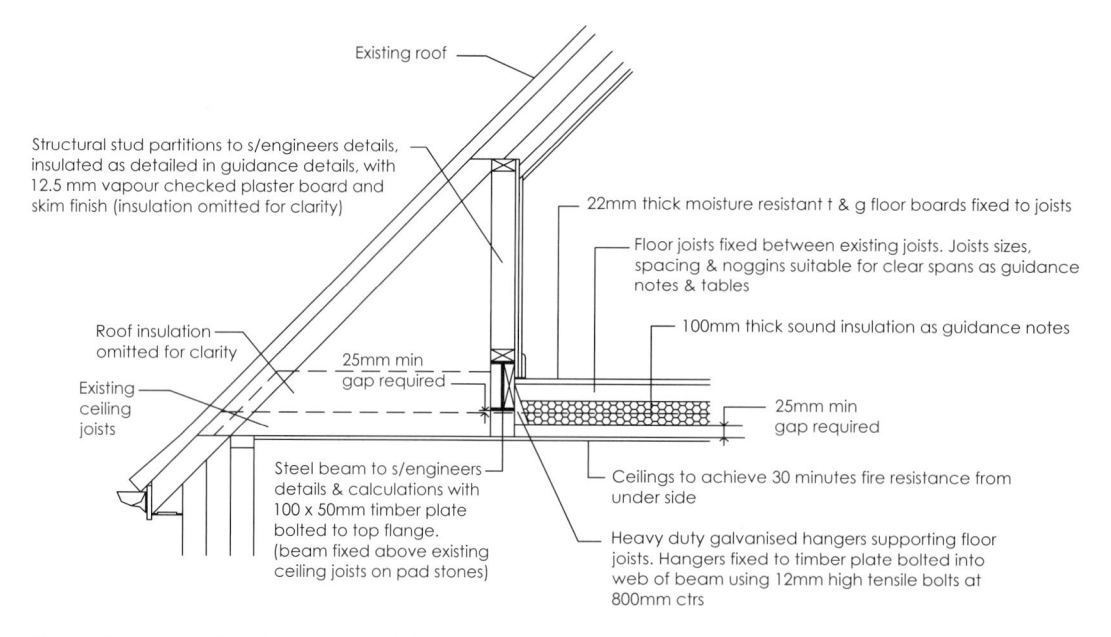

**Figure 4.6:** Typical section through existing roof indicating new loft floor construction *(not to scale)*

Means-of-escape windows are to be fitted with proprietary hinges to open to the minimum required clear width of 450 mm. Escape windows must have minimum clear opening casement dimensions of 0.33 m² and 450 mm (typically 450 mm wide × 750 mm high), located within 800–1100 mm above floor level to all bedrooms and habitable rooms at first-floor level and inner habitable rooms on the ground floor. Windows above the ground-floor storey and within 800 mm of floor level are to be provided with containment/guarding/ proprietary catches, which should be removable (but childproof) in the event of a fire. Where escape windows cannot be achieved, direct access to protected stairs (or a protected route to inner rooms) is acceptable in compliance with the guidance sections below and ADB1 Para 2.6 (a) or (b).

## Two-storey dwellings with loft conversion (or new third storey)

**Four options are available for means of escape:**

### Option 1: Protected stairway

The new and existing stairs, landings and hallway from the new third storey down to the ground floor must be protected and enclosed in 30-minute fire-resisting construction and the protected stairs must discharge directly to an external door, as in the guidance diagram below. 30 minutes' fire-resisting construction is required to the underside of the new upper-storey floor.

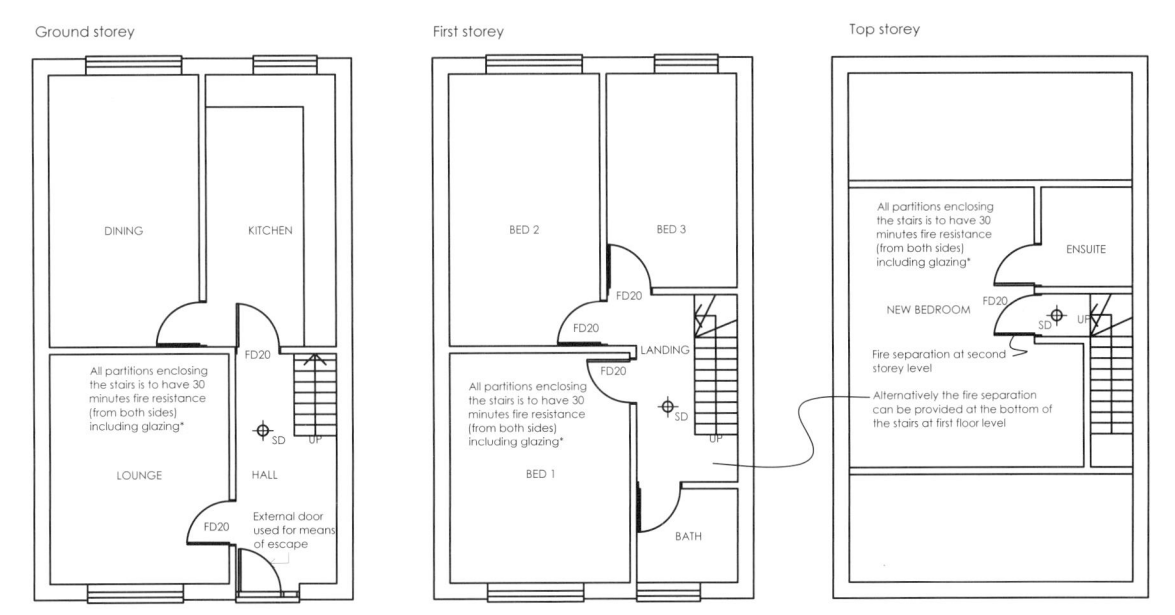

Notes: (i) 30 minutes fire resistance is required to underside of new storey floor. (ii) Cupboards within the stairway enclosure to be fitted with FD 20 fire doors, intumescent strips and fire resistant hinges.

**KEY TO ITEMS INDICATED ON LAYOUT** (to be read in conjunction with guidance details)
FD20   20 minute fire resisting door and frame fitted with intumescent strips and 3 fire resistant hinges (excludes toilet and bathroom doors- providing the partitions between the stairway and habitable room has 30 minutes fire resistance. (note: a self closing FD 30s fire door is required to a door opening between the dwelling and an attached garage - no glazing is permitted other than 30 minutes fire insulated glass)
SD/HD interconnected mains operated smoke/heat alarm with battery back up fitted at ceiling level
*       Fire resistant glazing to be 1.1m minimum above floor level in walls - (unlimited in doors - 100mm minimum above floor level in basement doors, no glazing is permitted in a fire door between an attached garage and the dwelling). These glazing limitations do not apply to glazed elements which satisfy the relevant fire insulation criterion in Table A1 of ADB1

### Key to works indicated on diagram above

**FD 20:** 20-minute fire-resisting doors to BS 476-22:1987, fitted with intumescent strips rebated around sides and top of door or frame; excludes toilets/bathrooms/en suite providing the partitions protecting the stairs have 30 minutes' fire resistance from both sides. Self-closers are not required to doors in domestic loft conversions. (Note: A self-closing FD30 fire door with intumescent strips, smoke seals and 100mm-high fire-resistant thresholds is required if there is a doorway between the dwelling and the garage.)

Existing solid hardwood/timber doors may achieve 20 minutes' fire resistance, or may be suitable for upgrading to achieve 20 minutes' fire resistance (as agreed with building control) with a proprietary intumescent paint/paper system, in accordance with the manufacturer's details. More details are available from: www.fireproof.co.uk, who can supply (and apply where required) an

### 4.14 Domestic loft conversions

intumescent paint/paper system, which must be applied in accordance with the manufacturer's details. A copy of the intumescent paint/paper manufacturer's certificate of purchase (and application where applicable) must be provided for building control on completion.

**SD/HD:** An interlinked, mains-operated smoke alarm/heat alarm with battery backup is to be fitted at ceiling level to BS 5446 and installed to BS 5839 pt 6.

**Important notes:** *Fire-resistant glazing to be 1.1 m minimum above floor level in walls (unlimited in doors – 100 mm minimum above floor level in basement doors, no glazing is permitted in a fire door between an attached garage and the dwelling). These glazing limitations do not apply to glazed elements which satisfy the relevant fire insulation criterion in Table A1 of ADB1. (i). Means of escape windows are not required where the stairs are protected as detailed in this guidance.

## Option 2: Protected stairway with alternative exits at ground-floor level

The new and existing stairs, landings and hallway from the new top storey down to the ground floor must be protected and enclosed in 30-minute fire-resisting construction, and the protected stairs must give access to two or more FD20 fire doors on the ground floor that discharge into different rooms that are separated from each other by 30-minute fire-resisting construction. Each must have external doors for escape, as in the guidance diagram below.

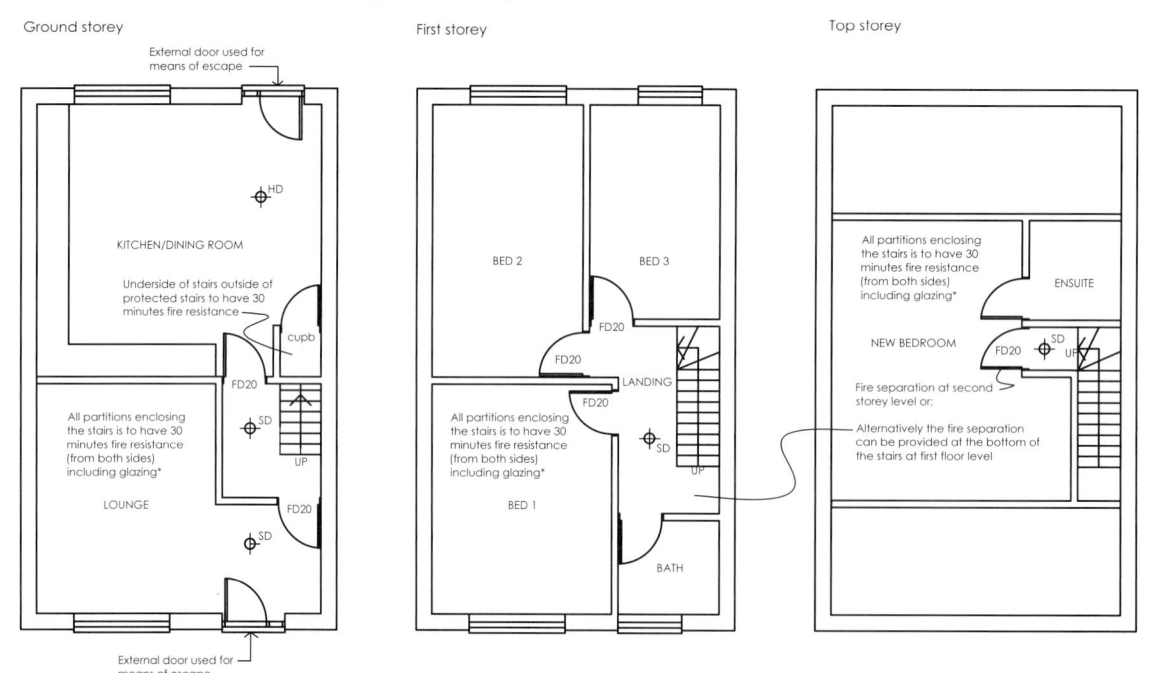

Notes: (i) 30 minutes fire resistance is required to underside of new storey floor. (ii) Cupboards within the stairway enclosure to be fitted with FD 20 fire doors, intumescent strips and fire resistant hinges.

**KEY TO ITEMS INDICATED ON LAYOUT** (to be read in conjunction with guidance details)
FD20  20 minute fire resisting door and frame fitted with intumescent strips and 3 fire resistant hinges (excludes toilet and bathroom doors- providing the partitions between the stairway and habitable room has 30 minutes fire resistance. (note: a self closing FD 30s fire door is required to a door opening between the dwelling and an attached garage - no glazing is permitted other than 30 minutes fire insulated glass)
SD/HD Interconnected mains operated smoke/heat alarm with battery back up fitted at ceiling level
*     Fire resistant glazing to be 1.1m minimum above floor level in walls - (unlimited in doors - 100mm minimum above floor level in basement doors, no glazing is permitted in a fire door between an attached garage and the dwelling). These glazing limitations do not apply to glazed elements which satisfy the relevant fire insulation criterion in Table A1 of ADB1

### Key to works indicated on diagram above

**FD 20:** 20-minute fire-resisting doors to BS 476-22:1987 fitted with intumescent strips rebated and top of door or frame; excludes toilets/bathrooms/en suite providing the partitions protecting the stairs have 30 minutes' fire resistance from both sides. Self-closers are not required to doors in domestic loft conversions. (Note: A self-closing FD30 fire door with intumescent strips, smoke seals and 100 mm-high fire-resistant thresholds is required if there is a doorway between the dwelling and the garage.)

Existing solid hardwood/timber doors may achieve 20 minutes' fire resistance, or may be suitable for upgrading to achieve 20 minutes' fire resistance (as agreed with building control) with a proprietary intumescent paint/paper system, in accordance with the manufacturer's details. More details are available from: www.fireproof.co.uk, who can supply (and apply where required) an intumescent paint/paper system, which must be applied in accordance with the manufacturer's details. A copy of the intumescent paint/paper manufacturer's certificate of purchase (and application where applicable) must be provided for building control on completion.

**SD/HD:** An interlinked, mains-operated smoke alarm/heat alarm with battery backup is to be fitted at ceiling level to BS 5446 and installed to BS 5839 pt 6.

**Important notes:** *Fire-resistant glazing to be 1.1 m minimum above floor level in walls (unlimited in doors – 100 mm minimum above floor level in basement doors; no glazing permitted in a fire door between an attached garage and the dwelling). These limitations do not apply to elements satisfying the relevant fire insulation criterion in Table A1 of ADB1. (i). Means-of-escape windows are not required where the stairs are protected as detailed in this guidance.

## Option 3: Fire-separated top storey with alternative external/internal fire exit

The new top storey should be separated from the lower storeys by 30-minute fire-resisting construction and provided with an alternative escape route leading to its own final exit, as illustrated on the guidance diagram below

The external stairs must not be within 1.8 m of any unprotected opening at the side of the stairs, and no openings are permitted below the stairs unless the opening is fitted with 30-minute fire-resisting glass and a proprietary bead system and is permanently sealed shut – subject to adequate ventilation. (See Figure 2.49 for full details.)

The external stairs may require planning permission before the works commence on site.

The alternative fire exit can also be formed internally as a protected stairway.

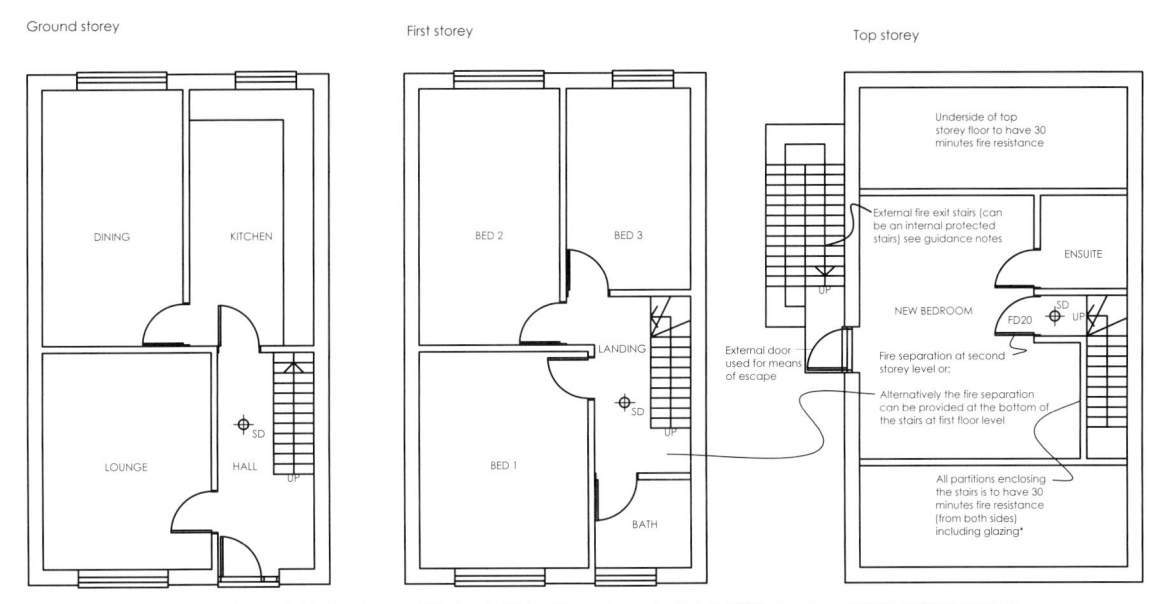

### 4.16 Domestic loft conversions

*Key to works indicated on diagram above*

**FD 20:** 20-minute fire-resisting doors to BS 476-22:1987 fitted with intumescent strips rebated around sides and top of door or frame, excludes toilets/bathrooms/en suite providing the partitions protecting the stairs have 30 minutes' fire resistance from both sides. Self-closers are not required to doors in domestic loft conversions. (Note: A self-closing FD30 fire door with intumescent strips, smoke seals and 100mm-high fire-resistant thresholds is required between the dwelling and the garage.)

Existing solid hardwood/timber doors may achieve 20 minutes' fire resistance, or may be suitable for upgrading to achieve 20 minutes' fire resistance (as agreed with building control) with a proprietary intumescent paint/paper system in accordance with the manufacturer's details. More details are available from: www.fireproof.co.uk, who can supply (and apply where required) an intumescent paint/paper system, which must be applied in accordance with the manufacturer's details. A copy of the manufacturer's certificate of purchase/ application must be provided for building control on completion.

**SD/HD:** An interlinked, mains-operated smoke alarm/heat alarm with battery backup is to be fitted at ceiling level to BS 5446 and installed to BS 5839 pt 6

**Important notes:** *Fire-resistant glazing to be 1.1m minimum above floor level in walls (unlimited in doors – 100mm minimum above floor level in basement doors; no glazing permitted in a fire door between an attached garage and the dwelling). These limitations do not apply to elements satisfying the relevant fire insulation criterion in Table A1 of ADB1. (i). Means-of-escape windows are not required where the stairs are protected as detailed in the guidance.

## Option 4: Residential sprinkler systems for means of escape

Where fire-safety requirements of the Building Regulations cannot be met for loft conversions, proposals for a fire-engineered solution, which may incorporate a sprinkler suppression system as part of the solution, can be allowed against the requirements of Approved Document B where a risk assessment has been carried out by a suitably qualified and experienced fire engineer and approved by building control before works commence on site. The residential sprinkler system is to be designed and installed by a suitably qualified specialist to BS 9251:2005, incorporating BAFSA technical guidance note No.1 June 2008, and must be approved by building control before works commence on site.

If the requirement not being met is a ground-floor open-plan arrangement and the stairs discharge into a habitable open-plan area in a three-storey building, then a partial sprinkler installation to the whole of all connected open-plan areas may be used. Fire separation of the route will be required between the upper floor and this open-plan area, with a 30-minute fire-resisting partition and FD20 fire door fitted with intumescent strips; means of escape would be via an escape window at first-floor level, in accordance with this guidance. Instead of the separation it might be possible to fully sprinkler the whole dwelling and retain the open-plan arrangement, with the agreement of building control.

With the agreement of the building control, it should be possible to reduce fire protection throughout the dwelling by 30 minutes with the introduction of a full sprinkler installation.

Contact: The British Automatic Sprinkler Association (BAFSA)- *Sprinklers for Safety: Use and Benefits of Incorporating Sprinklers in Buildings and Structures* (2006) ISBN: 0 95526 280 1. See also: www.bafsa.org.uk ISBN 0-9552628-3-6 technical guidance note no.1 and see also www.firesprinklers.org.uk.

### Important note

The above four options need not be followed if the dwelling house has more than one internal stairway, which afford effective alternative means of escape and are physically separated from each other (subject to building control approval).

Domestic loft conversions 4.17

## PART C: SITE PREPARATION AND RESISTANCE TO CONTAMINANTS AND MOISTURE

Follow the guidance details in Section 2 of this guidance for Domestic Extensions.

## PART D: CAVITY WALL FILLING WITH INSULATION

Follow the guidance details in Section 2 of this guidance for Domestic Extensions.

## PART E: RESISTANCE TO THE PASSAGE OF SOUND

Follow the guidance details in Section 2 of this guidance for Domestic Extensions.

## PART F: VENTILATION

Follow the guidance details in Section 2 of this guidance for Domestic Extensions.

## PART G: SANITATION, HOT-WATER SAFETY AND WATER EFFICIENCY

Follow the guidance details in Section 2 of this guidance for Domestic Extensions.

## PART H: DRAINAGE AND WASTE DISPOSAL

Follow the guidance details in Section 2 of this guidance for Domestic Extensions.

## PART J: COMBUSTION APPLIANCES AND FUEL STORAGE SYSTEMS

Follow the guidance details in Section 2 of this guidance for Domestic Extensions.

## PART K: PROTECTION FROM FALLING, COLLISION AND IMPACT

Follow the guidance details in Section 2 of this guidance for Domestic Extensions.

## PART L: CONSERVATION OF FUEL AND POWER IN CONVERSIONS

Follow the guidance details in Section 2 of this guidance for Domestic Extensions.

## PART M: ACCESS TO AND USE OF BUILDINGS FOR DISABLED

Follow the guidance details in Section 2 of this guidance for Domestic Extensions.

## PART N: SAFETY GLAZING, OPENING AND CLEANING

Follow the guidance details in Section 2 of this guidance for Domestic Extensions.

## PART P: ELECTRICAL SAFETY

Follow the guidance details in Section 2 of this guidance for Domestic Extensions.

# Section 5  Domestic garage and basement conversions into habitable rooms and conversion of barns and similar buildings into new dwellings

| | |
|---|---|
| Conversion of existing buildings | 5.5 |
| Material change of use | 5.5 |
|    Meaning of a material change of use | 5.5 |
|    Requirements relating to material change of use of the whole building (e.g. conversion of an agricultural barn into a dwelling) | 5.5 |
|    Requirements relating to material change of use of part of a building (e.g. garage or basement conversion into a habitable room) | 5.6 |
| Assessing the feasibility of your building for conversion | 5.6 |
|     1. Inspection of foundations to support additional loadings | 5.7 |
|     2. Structural repairs to cracking in masonry walls | 5.7 |
|     3. Leaning or bulging walls | 5.7 |
|     4. External and internal walls not bonded together | 5.7 |

*Guide to Building Control: For Domestic Buildings*, First Edition. Anthony Gwynne.
© 2013 John Wiley & Sons, Ltd. Published 2013 by John Wiley & Sons, Ltd.

| | |
|---|---|
| 5. Lintels and beams over openings and structural columns | 5.7 |
| 6. Ground-floor construction | 5.7 |
| 7. Timber floors above ground-floor level | 5.8 |
| 8. Roof structure | 5.8 |
| 9. Notches/holes in floor and ceiling joists and structural timbers | 5.9 |
| 10. Roof coverings | 5.9 |
| 11. Party walls and floors separating dwellings | 5.9 |
| 12. Remedial works to decay of timbers, including insect attack | 5.10 |
| 13. Non-traditional construction and framed buildings | 5.10 |
| 14. Fire safety and means of escape in event of a fire | 5.10 |
| 15. Resistance of the passage of contaminants (e.g. radon gas) | 5.10 |
| 16. Resistance to the passage of rising damp and penetrating damp | 5.11 |
|     16.1 Existing cavity walls | 5.11 |
|     16.2 Existing solid masonry walls | 5.11 |
| 17. Tanking systems | 5.11 |
| 18. Roof ventilation to prevent condensation | 5.12 |
| 19. Sound insulation | 5.12 |
| 20. Ventilation | 5.12 |
| 21. Sanitation, hot-water safety and water efficiency | 5.12 |
| 22. Foul- and surface-water drainage and disposal | 5.12 |
| 23. Heat-producing appliances and fuel storage | 5.13 |
| 24. Stairs and guarding | 5.13 |
| 25. Conservation of fuel and power, including upgrading of insulation standards | 5.13 |
| 26. Access to and use of buildings for disabled | 5.13 |
| 27. Safety glazing | 5.13 |
| 28. Electrical safety | 5.13 |
| 29. Asbestos | 5.13 |
| 30. Breathable buildings | 5.13 |
| Part A: Structure | 5.14 |
| A1: Underpinning works | 5.14 |
| Traditional underpinning | 5.14 |
| Sections through proposed underpinning *(not to scale)* | 5.15 |
|     Option 1: Mass concrete with weight of concrete providing final pinning | 5.15 |
|     Option 2: Mass concrete with packed mortar providing final pinning | 5.16 |
| A2: Single-wall garage conversions (or similar buildings) into habitable rooms (typical details indicated in Figures 5.2–5.12) | 5.16 |
| Upgrading pitched roofs | 5.16 |
| Upgrading flat roofs | 5.19 |
| Infilling of garage door opening | 5.20 |
| Upgrading single-skin external walls | 5.21 |
|     Option 1: Upgrading external walls with internal cavity and non-load-bearing timber-framed walls | 5.21 |
|     Option 2: Upgrading external single wall into non-load-bearing cavity wall construction – non-load-bearing inner wall | 5.23 |
|     Option 3: Upgrading external single wall into cavity wall construction – load-bearing inner wall | 5.24 |

| | |
|---|---|
| Option 4: Upgrading external walls with internal tanking and insulated timber-framed walls | 5.24 |
|     Suggested system | 5.24 |
|     Insulation systems | 5.25 |
| Upgrading garage ground floors (or similar) with upgraded enclosing single-skin walls | 5.26 |
|     Option 1: Floating floor laid over existing floor | 5.26 |
|     Option 2: Screed floor laid over existing floor | 5.26 |
|     Option 3: Take up existing floor and lay new concrete floor | 5.28 |
| A3: Cavity wall garage conversions (or similar buildings) into habitable rooms (typical details indicated in Figures 5.16–5.26) | 5.29 |
| Upgrading pitched roof | 5.29 |
| Infilling of garage door opening | 5.31 |
| Upgrading external cavity walls | 5.33 |
| Upgrading garage ground floors (or similar) with enclosing upgraded cavity walls | 5.34 |
|     Option 1: Floating floor laid over existing floor | 5.34 |
|     Option 2: Screed floor laid over existing floor | 5.35 |
|     Option 3: Take up existing floor and lay new concrete floor | 5.36 |
|     Option 4: Suspended timber floor laid over existing floor void | 5.37 |
| A4: Basement conversions into habitable rooms | 5.38 |
| Existing basements and tanking systems | 5.38 |
|     Site investigation and risk assessment | 5.38 |
|     Existing basement sub-structure | 5.39 |
|     Tanking systems | 5.40 |
|     Ventilation and means of escape | 5.40 |
|     Cavity closers | 5.40 |
| A5: Conversion of barns and similar buildings into new dwellings – Technical and practical guidance | 5.40 |
| Part A: Structure | 5.40 |
|     Upgrading pitched roofs | 5.41 |
|         Existing roof coverings and roofing felt | 5.41 |
|         Existing roof structure | 5.41 |
|         Thermal upgrade of insulation and ventilation levels | 5.42 |
|     Upgrading thermal insulation to existing solid masonry walls | 5.43 |
|     New upper floors supported by existing walls | 5.43 |
|     New upper floors supported by load-bearing ground-floor walls and foundations | 5.45 |
|         Upgrading existing ground floors | 5.45 |
|         Take up existing floor and lay new concrete floor | 5.46 |
|         Alternative options for upgrading ground floors | 5.47 |
|     Tanking of existing sub-structure walls with higher external ground levels | 5.47 |
|         Existing concrete floor slab | 5.48 |
|         Existing walls | 5.48 |
|         Tanking systems | 5.48 |
|         Floor-grade thermal insulation | 5.48 |
|         Floor screed | 5.48 |
|     First-floor walls built off existing ground-floor walls | 5.49 |
| Part B: Fire safety and means of escape | 5.51 |

| | |
|---|---|
| Part C: Site preparation and resistance to contaminants and moisture | 5.51 |
| Part D: Cavity wall filling with insulation | 5.51 |
| Part E: Resistance to the passage of sound | 5.51 |
| Performance standards | 5.51 |
|     Exemptions/relaxations for conversion of historic buildings into dwellings/flats | 5.51 |
|     Party walls and floors separating conversion of buildings into new dwellings/flats | 5.51 |
|     Pre-completion sound testing of party walls and floors | 5.52 |
|         Testing rate requirements | 5.52 |
|         Remedial works and retesting | 5.52 |
|     New internal walls and floors | 5.52 |
|     Existing internal walls and floors | 5.52 |
| Part F: Ventilation to new dwellings | 5.52 |
| Part G: Sanitation, hot-water safety and water efficiency | 5.52 |
| Part H: Drainage and waste disposal | 5.53 |
| Part J: Combustion appliances and fuel storage systems | 5.53 |
| Part K: Protection from falling, collision and impact | 5.53 |
| Part L: Conservation of fuel and power in conversions | 5.53 |
| Energy Performance Certificate (EPC) | 5.53 |
| Part M: Access to and use of buildings for disabled | 5.53 |
| Part N: Safety glazing, opening and cleaning | 5.54 |
| Part P: Electrical safety | 5.54 |

## Conversion of existing buildings

Converting an existing garage into a habitable room can be an easy and cost-effective way of increasing living accommodation in most houses. Guidance is also given on the conversion of barns into new dwellings. This guide to conversions will provide useful advice on how some of the more common technical and construction requirements can be achieved.

## Material change of use

A material change of use occurs in specified circumstances in which a building or part of a building that was previously used for one purpose will be used for another – for example, conversion of an agricultural barn into a dwelling. Where there is a material change of use the Building Regulations have set the following requirements that must be met before the building can be used for its new purpose:

### Meaning of a material change of use

Under Regulation 5 of the Building Regulations 2010, there is a material change of use where there is a change in the purposes for which, or the circumstances in which, a building is used, so that after that change- the building is used as a dwelling where previously it was not.

### Requirements relating to material change of use of the whole building (e.g. conversion of an agricultural barn into a dwelling)

Under Regulation 6 (1) of the Building Regulations 2010, where there is a material change of use of an existing building into a dwelling (of the whole building), such work, if any, shall be carried out as is necessary to ensure that the building complies with the applicable requirements of paragraphs of Schedule 1 of the Building Regulations 2010, as follows:

- A1 to A3 (structure –applies only to existing buildings in classes 1 to 6 in Schedule 2 of Building Regulations 2010, for example agricultural buildings);
- B1 (means of warning and escape);
- B2 (internal fire spread – linings);
- B3 (internal fire spread – structure);
- B4 (2) (external fire spread – roofs);
- B5 (access and facilities for the fire service);
- C1 (resistance to contaminates);
- C2 (resistance to moisture, interstitial and surface condensation);
- E 1 to E3 (resistance to the passage of sound);
- F1 (ventilation);
- G1 (cold water supply);
- G2 (water efficiency);
- G3 (1) to (4) (hot water supply and systems);
- G4 (sanitary conveniences and washing facilities);

- G5 (bathrooms);
- G6 (kitchens and food-preparation areas);
- H1 (foul-water drainage);
- H6 (solid waste storage);
- J1 to J4 (combustion appliances);
- L1 (conservation of fuel and power);
- P1 (electrical safety).

### Requirements relating to material change of use of part of a building (e.g. garage or basement conversion into a habitable room)

Under Regulation 6 (2) of the Building Regulations 2010, where there is a material change of part only of an existing building, such work, if any, shall be carried out as is necessary to ensure that part of the building complies with the applicable requirements of paragraphs of Schedule 1 of the Building Regulations 2010, as follows:

- A1 to A3 (structure – applies only to existing buildings in classes 1 to 6 in Schedule 2 of Building Regulations 2010);
- B1 (means of warning and escape);
- B2 (internal fire spread – linings);
- B3 (internal fire spread – structure);
- B4 (2) (external fire spread – roofs);
- B5 (access and facilities for the fire service);
- C1 (resistance to contaminates – contact your building control surveyor for their specific requirements in your area);
- C2 (resistance to moisture, interstitial and surface condensation);
- E1 to E3 (resistance to the passage of sound);
- F1 (ventilation);
- G1 (cold water supply);
- G2 (water efficiency);
- G3 (1) to (4) (hot water supply and systems);
- G4 (sanitary conveniences and washing facilities);
- G5 (bathrooms);
- G6 (kitchens and food-preparation areas);
- H1 (foul-water drainage);
- H6 (solid waste storage);
- J1 to J4 (combustion appliances);
- L1 (conservation of fuel and power);
- P1 (electrical safety).

## Assessing the feasibility of your building for conversion

Before commencing conversion works it is important to assess the feasibility of the project. This will involve inspection of the existing building prior to conversion, to assess the suitability of the following elements applicable to the project:

### 1. Inspection of foundations to support additional loadings

Existing foundations that will be built off or used to support new or modified loads should be exposed for inspection at the discretion of building control, and/or a suitably qualified and experienced property professional, to ensure that they are adequate and suitable.

Necessary remedial works, e.g. underpinning, should be carried out in compliance with the guidance details below and a structural engineer's details and calculations, which should be approved by building control before works commence on site.

### 2. Structural repairs to cracking in masonry walls

Any cracking evident in the walls should be inspected by a suitably qualified and experienced property professional. The root cause of the structural defect should be remedied and additional strengthening and/or underpinning works may be required to support new or modified loads, in compliance with the guidance below and a structural engineer's details and calculations, which should be approved by building control before works commence on site.

### 3. Leaning or bulging walls

Leaning or bulging walls should be inspected by a suitably qualified and experienced property professional and the root cause of the structural defect should be remedied. The wall may require rebuilding or additional strengthening works may be required, in compliance with a structural engineer's details and calculations, which should be approved by building control before works commence on site.

### 4. External and internal walls not bonded together

Lack of bonding causing light cracking between external walls and internal walls should be remedied by bonding the walls together (i.e. proprietary galvanised steel straps mechanically fixed to walls at 600 mm centres) or, where excessive movement has occurred, tie the building together in accordance with a structural engineer's details

### 5. Lintels and beams over openings and structural columns

Existing lintels, beams and columns that will be retained, built off or used to support new or modified loads should be exposed for inspection at the discretion of the building control surveyor and structural engineer, to ensure that they are adequate and suitable.

Defective lintels should be replaced with new manufactured proprietary lintels suitable for the proposed loads and clear spans, which should have the appropriate bearing as manufacturer's details and as detailed in this guidance. Defective, non-proprietary beams/columns should be replaced in compliance with a structural engineer's details and calculations, which should be approved by building control before works commence on site.

### 6. Ground-floor construction

Existing ground floors that will be built off or used to support new or modified loads may need to be exposed for inspection, at the discretion of the building control surveyor and/or a structural engineer, to ensure that they are adequate and suitable for the proposed conversion.

Ground-bearing concrete floor slabs should be of 100 mm minimum thickness and in good condition for the proposed upgrading works, as detailed in this guidance, and thicker if used to support non-load-bearing walls, as agreed with building control. If they do not appear to be adequate to support the proposed works, defective floors should be replaced with new floors, as detailed in this guidance.

### 7. Timber floors above ground-floor level

Existing timber floor joists should be inspected and retained providing they meet the following requirements:

- Joists should be of adequate sizes for the span (see Tables in this guidance; please contact building control if joists appear slightly undersized, as they may still be suitable if no deflection has occurred).
- Loads on floor should not be increased unless checked and confirmed as suitable by a structural engineer.
- Joists should not have not been weakened by excessive notching or drilling of holes, as detailed in this guidance.
- Existing timbers should be inspected and any defective timbers should be repaired and or replaced as necessary, and treated by a specialist against insect and fungal attack, with a warrant-backed guarantee. Major structural repairs and replacement works should be
  – in compliance with structural engineer's details and calculations, which should be
  – approved by building control before works commence on site.
- No masonry walls should be built off timber joists.
- Openings in floors, i.e. for stairs, hearths, etc. should be properly formed with trimming and trimmer joists, normally the same sizes as joists and doubled up to support trimmed joists. (Please contact building control if joists appear slightly undersized, as they may still be suitable if no deflection has occurred.)
- Joists should be strutted and strapped, as detailed in this guidance.
- New/replacement works should be carried out in compliance with details in this guidance.

### 8. Roof structure

Existing roof timbers should be inspected and retained, providing they meet the following requirements:

- Rafters, ceiling joists, purlins, hangers and binders, etc. should be of adequate sizes for the span (see tables in this guidance). Raised ceiling ties should be within the bottom third of rafters and connected together with bolted steel-toothed connectors. (Please contact building control if any timbers appear slightly undersized, as they may still be suitable if no deflection has occurred.)
- Principal roof trusses supporting purlins and trussed rafters are normally considered acceptable if no deflection has occurred; any defect or deflection should be remedied in compliance with the structural engineer's details and calculations, which should be approved by building control before works commence on site.
- Loads on roof should not be increased (unless checked and confirmed as suitable by a structural engineer); additional supports may be required for new water tanks, as detailed in this guidance.

- Ceiling joists should not have been weakened by excessive notching or drilling of holes, as detailed in this guidance.
- Existing timbers should be inspected and any defective timbers should be repaired and/or replaced as necessary and treated by a specialist against insect and fungal attack, with a warrant-backed guarantee. Major structural repairs and replacement works should be in compliance with the structural engineer's details and calculations, which should be approved by building control before works commence on site.
- Openings in roofs, i.e. for roof windows etc., should be properly formed with trimming and trimmer joists, normally the same sizes as joists and doubled up to support trimmed joists. (Please contact building control if joists appear slightly undersized, as they may still be suitable if no deflection has occurred.)
- Rafters and wall plates should be strutted and strapped, and trussed rafters should be braced, as detailed in this guidance.
- New/replacement works should be carried out in compliance with details in this guidance.

## 9. Notches/holes in floor and ceiling joists and structural timbers

Notches should not be deeper than 0.125 times the depth of the joists, should be not closer to the support than 0.07 times the span and not further away than 0.25 times the span. Holes should have a diameter not greater than 0.25 times the depth of joist and should be drilled at the joist centre line. They should be not less than 3 diameters (centre to centre) apart and should be located between 0.25 and 0.4 times the span from the support.

Rafters restrained by ceiling ties at eaves level can be birds-mouthed over supports to a depth not exceeding one-third of the rafter depth and should be mechanically fixed. Notches or holes exceeding the above requirements or cut into other structural members should be checked by a structural engineer.

## 10. Roof coverings

Existing roof coverings should be inspected, including associated support systems, i.e. battens, nails, felt, flat-roof deck, valleys, lead flashings, soakers, facia boards, soffit/barge boards and rainwater goods, etc., and retained if in a sound, wind- and watertight condition with a life span of at least 15 years (period of time as stated by some building warranty providers – but as agreed with building control). Defects should be repaired or replaced in materials to match the existing ones.

New/replacement works should be carried out in compliance with details in this guidance; roof coverings should be suitable for the roof pitch, be as the tile manufacturer's details and should not be heavier than the existing roof coverings; unless the roof structure is inspected by a suitably qualified and experienced property professional to ensure it is suitable to support the new loadings.

## 11. Party walls and floors separating dwellings

Existing party walls and or floors that separate dwellings should be upgraded to meet the required sound insulation values, as detailed in this guidance. New party walls and floors that separate dwellings should be constructed to meet the sound insulation values of the Building Regulations, as detailed in this guidance.

Factors to consider in meeting the sound insulation values are:

- Resistance to airborne sound
- Resistance to impact sound (floors only)
- Limiting flanking airborne sound
- Ensuring continuity of sound insulation at all junctions and service ducts, etc.
- Sealing of all gaps and services through the construction
- Ensuring discontinuity of structural elements.

### 12. Remedial works to decay of timbers, including insect attack

Existing timbers should be inspected and any defective timbers should be repaired and/or replaced as necessary and treated by a specialist against insect and fungal attack, with a warrant-backed guarantee. Major structural repairs and replacement works should be in compliance with a structural engineer's details and calculations, which should be approved by building control before works commence on site.

### 13. Non-traditional construction and framed buildings

An appraisal of the existing building should be carried out by a suitably qualified and experienced property professional and should take into consideration the following:

- Condition of the structure, including connections
- Proposals to increase loadings on the structure and foundations
- Alterations/modifications to the load paths
- Alterations/modifications to the stability systems
- Changes in environmental exposure
- Additional specialist reports and testing
- Suitability for upgrading of insulation.

Remedial works should be in compliance with a structural engineer's details and calculations, which should be approved by building control before works commence on site.

### 14. Fire safety and means of escape in event of a fire

Fire safety and means of escape in event of a fire should be carried out in compliance with details in this guidance for extensions and new dwellings.

### 15. Resistance of the passage of contaminants (e.g. radon gas)

Basic radon protection may be required as a minimum standard in the sub-structure to prevent radio-active gas present in the subsoil from entering the building through cracks in the construction, as detailed in this guidance, unless it is confirmed by a radon risk report from a specialist that radon protection is not required. Reports are available, at a cost, from www.ukradon.

**Note:** Full radon gas protection and/or methane gas protection may be required in the sub-structure by building control in certain geological areas; please contact building control for more advice.

## 16. Resistance to the passage of rising damp and penetrating damp

### 16.1 Existing cavity walls

Existing cavity walls should be inspected to ensure the following:

- Minimum 50 mm-wide clear cavity should extend the full height and width of external walls and extend at least 225 mm below damp-course level.
- Continuous horizontal damp-proof course should be in each wall leaf, 150 mm minimum above external ground level.
- Vertical damp-proof courses should be at all cavity closings to openings into the building and abutments to solid walls.
- Masonry walls should be sound, with wall ties and pointing in good condition on both sides.

### 16.2 Existing solid masonry walls

Existing solid masonry walls should be inspected to ensure the following:

- Brick or stonework should be in sound condition and at least 328 mm thick; dense concrete block work should be at least 250 mm thick; lightweight/aerated autoclaved concrete block work at least 215 mm thick; so as to resist passage of penetrating moisture/rain; and should have –
- pointing in good condition and have:–
- 20 mm thick render in two coats to exposed wall surfaces.

Resistance to rising damp should be prevented by a continuous horizontal damp-proof course in wall, 150 mm minimum above external ground level, or else should have a British Board of Agreement (BBA or other third-party accredited) damp-proof course installed and guaranteed (i.e. a chemically injected dpc) by a damp-proofing specialist.

Resistance to penetrating damp is normally prevented by the thickness of the wall, as detailed above; however, in situations where buildings are being converted for habitable use above ground and walls have insufficient thickness, they should be provided with a cavity to prevent the passage of moisture, as detailed in this guidance. (In certain circumstances where a cavity cannot be provided, a designed vapour-permeable tanking solution in accordance with the guidance below may be acceptable – subject to building control approval).

A basement conversion for habitable use will require a designed tanking system to prevent the passage of moisture into the building as detailed in the guidance below – subject to building control approval.

## 17. Tanking systems

Tanking systems providing either barrier, structural or drained protection to the building must be assessed, designed and installed for the particular project, in compliance with BS 8102: 2009 Code of Practice for Protection of Below Ground Structures against Water from the Ground. Tanking systems can be installed internally or externally in accordance with a tanking specialist's details.

The illustrated tanking section details in this guidance are suggested details only and actual details must be approved by building control before works commence on site. Forms of tanking include:

bonded sheet materials; liquid applied membranes; mastic asphalt, drained cavity membranes, and cementitious crystallisation and cementitious multi-coat renders.

Suitable tanking systems are to have British Board of Agreement (BBA or other approved third-party) accreditation and be individually assessed by a tanking specialist as suitable for the proposed situation. Tanking systems above ground should be vapour-permeable, to prevent condensation problems and mould growth.

Tanking systems must be designed/installed/applied by a tanking specialist for the particular project, in compliance with the tanking manufacturer's details, to resist the passage of water into the building and prevent condensation within the building and, where required, to prevent radon gas (and any other identified contaminants) from entering the building.

Tanking systems are to be properly connected to and made continuous with wall damp-proof courses/radon dpc trays. Perforation of the tanking system by service entry pipes etc. should be avoided or carried out strictly in accordance with the tanking manufacturer's details.

**Important note:** The risk of condensation with any tanking system should be assessed by a specialist; a condensation risk analysis should be carried out for the particular project; and the tanking and thermal insulation system should be designed and installed to prevent any potential condensation/interstitial condensation problems.

## 18. Roof ventilation to prevent condensation

Existing roofs should be upgraded and ventilated to prevent condensation and interstitial condensation, in compliance with details in this guidance.

## 19. Sound insulation

Sound insulation between internal walls and floors separating bedrooms, or a room containing a WC and other rooms, is to be carried out in accordance with the relevant details contained within this guidance. Walls separating dwellings should be constructed or upgraded in accordance with the relevant details within this guidance and sound tested if required by building control.

## 20. Ventilation

Purge (natural) ventilation and background ventilation to habitable rooms and mechanical ventilation should be carried out in compliance with details in this guidance.

## 21. Sanitation, hot-water safety and water efficiency

Sanitation, hot-water safety and water efficiency should be carried out in compliance with details in this guidance. (Hot-water safety and water efficiency apply only to a material change of use when creating a new dwelling.)

## 22. Foul- and surface-water drainage and disposal

Foul-water, waste-water and surface-water drainage and disposal should be carried out in compliance with details in this guidance. Where additional drainage effluent is to be connected to an existing septic tank/treatment system, it should be checked by a specialist and the sizes/condition of the tank/system is to be confirmed as suitable for treatment of the proposed additional effluent.

## 23. Heat-producing appliances and fuel storage

Heat-producing appliances and fuel storage should be carried out in compliance with details in this guidance.

## 24. Stairs and guarding

Stairs and guarding should be carried out in compliance with details in this guidance. Existing stairs may be retained – contact building control for more advice.

## 25. Conservation of fuel and power, including upgrading of insulation standards

New thermal elements, including external windows, doors and roof windows, should be carried out in compliance with this guidance.

Existing thermal elements that are to be retained and become part of the thermal envelope, or that are to be renovated and are uninsulated below the threshold values as detailed in this guidance, should be thermally renovated/upgraded to the U-values in compliance with details in this guidance.

Sap rating and an Energy Performance Certificate are required when creating a new dwelling by a material change of use, in compliance with details in this guidance. (Target emission rates do not have to be achieved.)

## 26. Access to and use of buildings for disabled

Approved Document M (ADM) does not apply to conversions into new dwellings formed by a material change of use.

## 27. Safety glazing

Safety glazing should be carried out in compliance with details in this guidance.

## 28. Electrical safety

Electrical safety should be carried out in compliance with details in this guidance.

## 29. Asbestos

Any suspected asbestos is to be inspected by a specialist, removed and disposed of off-site by a specialist licensed contractor, in compliance with the Control of Asbestos Regulations 2012.

## 30. Breathable buildings

The concept of naturally breathable buildings may be acceptable in certain circumstances where a sympathetic approach is required (for example, a heritage or similar existing building), as specified by a suitably qualified and experienced conservation specialist/local authority conservation officer and as agreed and approved by building control. Breathable building guidance is detailed in the section below.

## PART A: STRUCTURE

### A1: Underpinning works

Underpinning of the existing building/foundations may be required in the following circumstances:

- Existing building has moved and/or cracked.
- The existing foundations are inadequate (or non-existent).
- The proposals are to increase the loadings on the foundations.
- External or internal subsoil levels are to be lowered (i.e. when excavating for services or new ground-floor slab within building).

Underpinning works are normally carried out to extend foundations down to stable ground conditions, in compliance with BRE Digest 352 as amended, using one of the following systems:

- Traditional mass concrete underpinning for shallow depths (as detailed below); **or**
- Proprietary underpinning system carried out by an underpinning specialist (i.e. Angle Pile, Pile and Beam, Pier and Beam or other proprietary underpinning systems).

The system used depends on a number of factors, including the type of existing foundation, depth of suitable strata, position of water-table level, etc.

A full site survey and preliminary works should be implemented before any underpinning is carried out, including:

- All necessary requirements under the Party Wall Act
- Detailed survey, recording all defects/cracks with supporting photographs (dated)
- Health and safety risk assessment and provision of all necessary temporary protection, support, shoring and working platforms etc. (in compliance with current health and safety requirements and structural engineer's details), which are to be erected, maintained, certificated, dismantled and removed by a suitably qualified and insured specialist.

**All** underpinning works should be carried out in strict compliance with a suitably qualified and insured specialist's (i.e. a structural engineer's) details and calculations and should be approved by building control before works commence on site.

### Traditional underpinning

Traditional underpinning is carried out by excavating in a designed sequence of stages alongside and underneath the existing foundation to the required width and depth, casting a new foundation, building up to the underside of the existing foundation in mass concrete (or dense block work and mass concrete), and finally pinning between the old and new work, as detailed below.

The sequence and timing of underpinning should be carried out in compliance with an approved underpinning plan produced by a suitably qualified and insured specialist's (i.e. a structural engineer's) details and calculations, which should be approved by building control before works commence on site. Not less than 75 per cent of the existing wall should be supported at all times (no two bays should be worked together).

Maximum bay length should not exceed 1.0 m, reduced to 0.7 m if the wall is of poor quality. Excavations should be properly bottomed out and dewatered, using pumps where necessary. Size

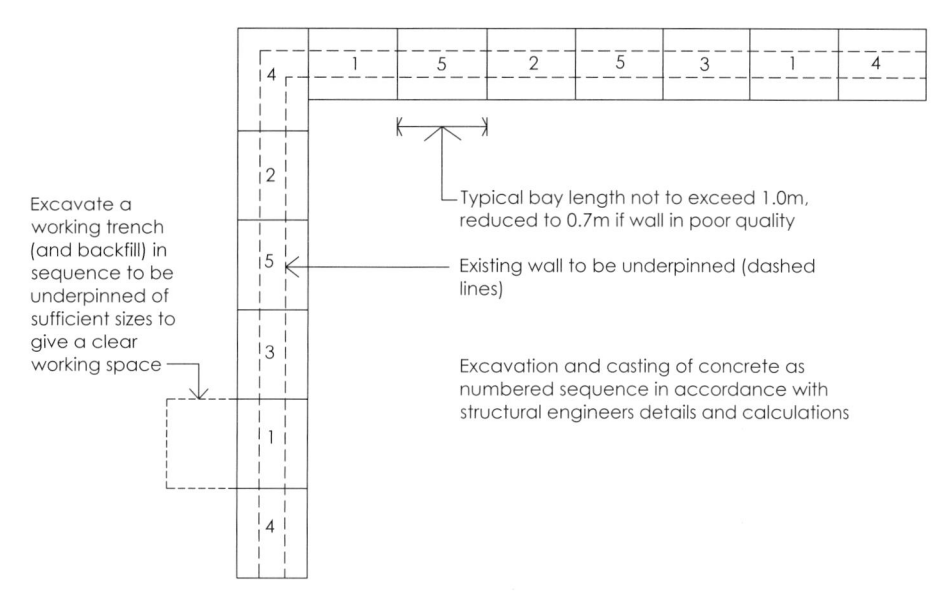

**Figure 5.1:** Typical plan layout of continuous underpinning works *(not to scale)*

Labels on figure:
- Excavate a working trench (and backfill) in sequence to be underpinned of sufficient sizes to give a clear working space
- Typical bay length not to exceed 1.0m, reduced to 0.7m if wall in poor quality
- Existing wall to be underpinned (dashed lines)
- Excavation and casting of concrete as numbered sequence in accordance with structural engineers details and calculations

and depth of foundations should be appropriate for the safe bearing capacity of the supporting subsoil, in compliance with a structural engineer's details.

## Sections through proposed underpinning *(not to scale)*

### Option 1: Mass concrete with weight of concrete providing final pinning

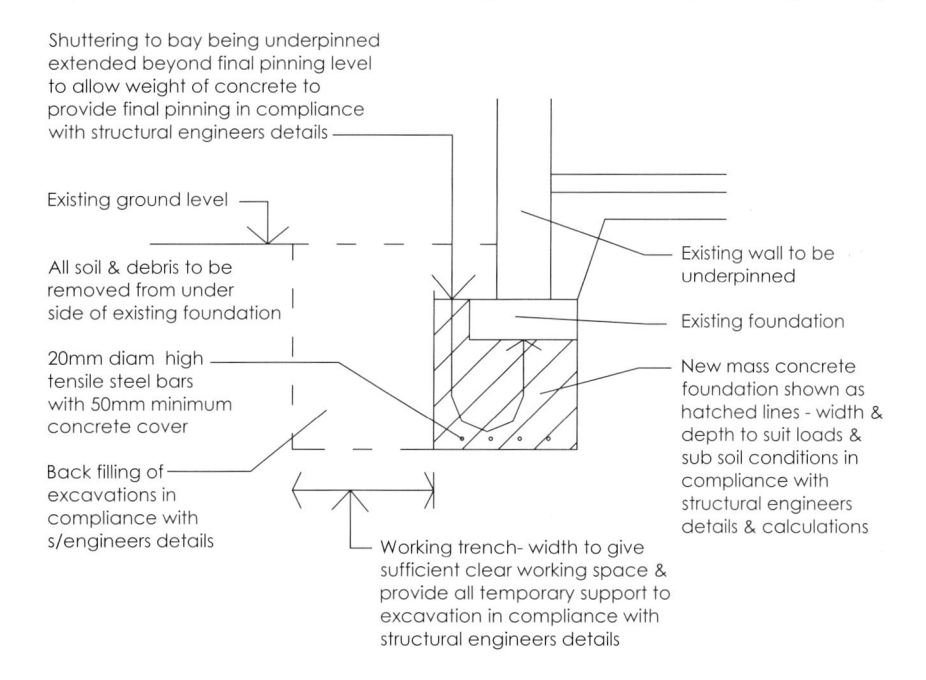

Labels on figure:
- Shuttering to bay being underpinned extended beyond final pinning level to allow weight of concrete to provide final pinning in compliance with structural engineers details
- Existing ground level
- All soil & debris to be removed from under side of existing foundation
- 20mm diam high tensile steel bars with 50mm minimum concrete cover
- Back filling of excavations in compliance with s/engineers details
- Existing wall to be underpinned
- Existing foundation
- New mass concrete foundation shown as hatched lines - width & depth to suit loads & sub soil conditions in compliance with structural engineers details & calculations
- Working trench- width to give sufficient clear working space & provide all temporary support to excavation in compliance with structural engineers details

## Option 2: Mass concrete with packed mortar providing final pinning

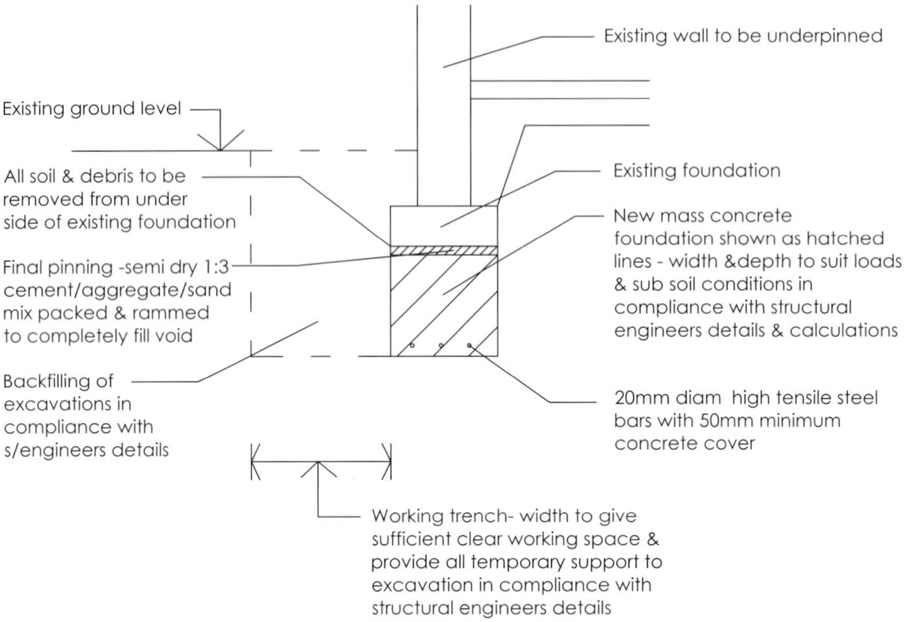

*Diagram labels:*
- Existing wall to be underpinned
- Existing ground level
- All soil & debris to be removed from under side of existing foundation
- Existing foundation
- New mass concrete foundation shown as hatched lines - width &depth to suit loads & sub soil conditions in compliance with structural engineers details & calculations
- Final pinning -semi dry 1:3 cement/aggregate/sand mix packed & rammed to completely fill void
- Backfilling of excavations in compliance with s/engineers details
- 20mm diam high tensile steel bars with 50mm minimum concrete cover
- Working trench- width to give sufficient clear working space & provide all temporary support to excavation in compliance with structural engineers details

## A2: Single-wall garage conversions (or similar buildings) into habitable rooms (typical details indicated in Figures 5.2–5.12)

Conversion of non-habitable garages into habitable rooms can be often complex and each case should be separately analysed, assessed and an appropriate method of conversion specified by a suitably qualified and experienced specialist. The details and diagrams below are for guidance only and the actual details must be approved by building control.

Suggested details and diagrams are in the guidance below.

### Upgrading pitched roofs (see Figure 5.6)

Existing roof coverings and roofing felt should be in a sound and weather-tight condition; any defective coverings/felt should be replaced with matching new/existing sound coverings and fixed as manufacturer's details.

In the existing roof structure, all existing timbers should be in a sound condition, and where necessary defective timber is to be replaced with new, in compliance with details and calculations carried out by a suitably qualified and experienced person. Existing timbers are to be inspected and where necessary treated against insect and fungal attack by a suitably qualified and experienced specialist.

Thermally upgrade insulation levels to an existing roof that has a threshold U-value worse than 0.35 to a U-value of 0.16 for a pitched roof, with insulation at ceiling level as follows: New thermal insulation is to be fixed between and over ceiling joists, as in domestic extension guidance details.

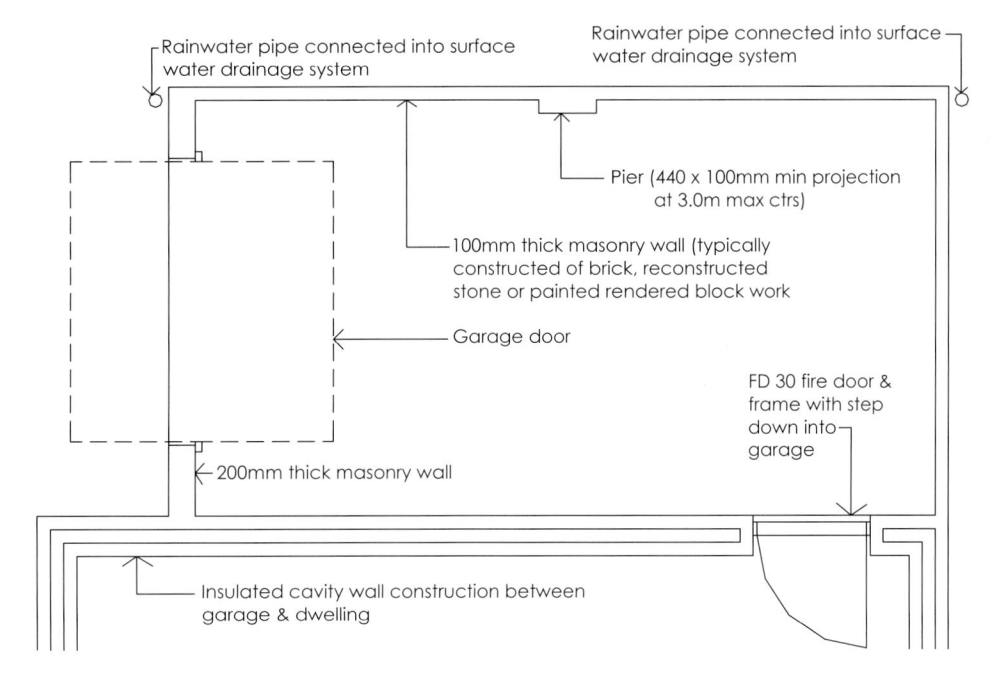

**Figure 5.2:** Typical plan layout of an existing attached single-skin garage *(not to scale)*

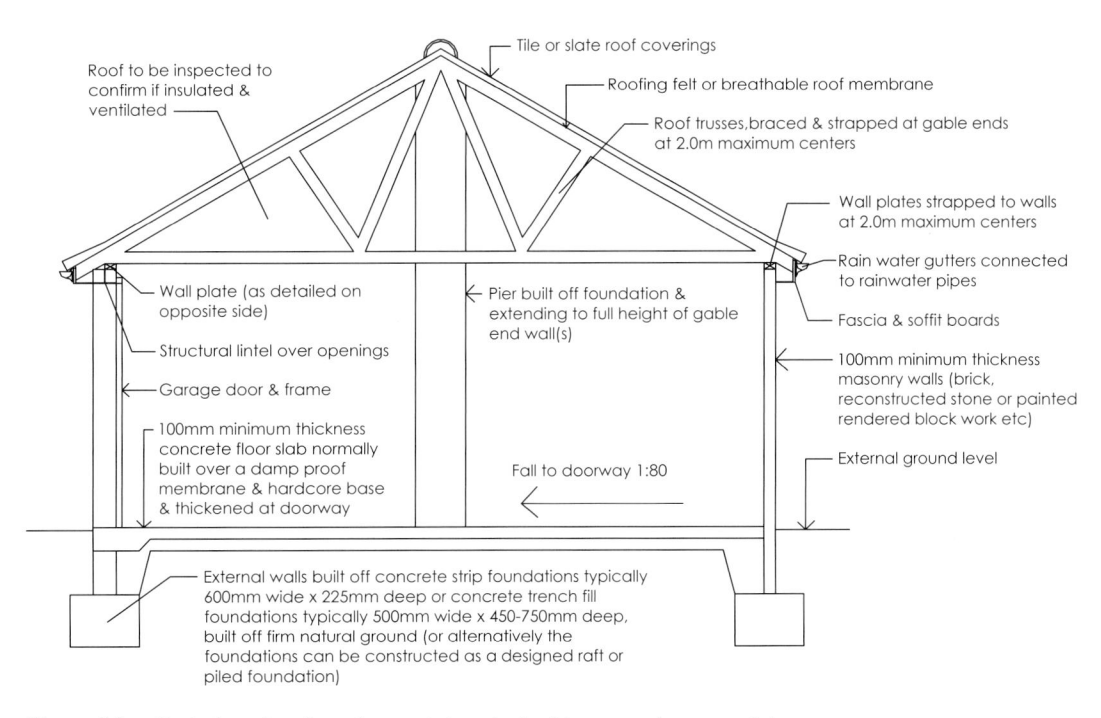

Labels (clockwise from top left):
- Rainwater pipe connected into surface water drainage system
- Rainwater pipe connected into surface water drainage system
- Pier (440 x 100mm min projection at 3.0m max ctrs)
- 100mm thick masonry wall (typically constructed of brick, reconstructed stone or painted rendered block work
- Garage door
- FD 30 fire door & frame with step down into garage
- 200mm thick masonry wall
- Insulated cavity wall construction between garage & dwelling

**Figure 5.3:** Typical section through an existing single-skin garage *(not to scale)*

Labels:
- Tile or slate roof coverings
- Roofing felt or breathable roof membrane
- Roof trusses, braced & strapped at gable ends at 2.0m maximum centers
- Wall plates strapped to walls at 2.0m maximum centers
- Rain water gutters connected to rainwater pipes
- Fascia & soffit boards
- 100mm minimum thickness masonry walls (brick, reconstructed stone or painted rendered block work etc)
- External ground level
- Roof to be inspected to confirm if insulated & ventilated
- Wall plate (as detailed on opposite side)
- Structural lintel over openings
- Garage door & frame
- 100mm minimum thickness concrete floor slab normally built over a damp proof membrane & hardcore base & thickened at doorway
- Pier built off foundation & extending to full height of gable end wall(s)
- Fall to doorway 1:80
- External walls built off concrete strip foundations typically 600mm wide x 225mm deep or concrete trench fill foundations typically 500mm wide x 450-750mm deep, built off firm natural ground (or alternatively the foundations can be constructed as a designed raft or piled foundation)

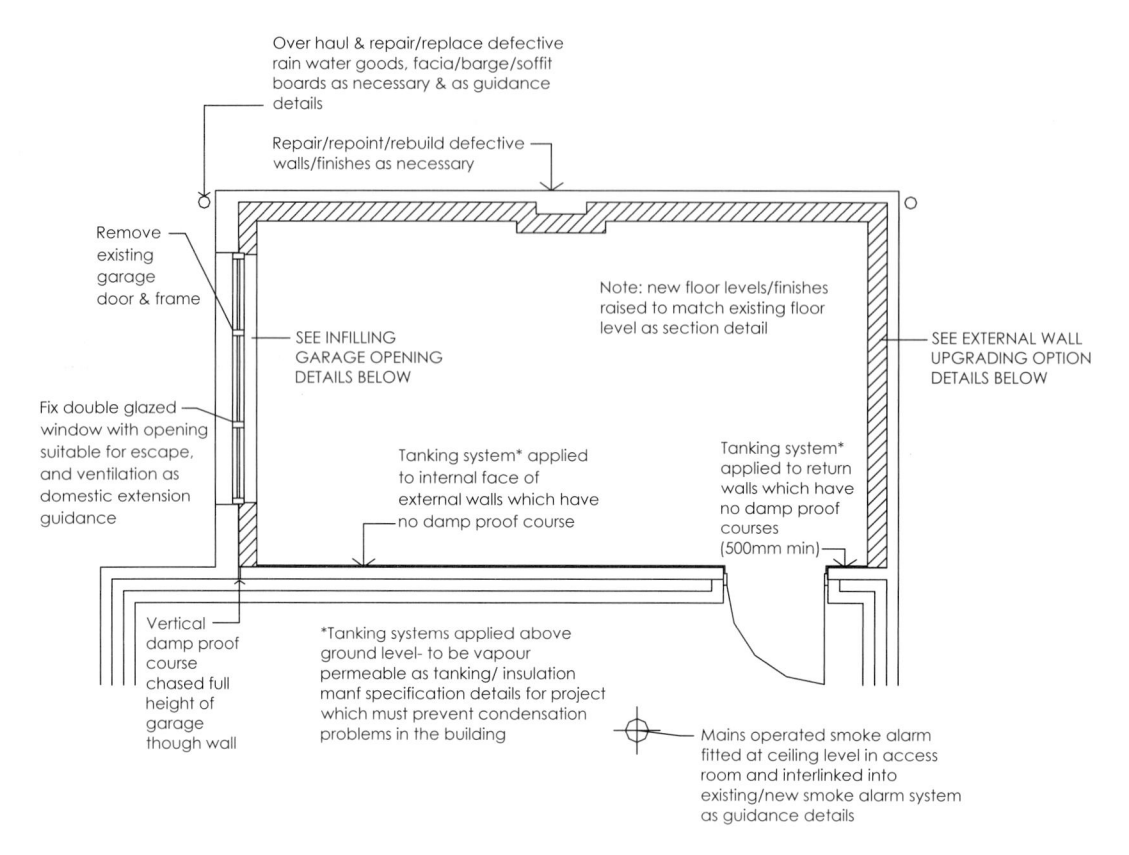

**5.18** Domestic garage and basement conversions into habitable rooms and new dwellings

**Figure 5.4:** Plan layout of a typical single-skin garage conversion *(not to scale)*

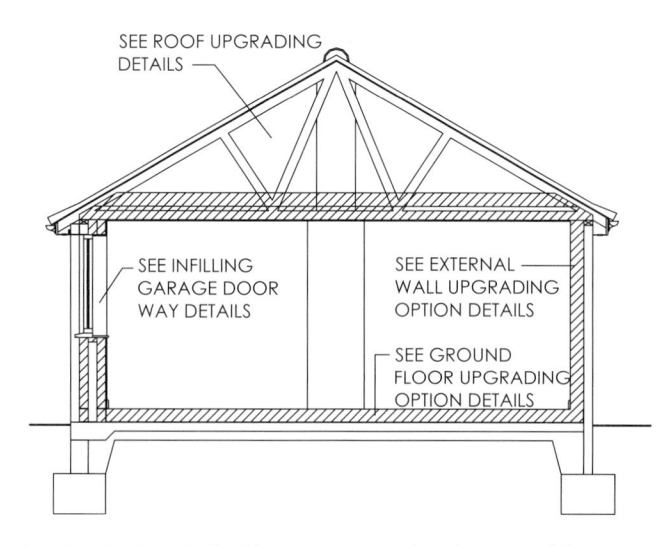

**Figure 5.5:** Section through a typical single-skin garage conversion *(not to scale)*

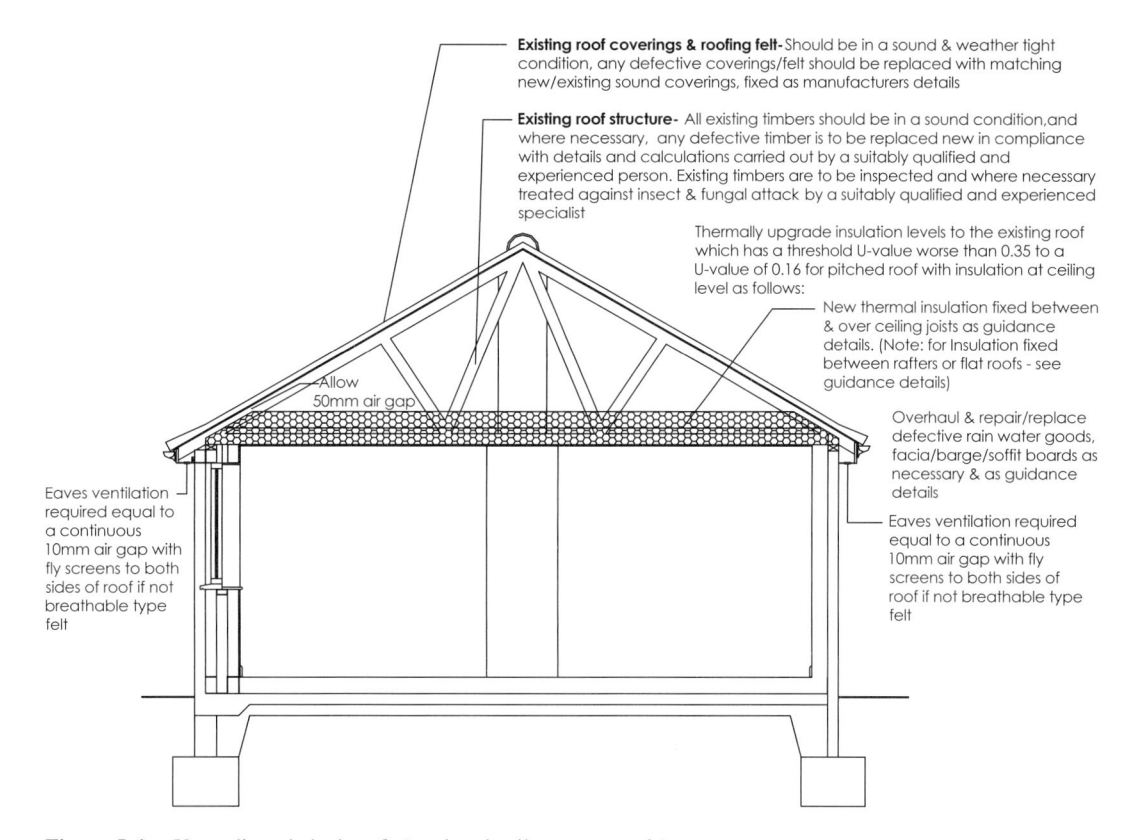

**Figure 5.6:** Upgrading pitched roofs (section detail – *not to scale*)

Install eaves ventilation where necessary, equal to a continuous 10 mm air gap with fly screens to both sides of roof if not breathable-type felt, using proprietary ventilators fixed in accordance with the manufacturer's details

Overhaul and repair/replace defective rainwater goods and facia/barge/soffit boards as necessary and as in the guidance details.

## Upgrading flat roofs (see Figure 5.7)

Existing roof coverings and roofing felt should be in a sound and weather-tight condition; any defective coverings/felt should be replaced with matching new/existing sound coverings, fixed by a flat-roofing specialist as manufacturer's details

In the existing roof structure, all existing timbers should be in a sound condition, and all defective timber is to be replaced with new in compliance with details and calculations carried out by a suitably qualified and experienced person. Existing timbers are to be inspected by a suitably qualified and experienced specialist and where necessary treated against insect and fungal attack.

Where the existing thermal insulation threshold U-value is worse than 0.35, upgrade the thermal insulation level to achieve a minimum U-value of 0.18 for flat roofs, as follows:

**5.20** Domestic garage and basement conversions into habitable rooms and new dwellings

**Existing roof coverings & roofing felt** - Should be in a sound & weather tight condition, any defective coverings/felt should be replaced with matching new/existing sound coverings, fixed by flat roofing specialist as manufacturers details

**Existing roof structure** - All existing timbers should be in a sound condition, all defective timber is to be replaced with new in compliance with details & calculations carried out by a suitably qualified and experienced person. Existing timbers are to be inspected & where necessary treated against insect & fungal attack by a suitably qualified & experienced specialist

Where the existing thermal insulation threshold U-value is worse than 0.35, upgrade thermal insulation level to achieve a minimum U-value of 0.18 for flat roofs as follows:

Friction fix thermal insulation tightly between existing flat roof joists & additional layer fixed under as guidance tables. Depth of joists to be increased with additional timbers glued & screwed to underside of joist to accommodate the required thickness of insulation where necessary.

50mm min continuous clear air gap required between top of insulation & underside of decking

1: 60 - 1: 80 Gradient minimum

Overhaul & repair/ replace defective rain water goods, facia/barge/soffit boards as necessary & as guidance details

Eaves ventilation required equal to a continuous 25mm air gap with fly screens on opposing (both) sides of roof continuous with 50mm air gap above insulation

treated timber battens to suit depth of insulation as guidance tables, fixed to underside of joists at 400mm ctrs

12.5mm vapour checked plaster board & skim fixed to underside of battens

Eaves ventilation required equal to a continuous 25mm air gap with fly screens on opposing (both) sides of roof continuous with 50mm air gap above insulation

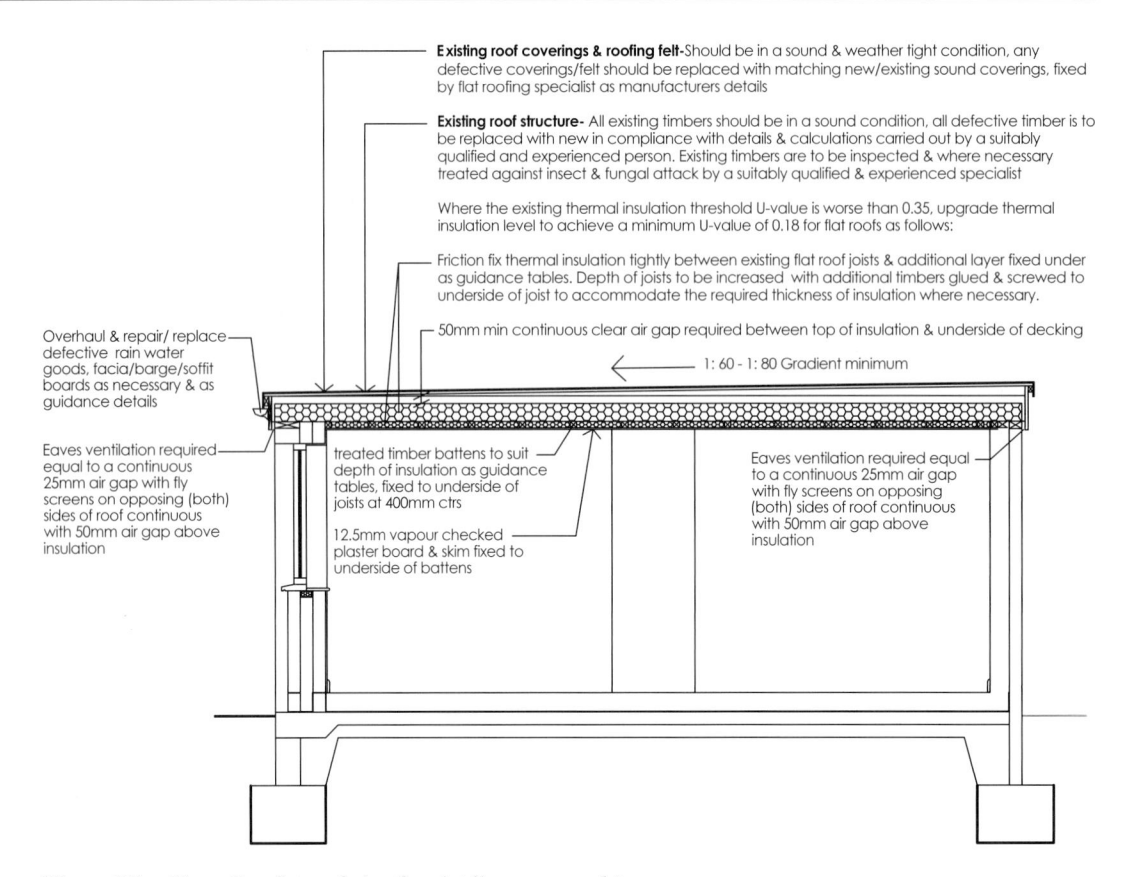

**Figure 5.7:** Upgrading flat roofs (section detail – *not to scale*)

Friction-fix thermal insulation tightly between existing flat-roof joists and fix an additional layer under joists, as in domestic extension guidance details. Depth of joists is to be increased with additional timbers glued and screwed to underside of joist to accommodate the required thickness of insulation where necessary.

A 50 mm minimum continuous clear air gap is required between top of insulation and underside of flat-roof decking.

Install eaves ventilation equal to a continuous 25 mm air gap, with fly screens on opposing (both) sides of roof continuous with 50 mm air gap above insulation.

Overhaul and repair/replace defective rainwater goods and facia/barge/soffit boards as necessary and as in the guidance details.

## Infilling of garage door openings (see Figure 5.8)

Remove existing garage door and frame and prepare walls/floor.

Check condition of existing lintel and replace where necessary, as in guidance details

Construct new foundations for infilling of garage doorway (foundation details to be in accordance with domestic extension guidance above); **or**

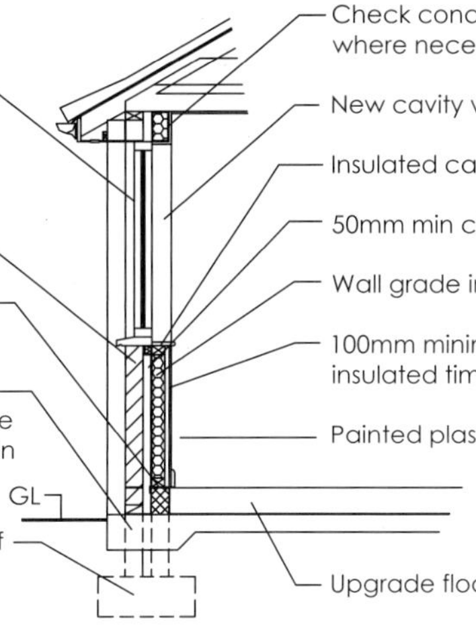
Remove existing garage door & frame
Fix double glazed window with opening suitable for ventilation & escape as guidance details
100mm minimum external wall as guidance details
Radon/DPC tray (225mm deep) & weep holes
New non load bearing cavity wall infilling garage doorway built off existing concrete floor slab 150mm minimum thickness and in good condition as agreed with building control, **or;**
alternatively build off existing foundation if suitable or construct new foundation as domestic extension guidance details as agreed with building control.
GL
Check condition of existing lintel (s) & replace where necessary as guidance details
New cavity wall construction tied into the existing:
Insulated cavity closer & dpc
50mm min clear cavity
Wall grade insulation as guidance table
100mm minimum width insulation block/ insulated timber frame as guidance details
Painted plaster finishes
Upgrade floor as guidance details
**Important note:** load bearing cavity walls (i.e. supporting lintels etc) to have new foundations in accordance with domestic extension guidance above

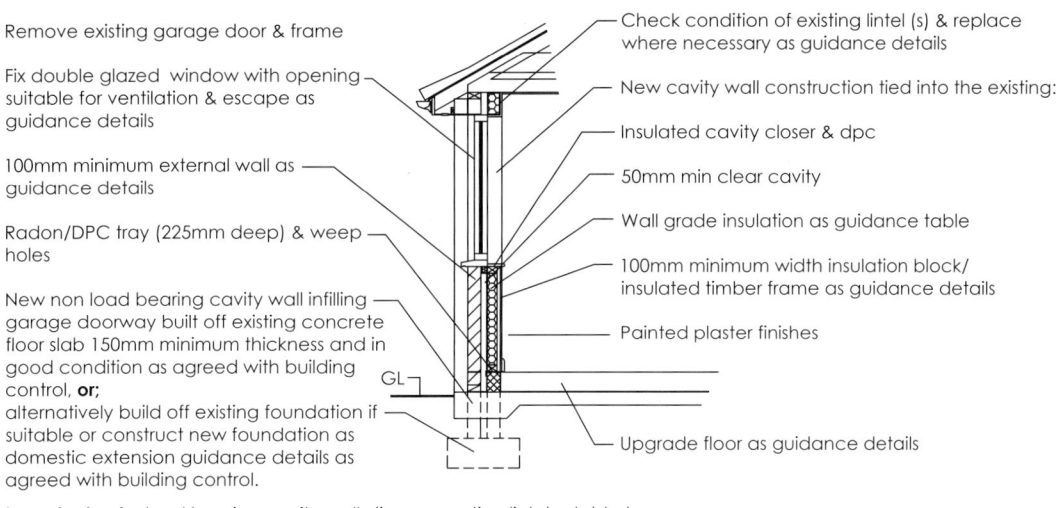

**Figure 5.8:** Infilling of garage door openings (section detail – *not to scale*)

Alternatively, it may be possible to build a **non**-load-bearing cavity wall directly off the existing floor slab, as detailed in Figure 5.8, providing the slab is of sufficient thickness (150 mm minimum/ reinforced) and in good condition (subject to inspection of the slab and approval by building control).

Construct new cavity wall (as in domestic extension guidance), tied into the existing structure, ensuring a continuous horizontal damp-proof course/radon gas tray built into the new cavity wall and sealed to damp-proof membrane, as detailed in Figure 5.8, ensuring vertical damp-proof courses/ tanking system is inserted/applied between the new and existing walls to prevent passage of moisture into the building.

## Upgrading single-skin external walls

**Four options are suggested.**

### Option 1: Upgrading external walls with internal cavity and non-load-bearing timber-framed walls (see Figures 5.9.1 and 5.9.2)

Thermally upgrade existing walls that have a threshold U-value of worse than 0.7, to a U-value of 0.3 for solid walls as follows:

Repair/re-point/rebuild existing walls and make good finishes as necessary. Construct a 100 mm-thick lightweight block work plinth built off existing concrete floor slab, which should be at least 100 mm thick and in good condition (subject to building control approval – or build off new foundation if required by building control, as in guidance detail below), to form a cavity at least 50 mm

**5.22** Domestic garage and basement conversions into habitable rooms and new dwellings

Upgrade roof as detailed in guidance

Repair/repoint/ existing walls/finishes as necessary

Thermally upgrade the existing walls which have a threshold U-value worse than 0.7, to a U-value of 0.3* for solid walls as follows:

Non load bearing independent treated timber frame- constructed and fixed into position using 50/75 x 50mm timber sole plates fixed over dpc to top of block work plinth, head plated fixed to under side of existing ceiling joists, with 50/75 x 50mm timber studs fixed vertically between plates at 400mm ctrs with proprietary British Board of Agreement (BBA or other third party accredited) breather membrane fixed to back of stud walls and forming drip over blockwork into cavity.

integral/12.5mm vapour checked plaster board & skim

Thermal insulation fixed between/ over studs as guidance table*

50mm min cavity

DPC 150mm min above ground level and sealed to DPM/Radon barrier

Proprietary application to seal cavity and prevent passage of radon gas into building if required by building control & weep holes formed in external wall above radon barrier to allow any moisture out

100mm insulation block forming min 50mm wide x 150mm deep cavity

Upgrade exisitng floor as options detailed in guidance

DPC 150mm minimum above ground level

Ground level

Inner wall built off exisitng floor slab (min 100mm thick in good condition) or new foundation as guidance (to be agreed with building control)

Existing foundations

*note: lesser standard is acceptable where reduces floor area by 5%

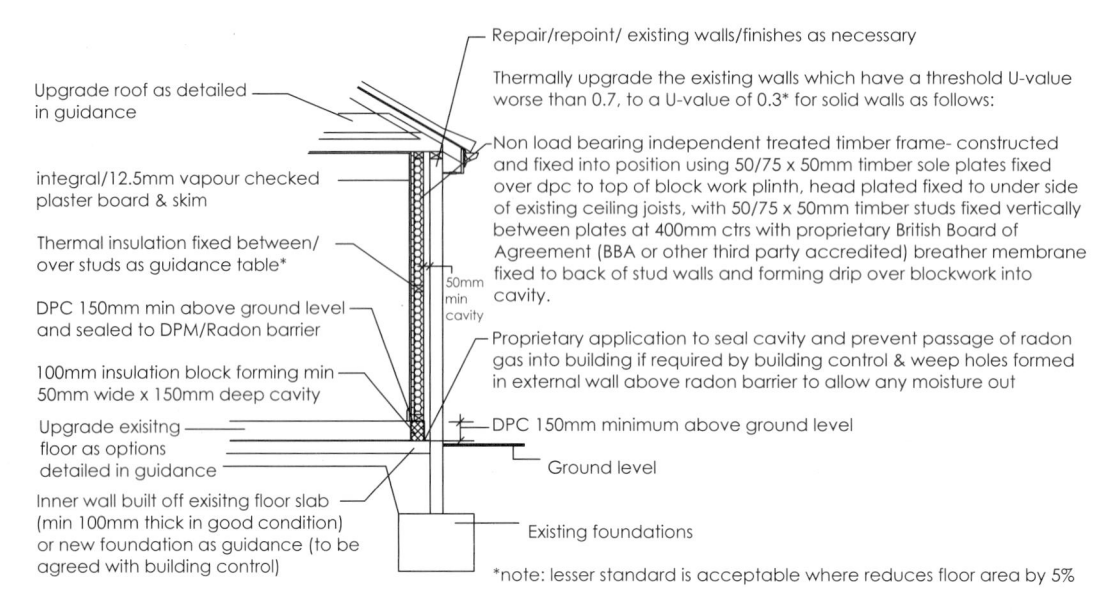

**Figure 5.9.1:** Upgrading external walls with internal cavity and non-load-bearing timber-framed walls (section detail – *not to scale*)

new 100mm thick external masonry wall as guidance

Remove existing garage door & frame

Existing garage wall repaired as necessary

Fix double glazed window with opening suitable for escape and ventilation as guidance

Proprietary s/steel wall ties fixed to existing wall at spacings as domestic extension guidance details

Existing garage masonry walls (repaired as necessary)

Vertical dpc & cavity closed & fire stopped

Insulated inner stud wall as detailed on opposite wall

50mm clear cavity

Vertical damp proof course chased full height of garage though wall

Insulation fixed between/over studs as guidance table

BBA (or other third party accredited) breather membrane fixed to back of stud walls and forming drip over blockwork into cavity.

Tanking system applied existing wall if it has no damp proof course, with plaster finishes

Integral plaster board/12.5mm plaster board with intergral vapor check or 500g vapour check

Non load bearing independent treated timber frame- 50/75 x 50mm timber studs fixed vertically between head & sole plates at 400mm ctrs (braced as necessary)

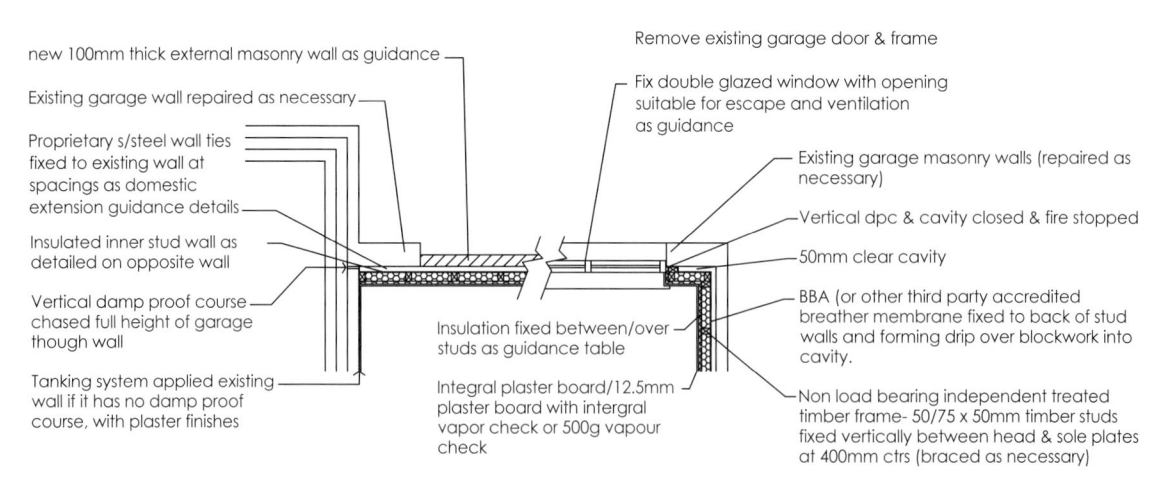

**Figure 5.9.2:** Upgrading external walls with internal cavity and non-load-bearing timber-framed walls (plan detail – *not to scale*)

wide and at least 150 mm deep to support the inner non-load-bearing timber frame (height of block work is to match existing internal floor level).

Fabricate and fix into position a non-load-bearing, lightweight, treated timber stud partition using 50/75 × 50 mm timber sole plates fixed over a dpc to top of new wall; head plates fixed to the underside of existing ceiling joists. Fix 50/75 × 50 mm studs vertically between head and sole plates at 400 mm centres (braced with external-quality plywood or similar approved material where necessary), and fix a proprietary BBA (British Board of Agreement or other third-party accredited) breathable membrane over back of stud work in cavity and extend 25 mm over face of new wall to form a drip into the cavity. Apply a proprietary radon gas-proof application in accordance with the manufacturer's details to seal the sub-floor cavity if required by building control, with a weep hole formed above in external wall as necessary.

Thermal insulation is to be fixed between/over stud walls as in Table 5.1 below; fix either 500 g vapour check/integral plasterboard or separate 12.5 mm vapour-checked plasterboard and skim internal finish. Note: A lesser insulation standard is acceptable where the construction reduces floor area by 5 per cent, as agreed with building control.

## Option 2: Upgrading external single wall into non-load-bearing cavity wall construction – non-load-bearing inner wall (see Figure 5.10)

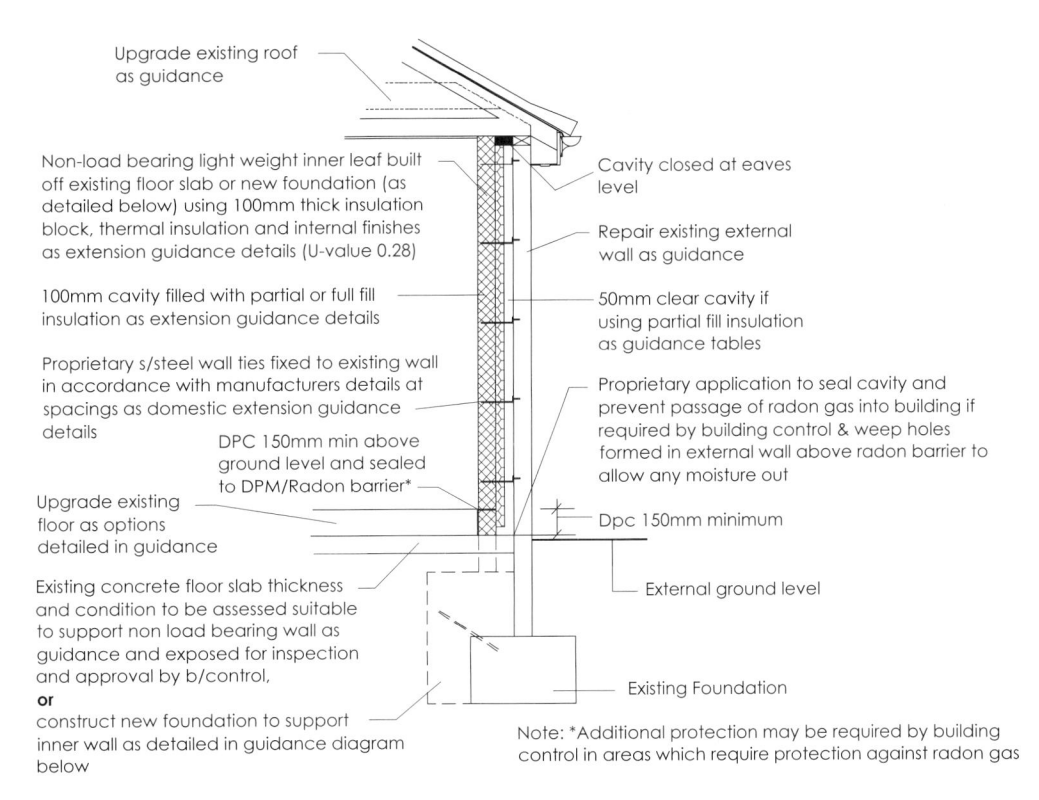

**Figure 5.10:** Upgrading external single wall into cavity wall construction – non-load-bearing inner wall (section detail – *not to scale*)

## Option 3: Upgrading external single wall into cavity wall construction – load-bearing inner wall (see Figure 5.11)

Upper floor/roof constructed as extension guidance details

Load bearing inner leaf built off new foundation using 100mm thick insulation block and internal finishes as extension guidance details

100mm cavity filled with partial or full fill insulation as extension guidance details (U-value 0.28)

Proprietary s/steel wall ties fixed to exisitng wall in accordance with manufacturers details at spacings as extension guidance details

DPC 150mm min above ground level and sealed to DPM/Radon barrier*

Upgrade existing floor as options detailed in guidance

Existing concrete floor slab to be exposed for inspection as guidance

Cut through existing floor slab & construct new foundation tied to existing to support new inner load bearing wall in accordance with structural engineers details. Make good disturbed construction to match existing, piece in new section of DPM sealed to existing, tie new section of concrete floor into existing with 300mm long x 12mm diam s/steel bars at 800mm ctrs, resin grouted 150mm into existing slab to s/engineers details

Build up new wall to match exisitng as extension guidance details

Repair existing external wall as guidance, existing wall & lintels to be exposed for inspection to support new loadings

50mm clear cavity if using partial fill insulation as guidance tables

Proprietary application to seal cavity and prevent passage of radon gas into building if required by building control & weep holes formed in external wall above radon barrier to allow any moisture out

Dpc 150mm minimum above ground level

External ground level

Back fill cavity with weak mix concrete within 225mm of dpc level

Existing foundations to be exposed for inspection by building control to ensure suitable to support new storey loadings

Note: *Additional protection may be required by building control in areas which require protection against radon gas

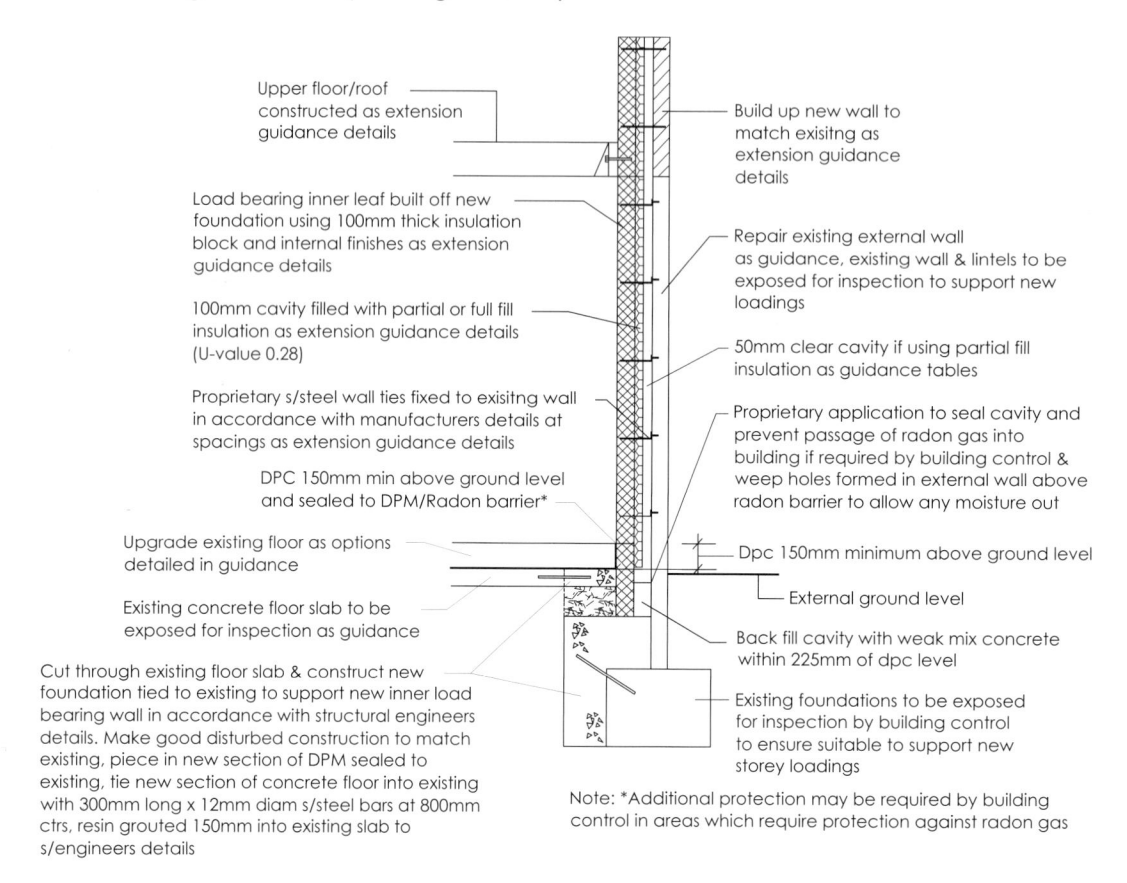

**Figure 5.11:** Upgrading external single wall into cavity wall construction – load-bearing inner wall (section detail – *not to scale*)

## Option 4: Upgrading external walls with internal tanking and insulated timber-framed walls (see Figure 5.12)

Resistance to penetrating damp is normally prevented in existing single-skin walls by forming a cavity to prevent the passage of moisture, typically as detailed in Options 1, 2 and 3 above.

In certain circumstances where a cavity cannot be provided, a designed tanking solution in accordance with Figure 5.12 below may be acceptable, subject to building control approval.

### *Suggested system*

A modified, vapour-permeable, cement-based waterproof tanking system must be assessed, designed and installed for the particular project by a tanking specialist, in compliance with BS 8102: 2009. The tanking system used must be suitable for the particular situation and be vapour-permeable above ground to prevent condensation/interstitial condensation problems within the building. Additional proprietary applications may be required to prevent radon gas from entering the building, if required by building control. (Note: Radon protection could affect the vapour permeability of the tanking system.) Tanking systems are to have a British Board of Agreement (BBA or other approved third-party) accreditation.

### Insulation systems

These must be assessed, designed and installed for the particular project, in compliance with an insulation manufacturer's details and Table 5.1 below. The risk of condensation with any tanking system should be assessed by a specialist; a condensation risk analysis should be carried out for the particular project; and the tanking and thermal insulation system should be designed and installed to prevent any potential condensation/interstitial condensation problems and mould growth within the converted building.

Tanking systems are to be properly connected to, and made continuous with, wall damp-proof courses/radon dpc trays. Perforation of the tanking system by service entry pipes etc. should be avoided or carried out strictly in accordance with the tanking manufacturer's details.

**Important notes**

1. The illustrated tanking section detail below is a suggested detail only and actual details must be in accordance with the tanking specialist's details for the particular project.
2. The existing external wall should be suitable for the proposed upgrading works and the application of any tanking system, and any such tanking application should not increase the risk of deterioration of the building fabric – subject to assessment by a suitably qualified/experienced specialist

Upgrade roof as detailed in guidance

Apply suitable tanking system in accordance with guidance details. Suggested system: 'SOVEREIGN' modified vapour permeable cement based water proof tanking system applied to prepared inner wall faces and top of concrete floor by a tanking specialist in accordance with manufacturer's details and specification for the particular project. Further information and individual specifications can be obtained from SOVEREIGN Technical: Contact Mark Gillen: 01229 870800 or by e mail at: mark.gillen@bostik.com (Note: only suggested details have been illustrated and actual details may differ from those shown)

22mm tongue & grooved moisture resistant flooring sheets with staggered joints mechanically pined & glued with water proof glue (or alternatively 75mm thick cement/sand screed/separating layer over insulation as guidance

Floor grade thermal insulation to minimum thickness as guidance table, made up to required floor level, with all joints staggered. Note: lesser insulation standard is acceptable where new work affects existing adjoining floor levels

Existing concrete floor to be exposed for inspection (minimum 100mm thick and in good sound condition), prepared and levelled up as necessay using proprietary levelling compounds suitable for tanking

Apply tanking to internal wall & floor surfaces as detailed above

Thermally upgrade the existing walls which have a threshold U-value worse than 0.7, to a U-value of 0.3* for solid walls in accordance with insulation manufacture's details for the particular project, typically as follows:

Treated timber frame, consisting of 50/75 x 50mm timber sole plates fixed to top of new floor & timber head plates fixed to under side of existing ceiling joists, with studs fixed vertically between plates at 400mm ctrs.

Thermal insulation fixed between/over studs as guidance table* with 500g vapour check and integral plater board finish or 12.5mm vapour checked plaster board & skim

Note: Timber frame position to tanking to be in accordance with the insulation & tanking manufactures details for the particular project to prevent potential condensation/interstitial condensation problems within the building

Solid masonry wall (Repair/ repoint/rebuild defective walls/ and finishes as necessary)

Ground level

*note: lesser standard is acceptable where reduces floor area by 5%

Existing foundations

Note: Additional proprietary radon gas proof applications may be required by building control in areas which require protection against radon gas, applied in accordance with the tanking manufacturer's details for a particular project, which could effect the vapour permeability of the wall.

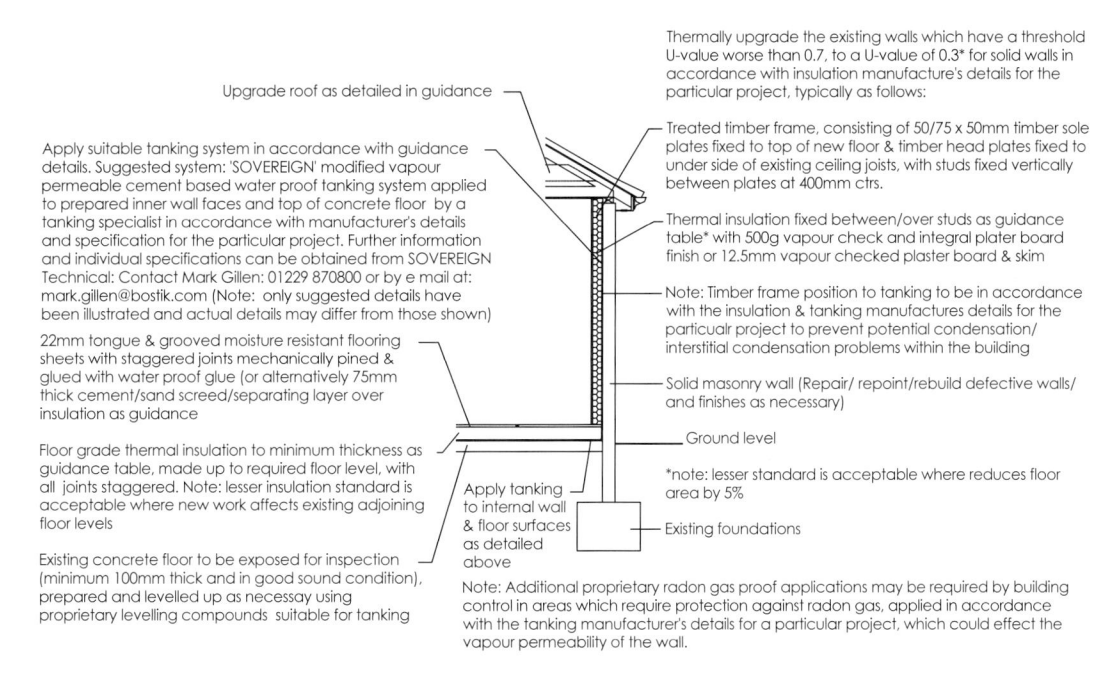

**Figure 5.12:** Upgrading external walls with internal tanking and insulated timber-framed walls (section detail – *not to scale*)

**Table 5.1:** Upgrading thermal insulation to existing 100/215 mm-thick, single-skin, solid masonry walls (dense block work or brickwork)

| Insulation product | K value | Minimum thickness (mm) |
|---|---|---|
| **U-value 0.3 W/m²k** | | |
| Kingspan Kootherm K18 | 0.020 | 72.5 mm fixed across studs with integral vapour-checked plasterboard and skim finish |
| Celotex FR5000 and Celotex PL4000 | 0.021 0.022 | 50 mm between 50 mm deep studs and 25 mm fixed across studs with integral vapour-checked plasterboard and skim finish |
| Celotex FR5000 and Celotex PL4000 | 0.021 0.022 | 40 mm between 75 mm deep studs and 25 mm fixed across studs with integral vapour-checked plasterboard and skim finish |

## Upgrading garage ground floors (or similar) with upgraded enclosing single-skin walls

**Three options are suggested.**

### Option 1: Floating floor laid over existing floor (see Figures 5.13.1 and 5.13.2)

Prepare and level up existing floor slab as necessary, using proprietary levelling products in compliance with manufacturer's details (existing walls to be upgraded in accordance with the guidance above, and floor slab to be minimum 100 mm-thick concrete in good condition). Lay 1200 g damp-proof and basic radon gas-proof membrane or apply suitable British Board of Agreement (BBA or other third-party accredited) tanking system to floor surfaces up to dpc level and seal to prevent passage of moisture (and passage of radon gas as necessary in radon gas-affected areas), as manufacturer's details.

Lay floor-grade thermal insulation to minimum thickness as domestic extension guidance table, made up to required floor level, with all joints staggered. Note: A lesser insulation standard may be acceptable where new work affects existing adjoining floor levels, subject to approval by building control.

Fix 22 mm-thick, moisture-resistant, tongue-and-grooved timber floorboards laid with joints staggered; all joints are to be glued (using waterproof glue) and pinned, in accordance with the floorboard manufacturer's details and current BS EN standards. Allow an expansion gap around wall perimeters as manufacturer's details (typically 10–15 mm); all floorboards are to be secured at perimeters with skirting boards.

**Note:** Where full radon gas protection is required by building control, additional radon protection is to be carried out, in accordance with a radon gas specialist's details.

### Option 2: Screed floor laid over existing floor (see Figure 5.14)

Prepare and level up existing floor slab as necessary using proprietary levelling products in compliance with manufacturer's details (existing walls to be upgraded in accordance with the guidance above and floor slab to be minimum 100 mm-thick concrete in good condition), Lay 1200 g

22mm tongue & grooved moisture resistant flooring sheets with staggered joints mechanically pined & glued with water proof glue (allow 10-15mm expansion gap around floor perimeter)

Upgrade walls as guidance details for single skin walls

Floor grade thermal insulation to minimum thickness as guidance table, made up to required floor level, with all joints staggered. Note: lesser insulation standard is acceptable where new work affects existing adjoining floor levels

10-15mm expansion gap around floor perimeter

DPC/radon barrier

Ground level

1200g damp proof membrane/radon gas barrier laid /tanking system as specified over existing floor slab and sealed to DPC/radon barrier

Prepare and level up existing floor as necessary using proprietary levelling products ready for dpm/tanking. Existing floor to be 100mm minimum thickness and in good condition

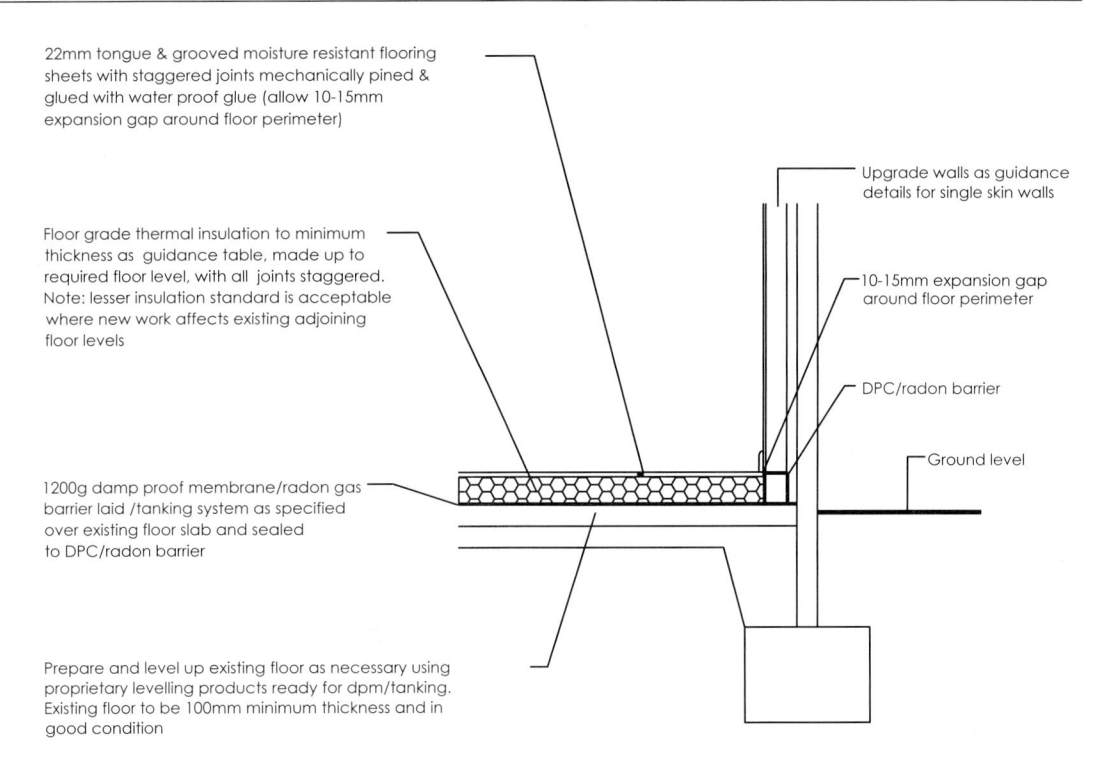

**Figure 5.13.1:** Floating floor laid over existing floor (section detail – *not to scale*) (Also see Figure 5.13.2 below for levelling of sloping floors with floating floor finish.)

22mm tongue & grooved moisture resistant flooring sheets glued with water proof glue and screwed to joists as guidance details:

Upgrade cavity walls as guidance diagram above

Level up existing slope in floor using 50mm wide treated timber floor joists to required level, joists supported onto dpc laid over dpm/tanking and existring floor slab at 400mm ctrs, with noggins fixed between joists ends and within spans to prevent twisting as guidance details for floors

Floor grade thermal insulation fixed tightly between joists, thickness as guidance table, made up to required floor level, with all joints staggered. Note: lesser insulation standard is acceptable where new work affects existing adjoining floor levels

1200g damp proof membrane/ radon gas barrier laid over ex floor & 150mm up ex walls to dpc level, chased 25mm into walls and sealed to dpc

Existing floor slab as guidance diagram

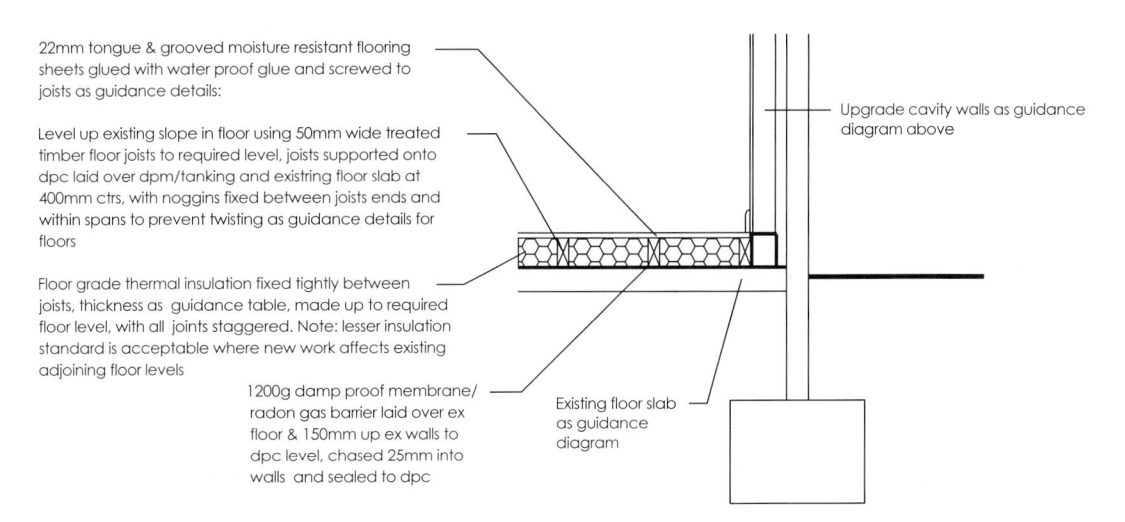

**Figure 5.13.2:** Timber joists used to level up existing sloping floors (section detail – *not to scale*)

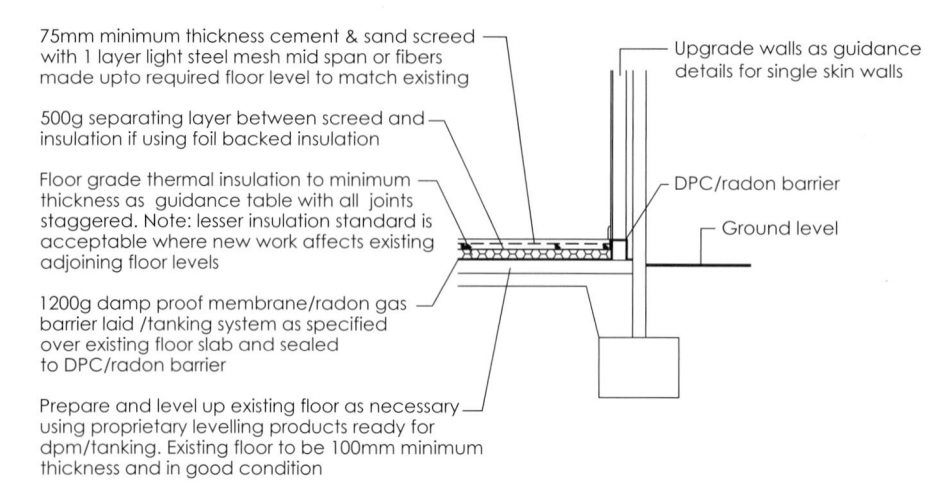

**5.28** Domestic garage and basement conversions into habitable rooms and new dwellings

**Figure 5.14:** Screed floor laid over existing floor (section detail – *not to scale*)

Callouts:
- 75mm minimum thickness cement & sand screed with 1 layer light steel mesh mid span or fibers made upto required floor level to match existing
- 500g separating layer between screed and insulation if using foil backed insulation
- Floor grade thermal insulation to minimum thickness as guidance table with all joints staggered. Note: lesser insulation standard is acceptable where new work affects existing adjoining floor levels
- 1200g damp proof membrane/radon gas barrier laid /tanking system as specified over existing floor slab and sealed to DPC/radon barrier
- Prepare and level up existing floor as necessary using proprietary levelling products ready for dpm/tanking. Existing floor to be 100mm minimum thickness and in good condition
- Upgrade walls as guidance details for single skin walls
- DPC/radon barrier
- Ground level

damp-proof and radon gas-proof membrane or apply suitable British Board of Agreement (BBA or other third-party accredited) tanking system to floor surfaces up to dpc level and seal to prevent passage of moisture (and passage of radon gas as necessary in radon gas-affected areas) as manufacturer's details

Lay floor-grade thermal insulation to minimum thickness as domestic extension guidance table, made up to required floor level, with all joints staggered. Note: A lesser insulation standard may be acceptable where new work affects existing adjoining floor levels, subject to approval by building control.

Lay 75 mm minimum thickness cement and sand screed with one layer light anti-crack mesh mid span (or anti-crack fibres), made up to required floor level to match existing. (Note: 500 g polythene separating layer is to be installed between the screed and insulation if using a foil-faced polyurethane/PIR-type insulation board.) Allow 10–15 mm minimum expansion gap around wall perimeters.

**Note:** Where full radon gas protection is required by building control, additional radon protection is to be carried out in accordance with a radon gas specialist's details.

## Option 3: Take up existing floor and lay new concrete floor (see Figure 5.15)

Take up existing concrete floor slab and remove off-site; lay sand-blinded hardcore, mechanically compacted in 150 mm layers (floor slab is to be designed by a suitably qualified person where depth of fill exceeds 600 mm).

Lay 100 mm minimum thickness ST2 or Gen1 concrete ground-bearing floor slab (concrete thickened to 150 mm under inner non load bearing wall) over mechanically compacted hardcore and finish with a trowel-smooth top surface. Lay 1200 g damp-proof and radon gas-proof membrane over floor slab, dressed up and sealed into walls at dpc level; or alternatively, apply suitable British Board of Agreement (BBA or other third-party accredited) tanking system to top of slab, up wall surfaces and seal to dpc to prevent passage of moisture into the building (and passage of radon gas as necessary in radon gas-affected areas), as manufacturer's details.

Domestic garage and basement conversions into habitable rooms and new dwellings       5.29

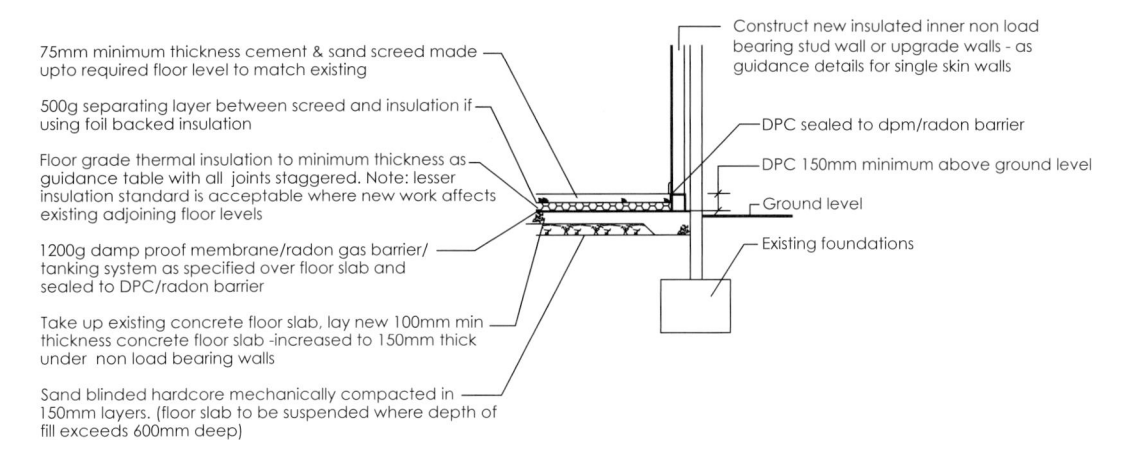

**Figure 5.15:** Take up existing floor and lay new concrete floor (section detail – *not to scale*)

Lay floor-grade thermal insulation to minimum thickness as domestic extension guidance table, with all joints staggered. Note: A lesser insulation standard is acceptable where new work affects existing adjoining floor levels.

Lay 75 mm minimum thickness cement and sand screed made up to required floor level to match existing. (Note: 500 g polythene separating layer is to be installed between the screed and insulation if using a foil-faced polyurethane/PIR-type insulation board). Allow 10–15 mm minimum expansion gap around wall perimeters.

**Note:** Where full radon gas protection is required by building control, additional radon protection to be carried out in accordance with a radon gas specialist's details.

## A3: Cavity wall garage conversions (or similar buildings) into habitable rooms (typical details indicated in Figures 5.16–5.26)

### Upgrading pitched roof (see Figure 5.20)

Existing roof coverings and roofing felt should be in a sound and weather-tight condition; any defective coverings/felt should be replaced with matching new/existing sound coverings, fixed as manufacturer's details.

In the existing roof structure, all existing timbers should be in a sound condition, and where necessary, defective timber is to be replaced with new in compliance with details and calculations carried out by a suitably qualified and experienced person. Existing timbers are to be inspected and where necessary treated against insect and fungal attack by a suitably qualified and experienced specialist.

Thermally upgrade insulation levels to an existing pitched roof that has a threshold U-value worse than 0.35, to a U-value of 0.16 for insulation at ceiling level (as detailed below), and 0.18 for insulation following slope of rafters – see domestic extension guidance details.

Install eaves ventilation equal to a continuous 10 mm air gap with fly screens to both sides of roof if not breathable-type felt, using proprietary ventilators, fixed in accordance with the manufacturer's details. Overhaul and repair/replace defective rainwater goods and facia/barge/soffit boards as necessary and as guidance details

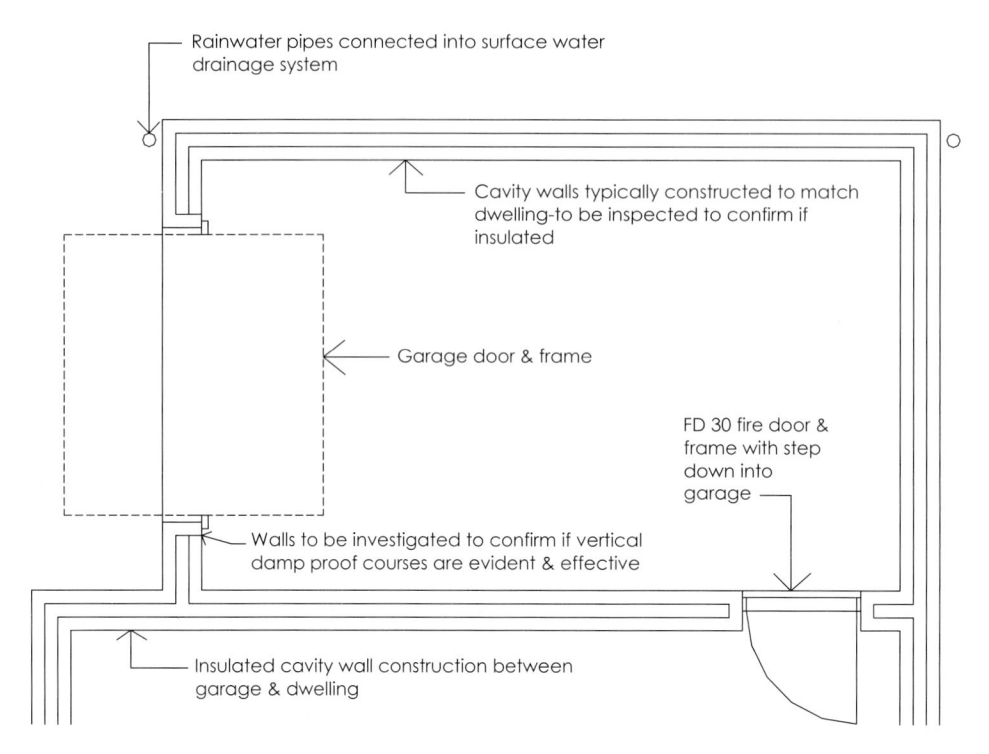

**5.30** Domestic garage and basement conversions into habitable rooms and new dwellings

**Figure 5.16:** Plan layout of existing attached garage *(not to scale)*

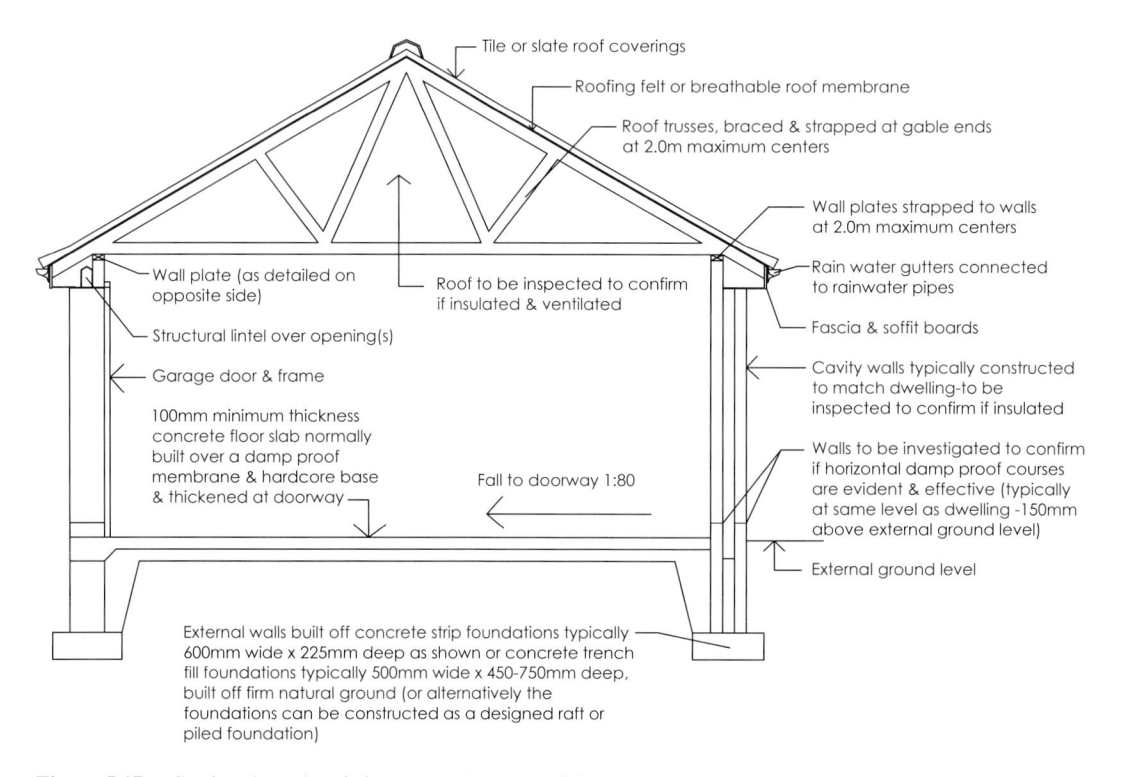

**Figure 5.17:** Section through existing garage *(not to scale)*

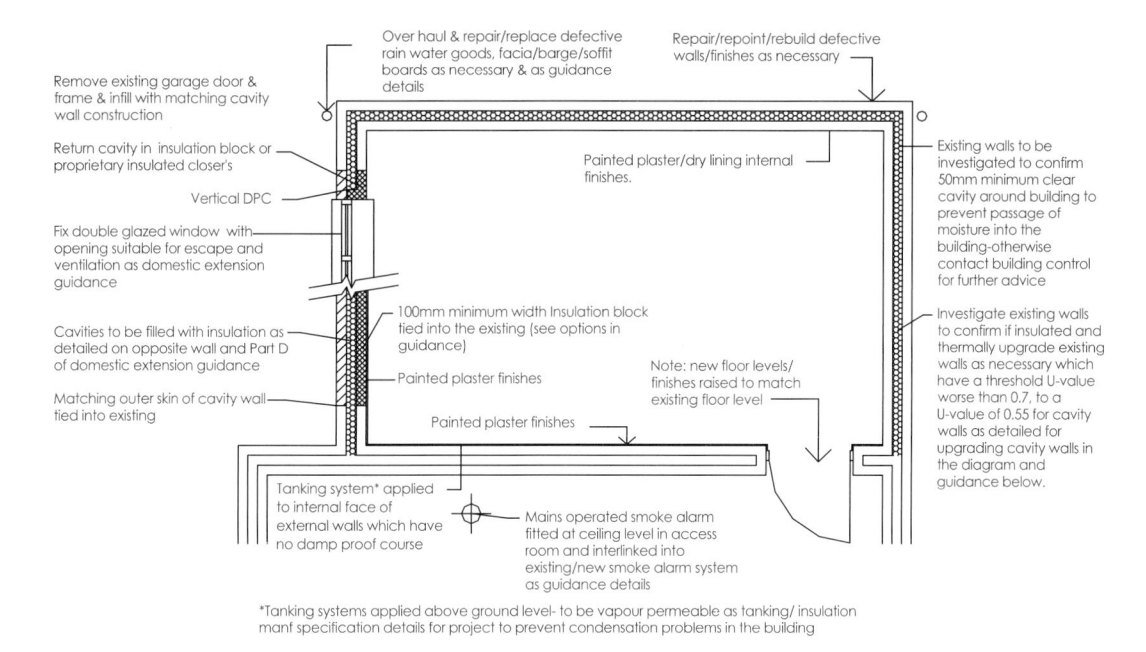

**Figure 5.18:** Plan layout of proposed garage conversion *(not to scale)*

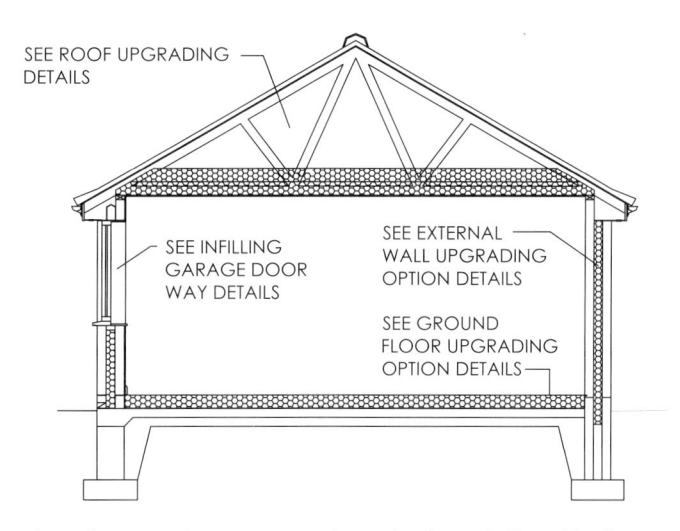

**Figure 5.19:** Section through proposed garage conversion – details are indicated in diagrams below *(not to scale)*

## Infilling of garage door opening (see Figure 5.21)

Remove existing garage door and frame and prepare walls/floor.
 Check condition of existing lintel and replace where necessary, as guidance details.
 Construct new foundations for infilling of garage doorway (foundation details to be in accordance with domestic extension guidance above), **or**

**5.32** Domestic garage and basement conversions into habitable rooms and new dwellings

**Existing roof coverings & roofing felt-** Should be in a sound & weather tight condition, any defective coverings/felt should be replaced with matching new/existing sound coverings, fixed as manufacturers details

**Existing roof structure-** All existing timbers should be in a sound condition, and where necessary, any defective timber is to be replaced new in compliance with details and calculations carried out by a suitably qualified and experienced person. Existing timbers are to be inspected and where necessary treated against insect & fungal attack by a suitably qualified and experienced specialist

Thermally upgrade insulation levels to the existing roof which has a threshold U-value worse than 0.35 to a U-value of 0.16 for pitched roof with insulation at ceiling level as follows:

New thermal insulation fixed between & over ceiling joists as guidance details. (Note: for Insulation fixed between rafters or flat roofs - see guidance details)

Overhaul & repair/replace defective rain water goods, facia/barge/soffit boards as necessary & as guidance details

Eaves ventilation required equal to a continuous 10mm air gap with fly screens to both sides of roof if not breathable type felt

Eaves ventilation required equal to a continuous 10mm air gap with fly screens to both sides of roof if not breathable type felt

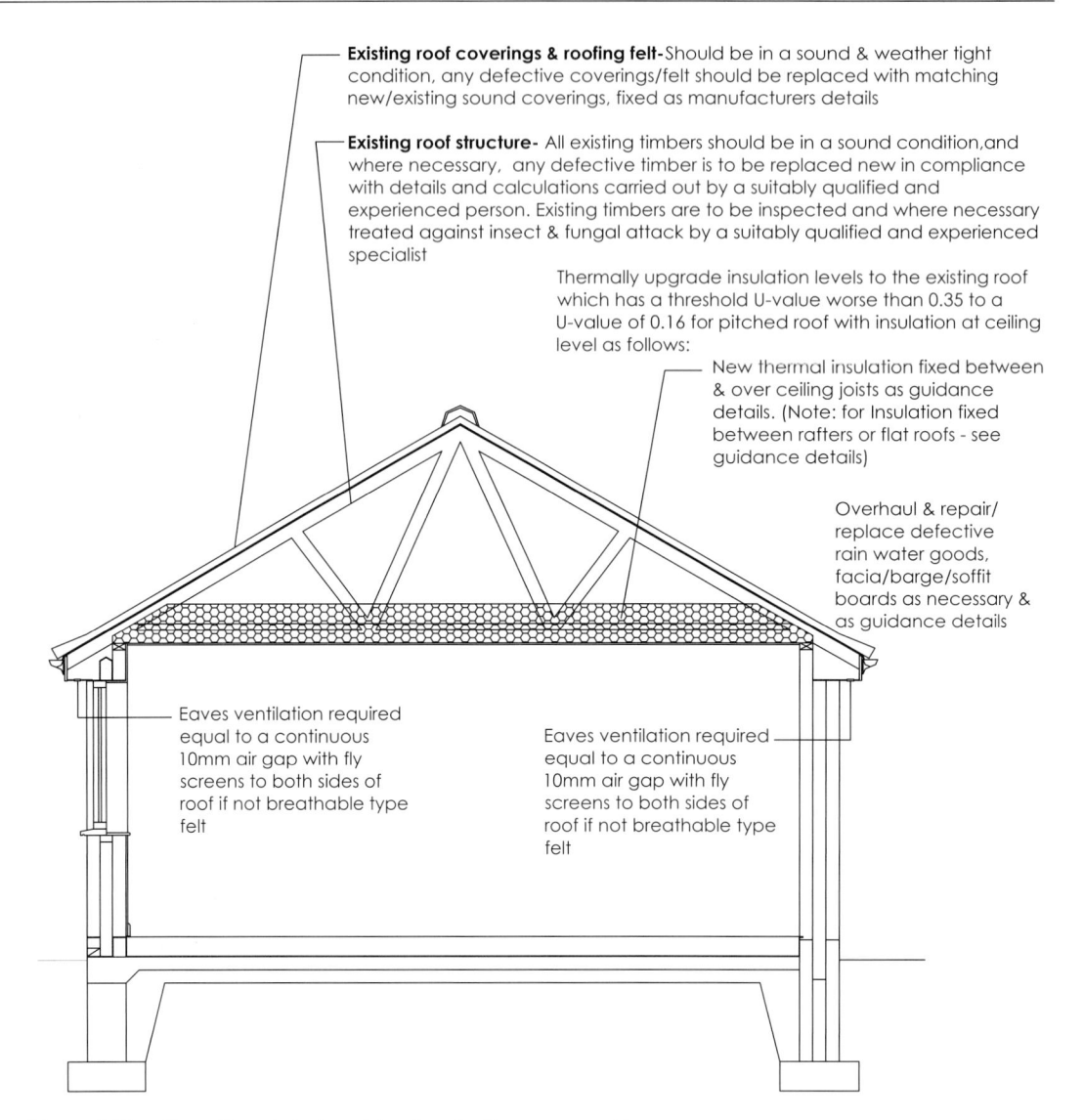

**Figure 5.20:** Upgrading pitched roof (section detail – *not to scale*)
**Note:** For upgrading of flat roofs, see Figure 5.7 above.

Alternatively, it may be possible to build a **non**-load-bearing cavity wall directly off the existing floor slab, as detailed in Figure 5.21 below, providing if it is of sufficient thickness (150 mm minimum/reinforced) and in good condition (subject to inspection of the slab and approval by building control).

Construct new cavity wall (see domestic extension guidance for details), tied into the existing structure, ensuring a continuous horizontal damp-proof course/radon gas tray built into new cavity wall and sealed to damp-proof membrane, as detailed in Figure 5.21 below, ensuring vertical damp-proof courses/tanking system is inserted/applied between the new and existing walls to prevent passage of moisture into the building.

## Figure 5.21

**Labels (left side):**
- Remove existing garage door & frame
- Fix double glazed window with opening suitable for ventilation & escape as guidance details
- New non load bearing cavity wall infilling garage doorway built off existing concrete floor slab 150 mm minimum thickness and in good condition as agreed with building control, **or**; alternatively build off existing foundation if suitable or construct new foundation as domestic extension guidance details as agreed with building control.

**Labels (right side):**
- Check condition of existing lintel & replace where necessary as guidance details
- Remove existing garage door & frame, build up opening in cavity wall construction as detailed on plan layout.
- Install injected cavity wall insulation as main wall guidance details
- Radon/DPC tray & weep holes
- Upgrade existing floor as guidance

**Important note:** load bearing cavity walls (i.e. supporting lintels etc) to have new foundations in accordance with domestic extension guidance above

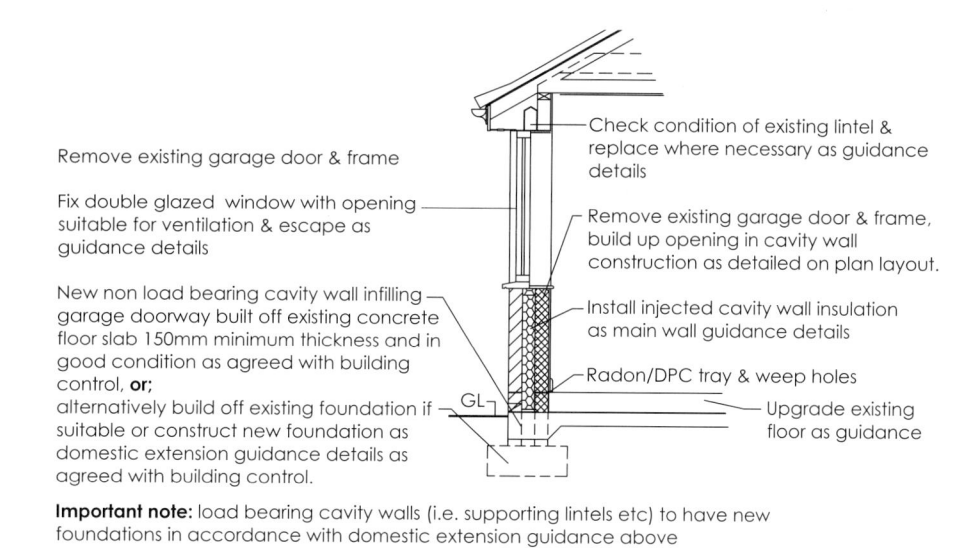

**Figure 5.21:** Infilling of garage door opening (section detail – *not to scale*)

## Figure 5.22

**Labels (left side):**
- Upgrade roof as detailed in guidance
- Existing walls should have an effective existing damp proof course-alternatively specialist to install a chemically inject a damp proof course 150 mm above external ground level
- Upgrade floor as detailed in guidance
- Radon gas proof tanking of all walls & floors may be required by building control in areas which require protection against radon gas

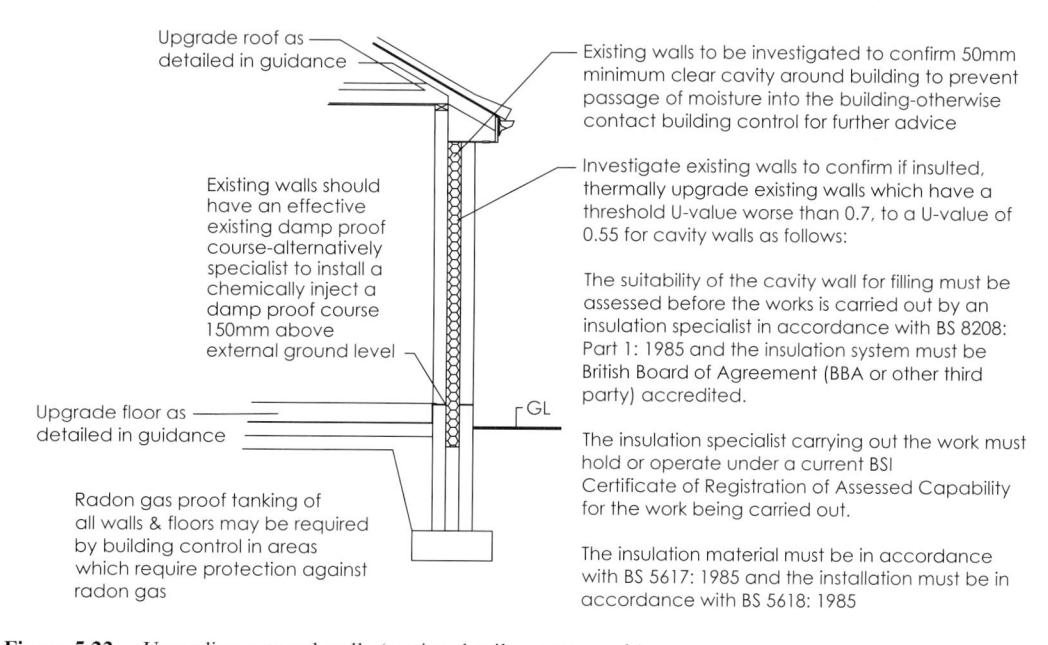

**Labels (right side):**
- Existing walls to be investigated to confirm 50 mm minimum clear cavity around building to prevent passage of moisture into the building-otherwise contact building control for further advice
- Investigate existing walls to confirm if insulted, thermally upgrade existing walls which have a threshold U-value worse than 0.7, to a U-value of 0.55 for cavity walls as follows:
- The suitability of the cavity wall for filling must be assessed before the works is carried out by an insulation specialist in accordance with BS 8208: Part 1: 1985 and the insulation system must be British Board of Agreement (BBA or other third party) accredited.
- The insulation specialist carrying out the work must hold or operate under a current BSI Certificate of Registration of Assessed Capability for the work being carried out.
- The insulation material must be in accordance with BS 5617: 1985 and the installation must be in accordance with BS 5618: 1985

**Figure 5.22:** Upgrading external walls (section detail – *not to scale*)

## Upgrading external cavity walls (see Figure 5.22)

Repair/re-point/rebuild defective walls/finishes as necessary, Prepare existing walls and floor slab (existing walls to be minimum 100 mm-thick masonry construction in good condition, and floor slab to be minimum 100 mm-thick concrete in good condition).

Thermally upgrade the existing walls that have a threshold U-value worse than 0.7, to a U-value of 0.55 by infilling of existing cavity walls, using a British Board of Agreement (BBA or other third-party accredited) cavity wall fill system, installed by specialists; **or** alternatively, where existing

Table 5.2: Upgrading thermal insulation to existing cavity masonry walls

| Insulation product | K value | Minimum thickness (mm) |
|---|---|---|
| **U-value 0.55 W/m²k** | | |
| Cavity wall filled using a British Board of Agreement (BBA or other third-party accredited) cavity wall insulation system, as detailed in ADE and Part D of domestic extensions in this guidance | | To be confirmed by cavity wall insulation specialist (cavity wall must be assessed by specialist to ensure it is suitable for cavity wall filling – see Part D in Section 2 of this guidance for domestic extensions) |

walls are not suitable for cavity wall filling, upgrade as a solid wall to a U-value of 0.3 using insulated stud partitions fixed against the inner wall face, as detailed in guidance above for single-wall garages.

## Upgrading garage ground floors (or similar) with enclosing upgraded cavity walls

**Four options are available.**

### Option 1: Floating floor laid over existing floor (see Figures 5.23.1 and 5.23.2)

Prepare and level up existing floor slab as necessary, using proprietary levelling products in compliance with manufacturer's details (existing floor slab to be minimum 100 mm-thick concrete and in good condition); lay 1200 g damp-proof and radon gas-proof membrane over existing floor slab, dressed up walls, overlapping and trimmed off above new floor level, chased and sealed into walls at dpc level in walls in radon-affected areas); or apply suitable British Board of Agreement (BBA or other third-party accredited) tanking system to floor surfaces up to dpc level and seal to prevent passage of moisture (and passage of radon gas as necessary in radon gas-affected areas), as manufacturer's details

Walls without a damp-proof course are to have a British Board of Agreement (BBA or other third-party accredited) chemically injected damp-proof course installed by a specialist, at least 150 mm above external ground level.

Lay floor-grade thermal insulation to minimum thickness as domestic extension guidance table, made up to required floor level, with all joints staggered. Note: A lesser insulation standard may be acceptable where new work affects existing adjoining floor levels, subject to approval by building control.

Fix 22 mm-thick, moisture-resistant, tongue-and-grooved timber floorboards laid with joints staggered; all joints are to be glued (using waterproof glue) and pinned, in accordance with floorboard manufacturer's details and current BS EN standards. Allow an expansion gap around wall perimeters as manufacturer's details (typically 10–15 mm). Floorboards are to be secured at perimeters with skirting boards.

**Note:** Where basic radon gas protection is required by Building Control, apply a British Board of Agreement (BBA or other third-party accredited) proprietary application to prevent the passage of radon gas into the building through the walls and floor, as manufacturer's details. Where full radon gas protection is required by building control, additional radon protection is to be carried out in accordance with a radon gas specialist's details.

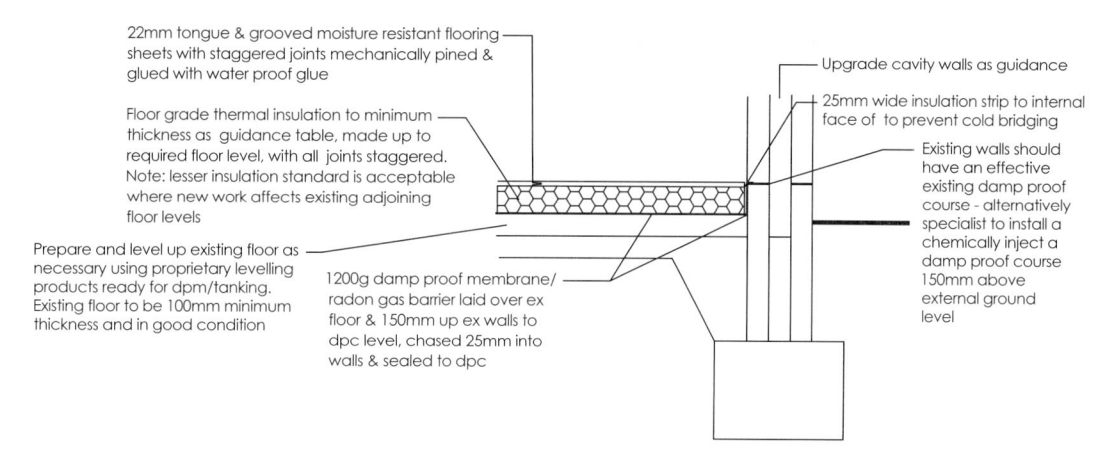

**Figure 5.23.1:** Floating floor laid over existing floor (section detail – *not to scale*) (Also see Figure 5.23.2 for levelling of sloping floors with a floating floor finish.)

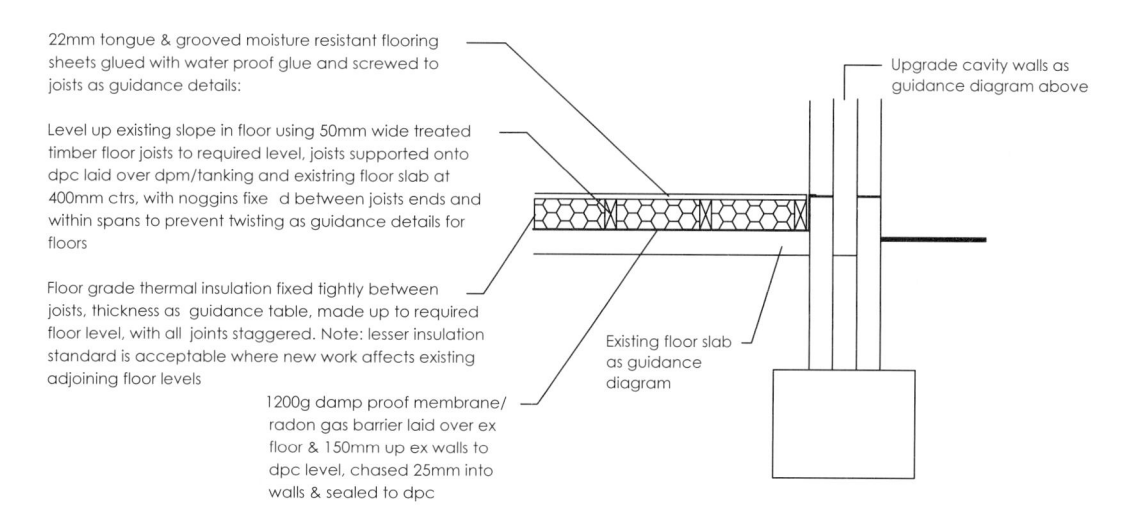

**Figure 5.23.2:** Timber joists used to level up existing sloping floors (section detail – *not to scale*)

## Option 2: Screed floor laid over existing floor (see Figure 5.24)

Prepare and level up existing floor slab as necessary using proprietary levelling products, in compliance with manufacturer's details (existing floor slab to be minimum 100 mm-thick concrete and in good condition); lay 1200 g damp-proof and radon gas-proof membrane over existing floor slab, dressed up walls, overlapping and trimmed off above new floor level, chased and sealed into walls at dpc level in walls in radon-affected areas); or apply suitable British Board of Agreement (BBA or other third-party accredited) tanking system to floor surfaces up to dpc level and seal to prevent passage of moisture (and passage of radon gas as necessary in radon gas-affected areas), as manufacturer's details.

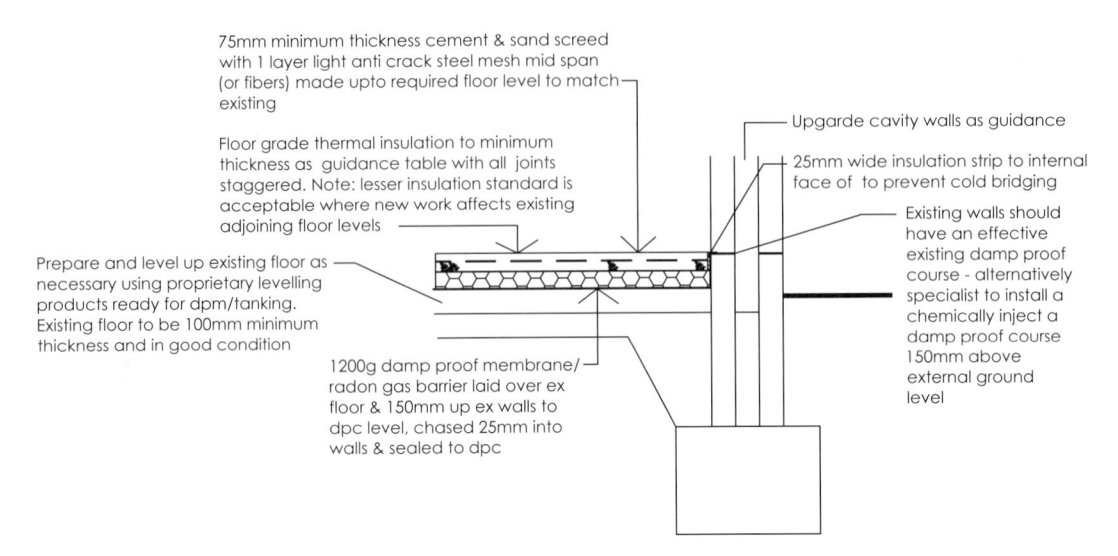

**5.36** Domestic garage and basement conversions into habitable rooms and new dwellings

**Figure 5.24:** Screed floor laid over existing floor (section detail – *not to scale*)

Walls without a damp-proof course are to have a British Board of Agreement (BBA or other third-party accredited) chemically injected damp-proof course installed by a specialist, at least 150 mm above external ground level.

Lay floor-grade thermal insulation to minimum thickness as domestic extension guidance table, made up to required floor level, with all joints staggered. Note: A lesser insulation standard may be acceptable where new work affects existing adjoining floor levels, subject to approval by building control.

Lay 75 mm minimum thickness cement and sand screed with one layer light anti-crack mesh mid span (or anti-crack fibres), made up to required floor level to match existing. (Note: 500 g polythene separating layer is to be installed between the screed and insulation if using a foil-faced polyurethane/PIR-type insulation board). Allow 10–15 mm minimum expansion gap around wall perimeters.

**Note:** Where basic radon gas protection is required by Building Control, apply a British Board of Agreement (BBA or other third-party accredited) proprietary application to prevent the passage of radon gas into the building through the walls and floor, as manufacturer's details. Where full radon gas protection is required by building control, additional radon protection is to be carried out in accordance with a radon gas specialist's details.

### Option 3: Take up existing floor and lay new concrete floor (see Figure 5.25)

Take up existing concrete floor slab and remove off-site; lay sand-blinded hardcore, mechanically compacted in 150 mm layers (floor slab to be designed by a suitably qualified person where depth of fill exceeds 600 mm).

Lay minimum 100 mm-thick ST2, or Gen1 concrete ground-bearing floor slab over hardcore and finish with a trowel-smooth top surface; lay 1200 g damp-proof and radon gas-proof membrane over floor slab, dressed up and sealed into walls at dpc level; **or** alternatively, apply suitable British Board of Agreement (BBA or other third-party accredited) tanking system to top of slab, up wall surfaces and seal to dpc to prevent passage of moisture into the building (and passage of radon gas as necessary in radon gas-affected areas), as manufacturer's details.

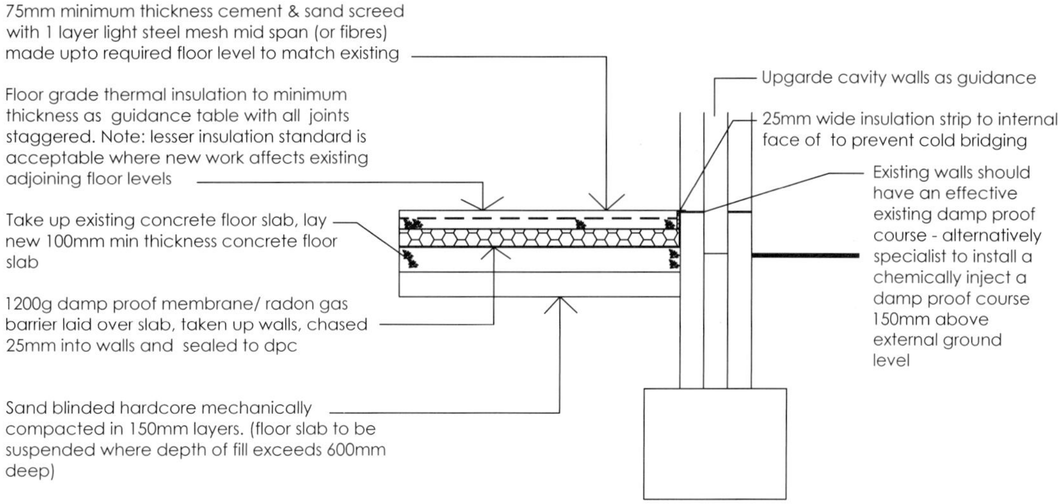
75mm minimum thickness cement & sand screed with 1 layer light steel mesh mid span (or fibres) made upto required floor level to match existing
Floor grade thermal insulation to minimum thickness as guidance table with all joints staggered. Note: lesser insulation standard is acceptable where new work affects existing adjoining floor levels
Take up existing concrete floor slab, lay new 100mm min thickness concrete floor slab
1200g damp proof membrane/ radon gas barrier laid over slab, taken up walls, chased 25mm into walls and sealed to dpc
Sand blinded hardcore mechanically compacted in 150mm layers. (floor slab to be suspended where depth of fill exceeds 600mm deep)
Upgarde cavity walls as guidance
25mm wide insulation strip to internal face of to prevent cold bridging
Existing walls should have an effective existing damp proof course - alternatively specialist to install a chemically inject a damp proof course 150mm above external ground level

**Figure 5.25:** New concrete floor (section detail – *not to scale*)

Walls without a damp-proof course are to have a British Board of Agreement (BBA or other third-party accredited) chemically injected damp-proof course installed by a specialist, at least 150 mm above external ground level.

Lay floor-grade thermal insulation to minimum thickness as domestic extension guidance table, with all joints staggered. Note: A lesser insulation standard is acceptable where new work affects existing adjoining floor levels.

Lay 75 mm minimum thickness cement and sand screed with one layer light anti-crack mesh mid span (or anti-crack fibres), made up to required floor level to match existing. (Note: 500 g polythene separating layer is to be installed between the screed and insulation if using a foil-faced polyurethane/PIR-type insulation board). Allow 10–15 mm minimum expansion gap around wall perimeters.

**Note:** Where basic radon gas protection is required by Building Control, apply a British Board of Agreement (BBA or other third-party accredited) proprietary application to prevent the passage of radon gas into the building through the walls and floor, as manufacturer's details. Where full radon gas protection is required by building control, additional radon protection is to be carried out in accordance with a radon gas specialist's details.

## Option 4: Suspended timber floor laid over existing floor void (see Figure 5.26)

Where new floor levels are to be made up over an existing deep floor void, a suspended floor can be constructed, as in Figure 5.26 below and as follows:

Existing walls to be minimum 100 mm-thick masonry construction in good condition, and floor slab to be minimum 100 mm-thick concrete in good condition

Suspended timber floor is to be constructed in accordance with domestic extensions guidance and span tables suitable for the clear spans. Joists are to be supported, typically by heavy-duty proprietary galvanised metal restraint joist hangers fixed to treated timber wall plates (same sizes as joists), resin-bolted 75–100 mm minimum into sound walls at 600– 800 mm centres using approved 12–16 mm-diameter stainless-steel fixings – to be agreed with building control. (Provide damp-proof course/tanking system and water plug to seal between bolted timber wall plates and

**5.38** Domestic garage and basement conversions into habitable rooms and new dwellings

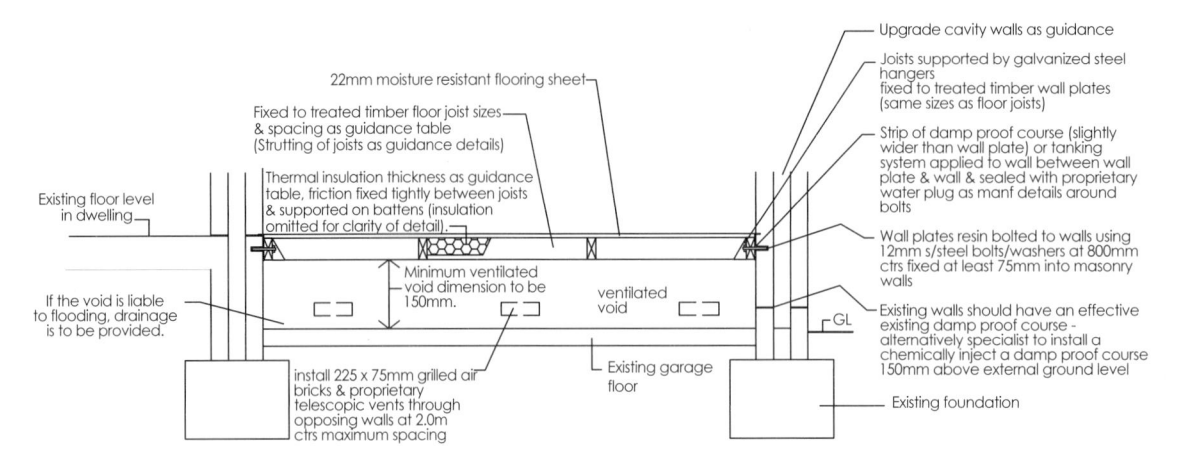

**Figure 5.26:** Suspended timber floor laid over existing floor void (section detail – *not to scale*)

masonry walls where there is no damp-proof course.) Floor joists are to be provided with one row of 38 × ¾ depth solid strutting at ends between joist hangers, or proprietary galvanised struts to BS EN 10327 fixed as manufacturer's details, at mid-span for 2.5–4.5 m spans, and two rows at 1/3 centres for spans over 4.5 m. Joists are to be doubled up and bolted together for trimmers, under partitions and baths.

Thermal insulation is to be friction-fixed tightly between joists (supported on timber battens fixed to sides of joists), as domestic extension guidance table. Note: A lesser insulation standard may be acceptable where new work affects existing adjoining floor levels, subject to approval by building control.

Fix 22 mm-thick, moisture-resistant, tongue-and-grooved timber floorboards laid with joints staggered, long edge fixed across the joists and all joints positioned over joists/noggins. All boards are to be glued and screwed to floor joists with all joints glued (using waterproof glue) and pinned, in accordance with the floorboard manufacturer's details and current BS EN standards. Allow an expansion gap around wall perimeters as manufacturer's details (typically 10–15 mm).

**Note:** Where basic radon gas protection is required by Building Control, apply a British Board of Agreement (BBA or other third-party accredited) proprietary application to prevent the passage of radon gas into the building through the walls and floor, as manufacturer's details. Where full radon gas protection is required by building control, additional radon protection is to be carried out in accordance with a radon gas specialist's details.

## A4: Basement conversions into habitable rooms

## Existing basements and tanking systems

### Site investigation and risk assessment

These are to be carried out before works commence to establish: ground conditions and water-table levels, presence of any contaminates and radon gas, including location of drains and services, etc.

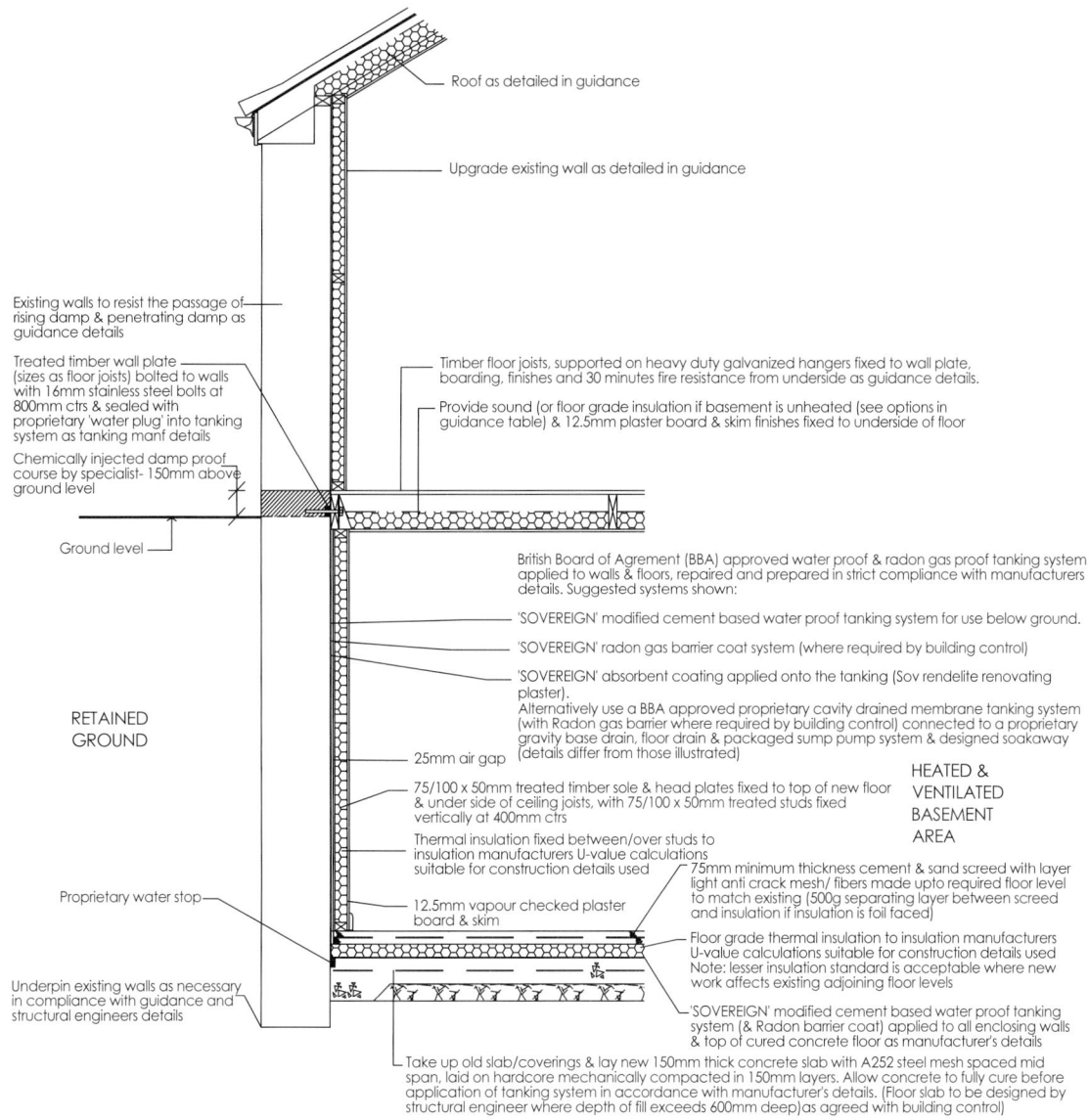

**Figure 5.27:** Tanking of existing basement (section detail – *not to scale*)

### Existing basement sub-structure

This is to be checked for structural stability by a suitably qualified specialist (i.e. a structural engineer) and any remedial works are to be carried out as necessary, in compliance with a suitably qualified specialist's details before tanking/conversion works commence.

### Tanking systems

Tanking systems providing either barrier, structural or drained protection to the building must be assessed, designed and installed in compliance with BS 8102: 2009 Code of Practice for Protection of Below Ground Structures against Water from the Ground.

The illustrated tanking section details in Figure 5.27 above are suggested details only, and actual details must be approved by building control before works commence on site. Other forms of barrier protection tanking must be suitable for existing basement situations.

Suitable tanking systems are to have BBA or other approved accreditation and must be assessed by a tanking specialist as suitable for the proposed situation.

Tanking systems must be designed/installed/applied by a tanking specialist, in compliance with the tanking manufacturer's details and specification for a particular project, so as to resist the passage of water and be vapour-permeable when applied above ground to prevent condensation/interstitial condensation within the building. Additional proprietary protective application to prevent radon gas from entering the building will be required in radon-affected areas (this may affect the vapour permeability of the tanking system). Tanking systems are to be properly connected to, and made continuous with, wall damp-proof courses/radon dpc trays. Perforation of the tanking system by service entry pipes etc. should be avoided, or else carried out strictly in accordance with the tanking manufacturer's details.

**Important note:** The risk of condensation with any tanking system should be assessed by a specialist, a condensation risk analysis should be carried out, and the tanking system should be designed and installed to prevent any potential condensation and mould problems.

### Ventilation and means of escape

These are to be carried out in compliance with domestic extension guidance details.

### Cavity closers

Proprietary British Board of Agreement (BBA or other third-party accredited) acoustic/thermally insulated/fire-resistant cavity closers, or similar, are to be provided to all cavity openings/closings, tops of walls and junctions with other properties, in accordance with the manufacturer's details.

Tops of cavity walls can be closed to prevent the passage of fire using a proprietary British Board of Agreement (BBA or other third-party accredited) 30-minutes fire-resistant rigid board, fixed in accordance with the manufacturer's details.

## A5: Conversion of barns and similar buildings into new dwellings – Technical and practical guidance (see Figures 5.28–5.36)

## Part A: Structure

(i) Assessing the feasibility of your building for conversion should be carried out in accordance with the details given in the guidance details above on domestic garages, basements and barn conversions.

(ii) Conversion works carried out to the building should be in accordance with the guidance details below.

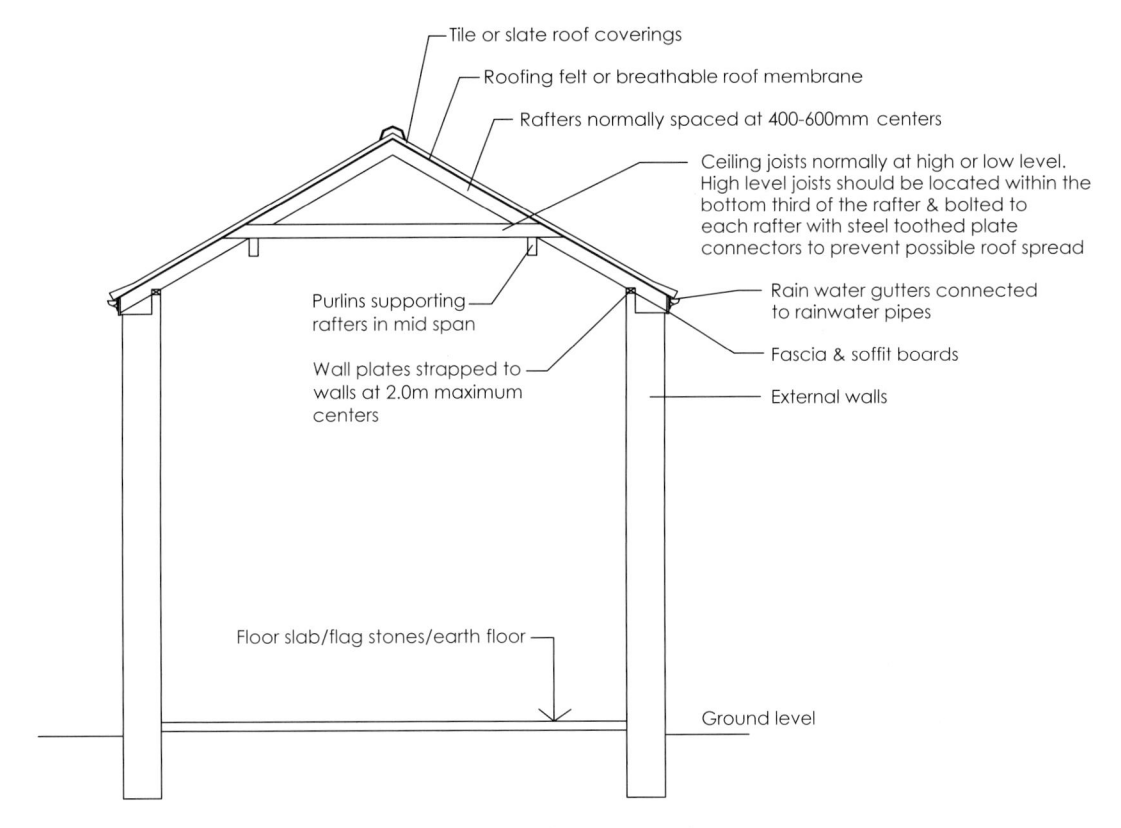

**Figure 5.28:** Typical section through an existing building *(not to scale)*

Labels in figure:
- Tile or slate roof coverings
- Roofing felt or breathable roof membrane
- Rafters normally spaced at 400-600mm centers
- Ceiling joists normally at high or low level. High level joists should be located within the bottom third of the rafter & bolted to each rafter with steel toothed plate connectors to prevent possible roof spread
- Purlins supporting rafters in mid span
- Rain water gutters connected to rainwater pipes
- Fascia & soffit boards
- Wall plates strapped to walls at 2.0m maximum centers
- External walls
- Floor slab/flag stones/earth floor
- Ground level

(iii) Any new building works should be carried out in accordance with the guidance for domestic extensions above.

## Upgrading pitched roofs

### *Existing roof coverings and roofing felt*

These should be in a sound and weather-tight condition; any defective coverings/felt should be replaced with matching new/existing sound coverings, in compliance with Section 2 of these guidance details and the manufacturer's details.

### *Existing roof structure*

All existing timbers should be in a sound condition, and where necessary defective timber is to be replaced with new in compliance with Section 2 of these guidance details. Existing timbers are to be inspected and where necessary treated against insect and fungal attack by a suitably qualified and experienced specialist.

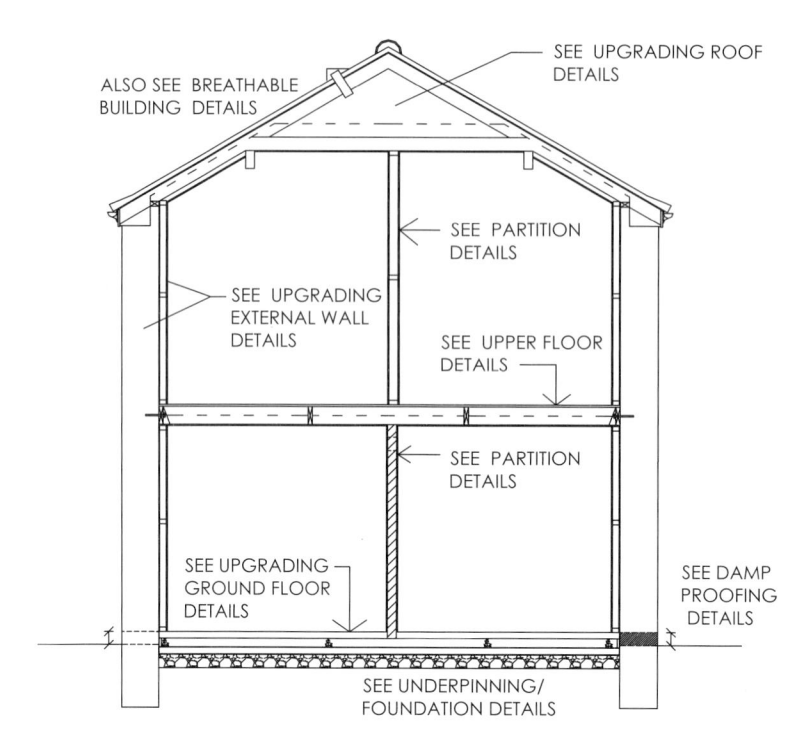

**Figure 5.29:** Section through proposed building conversion – details are indicated in diagrams below *(not to scale)*

## *Thermal upgrade of insulation and ventilation levels*

This is to be made to the existing roof as follows:

*Insulation at ceiling level* that has a threshold U-value worse than 0.35 should be improved to a U-value of 0.16 for pitched roof with insulation at ceiling level as follows: New thermal insulation is to be fixed between and over ceiling joists as domestic extension guidance details. Install eaves ventilation (low-level ventilation) equal to a continuous 10 mm air gap to both sides of roof; if not, British Board of Agreement (BBA or other third-party accredited) approved breathable-type felt, using proprietary ventilators fitted with insect screens, fixed in accordance with the manufacturer's details.

*Insulation at rafter level* that has a threshold U-value worse than 0.35 should be improved to a U-value of 0.18 for insulation fixed at rafter level as follows: New thermal insulation is to be fixed between and under rafters as Section 2 of these guidance details.

*Eaves ventilation* (low-level ventilation) should be installed to both sides of the roof equal to a continuous 25 mm air gap for the length of the roof, and also at ridge level (high-level ventilation) equal to a continuous 5 mm air gap for the length of the roof, using proprietary ventilators fitted with insect screens, fixed in accordance with the manufacturer's details; or, alternatively, replace roofing felt using a British Board of Agreement (BBA or other third-party accredited) approved breathable-type felt, fixed and ventilated as the manufacturer's details.

Overhaul and repair/replace defective rainwater goods and facia/barge/soffit boards as necessary and as guidance details.

Proprietary high level roof vents to be installed where insulation follows slope of roof- equal to a continuous 5mm air gap with insect screen. Note: high level vents is not required where insulation is at ceiling level (or where breathable roof membrane is used- subject to manufacturers details)

Roof insulation fixed at ceiling level see options in guidance

Roof insulation fixed at rafter level see options in guidance

50mm clear air gap (or 25mm air gap if using a vapour permeable membrane)

Renew rain water goods & facia boards etc as necessary as detailed in guidance

Install eaves ventilation equal to a continuous 25mm air gap with insect screen reduced to 10mm gap if insulation is at ceiling level. Note: eaves vents may not be required where a breathable roof membrane is used- subject to manufacturers details

Existing roof coverings should be inspected and retained providing they meet the requirements of this guidance. New/replacement works should be carried out in compliance with details in this guidance

Existing roof structure should be inspected and retained providing it meets the requirements of this guidance. New/ replacement works should be carried out in compliance with details in this guidance

Thermally upgrade existing roof which have a threshold U-value worse than 0.35, to a U-value of 0.16 for pitched roofs with insulation at ceiling level and 0.18 for pitched roofs with insulation at rafter level, and flat roofs with insulation at ceiling level as guidance details.

Thermally upgrade existing walls which have a threshold U-value worse than 0.7, to a U-value 0.3* for solid walls (or 0.55 for cavity walls) as follows :

Continuous 25mm minimum ventilated air gap provided between masonry wall and insulation to allow evaporation of any moisture (subject to any fire stopping requirement) as agreed with building control.

50/75 x 50mm timber sole & head plates fixed to top of new floor & under side of existing ceiling joists, with 50/75 x 50mm studs fixed vertically at 400mm ctrs

Thermal insulation fixed between/over studs as guidance table*

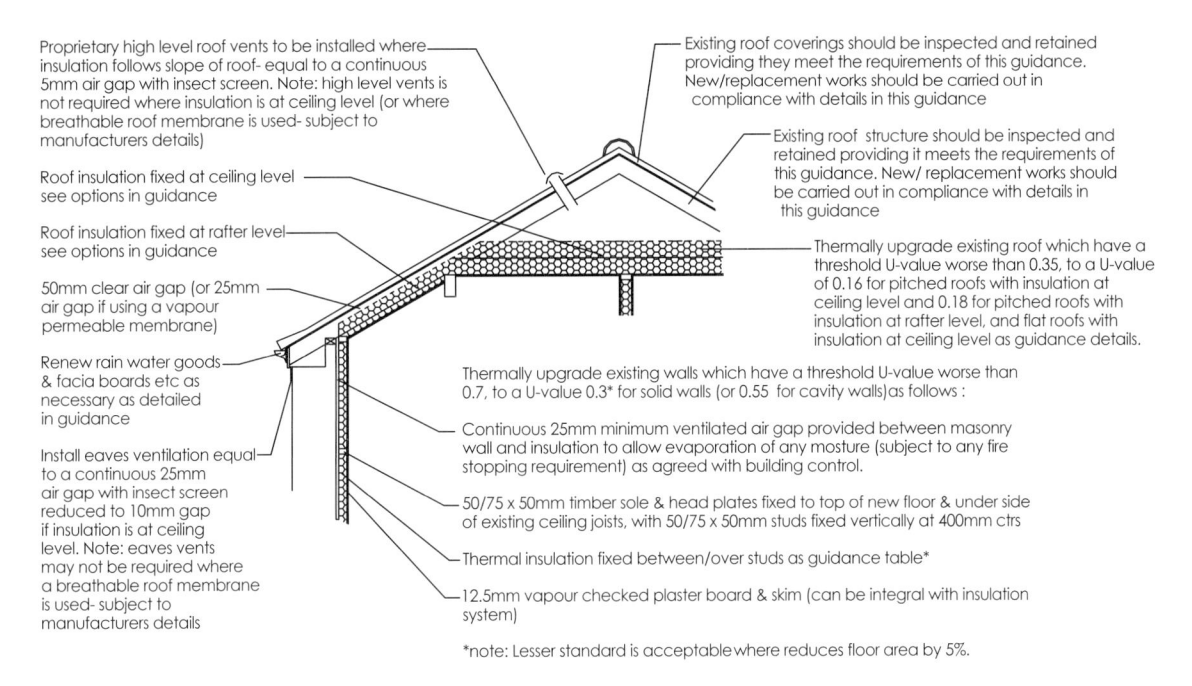

12.5mm vapour checked plaster board & skim (can be integral with insulation system)

*note: Lesser standard is acceptable where reduces floor area by 5%.

**Figure 5.30:** Upgrading pitched roof and external walls (section detail – *not to scale*).
Note: For upgrading of flat roofs, see Figure 5.7 above.

## Upgrading thermal insulation to existing solid masonry walls

Thermally upgrade existing solid masonry walls that have a threshold U-value worse than 0.7 to a U-value of 0.3, as follows:

- 50/75 (depth to suit actual thickness of insulation used) × 50 mm treated timber sole and head plates to be fixed to top of new/existing floor and underside of existing/new ceiling joists, infilled with 50/75 (depth to suit actual thickness of insulation used) × 50 mm treated timber studs fixed vertically at 400 mm centres.
- The new inner wall construction should allow for the passage of a 25 mm minimum ventilated air space (with fire-stopping where necessary) between the existing solid masonry walls and the face of the new stud wall/wall insulation to allow evaporation of any moisture, using details specified by a suitably qualified and experienced person or as agreed with building control.
- Thermal insulation is to be fixed between/over stud partition in accordance with the manufacturer's details and as in Table 5.3 below. *Note: A lesser insulation standard may be acceptable where the floor area is reduced by 5 per cent, subject to approval by building control.
- Insulation and vapour check integral with plasterboard or 12.5 mm-thick vapour-checked plasterboard finish is to have all junctions taped with scrim and finished with 3–5 mm board finish plaster, in accordance with the manufacturer's details.
- Upgrading of cavity walls is detailed in Figure 5.22.

## New upper floors supported by existing walls (see Figure 5.31)

Floor is to be constructed of kiln-dried structural-grade timber joists with sizes and spacing suitable for the proposed clear spans, in compliance with guidance span tables (Table 2.20). Joists are to be

**Table 5.3:** Upgrading thermal insulation to existing solid masonry walls

| Insulation product | K value | Minimum thickness (mm) |
|---|---|---|
| **U-value 0.3 W/m²k** | | |
| Kingspan Kootherm K18 | 0.020 | 72.5 mm fixed across studs with integral vapour-checked plaster board and skim finish |
| Celotex FR5000 and Celotex PL4000 | 0.021 / 0.022 | 50 mm between 50 mm-deep studs and 25 mm fixed across studs with integral vapour-checked plasterboard and skim finish |
| Celotex FR5000 and Celotex PL4000 | 0.021 / 0.022 | 40 mm between 75 mm-deep studs and 25 mm fixed across studs with integral vapour-checked plasterboard and skim finish |
| **Natural breathable insulations/ boards and finishes*** | | See 'breathable buildings' detailed in guidance. |

Note: *Natural breathable insulations, wall boards and finishes must be specified by a suitably qualified and experienced conservation specialist and subject to building control approval. Some natural breathable insulations/boards and finishes can also be applied externally. For further information, product advice, specification and supply of products, contact Ty-Mawr at: www.lime.org.uk.

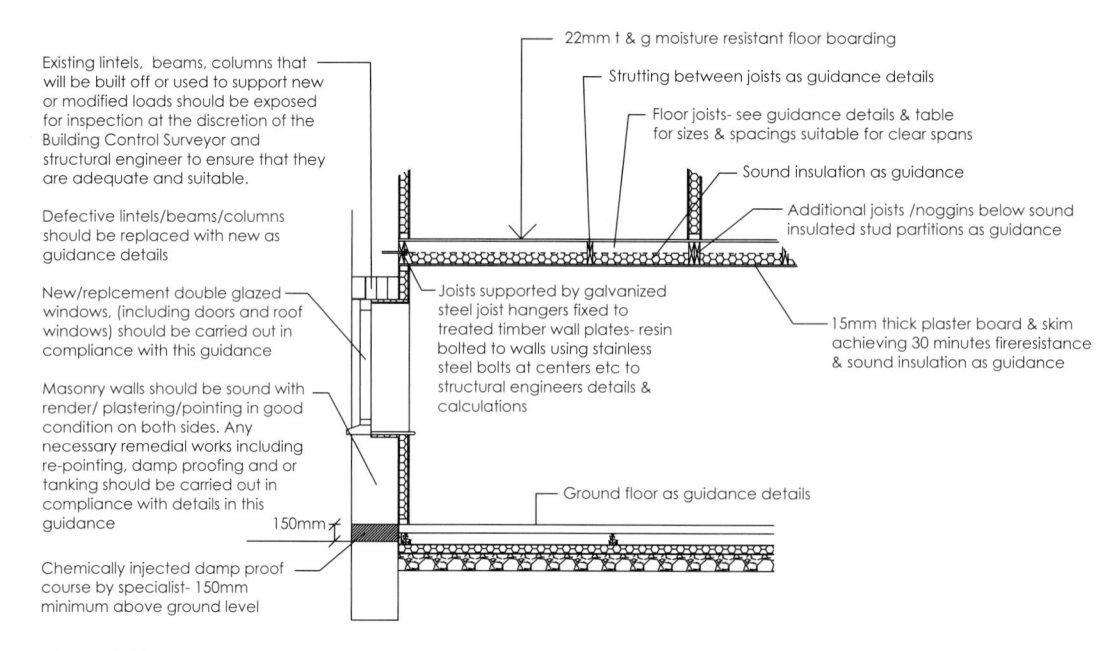

**Figure 5.31:** New upper floors supported by wall plates and existing walls (section detail – *not to scale*)

supported by heavy-duty proprietary galvanised metal restraint joist hangers built into walls or fixed to treated timber wall plates (wall plate sizes to be the same sizes as joists), resin-bolted 100–150 mm minimum into sound walls at 600–800 mm centres using approved 16 mm-diameter stainless-steel fixings: as agreed with building control. Joists are to be doubled up and bolted together for trimmers, under partitions and baths.

Floor void between joists is to be insulated with a minimum thickness of 100 mm of 10 kg/m³ proprietary sound insulation quilt; ceiling is to be minimum 15 mm plasterboard and skim.

Fix 22 mm-thick, moisture-resistant, tongue-and-grooved timber floorboards laid with joints staggered, long edge fixed across the joists and all joints positioned over joists/noggins. All boards are to be glued and screwed to floor joists, with all joints glued (using waterproof glue) and pinned, in accordance with the floorboard manufacturer's details and current BS EN standards. Allow an expansion gap around wall perimeters as manufacturer's details (typically 10–15 mm).

Floor joists are to be provided with one row of 38 × ¾ depth solid strutting at ends between joist hangers or proprietary galvanised struts to BS EN 10327, fixed as manufacturer's details, at midspan for 2.5–4.5 m spans, and two rows at 1/3 centres for spans over 4.5 m.

**Note:** Where required by building control, the new floor construction should allow for the passage of a ventilated air space (with fire-stopping where necessary) between the existing solid masonry walls and new wall insulation to allow evaporation of any moisture, using details specified by a suitably qualified and experienced person and as agreed with building control.

## New upper floors supported by load-bearing ground-floor walls and foundations (see Figure 5.32)

Construct new concrete foundation and integral ground floor (as Figure 5.32 below) in compliance with a structural engineer's details and calculations, to support 100 mm minimum width load-bearing insulation block wall with thermal insulation, stainless-steel wall ties, 50 mm minimum clear cavity and internal finishes as guidance details for new cavity walls, to support upper floor joists.

Upper floor is to be constructed of kiln-dried structural-grade timber joists with sizes and spacing suitable for the proposed clear span, in compliance with domestic extension guidance table (Table 2.20). Joists are to be built into walls using approved proprietary sealed joist caps, or sealed with silicon sealant to provide an airtight seal – to be agreed with building control. Joists are to be doubled up and bolted together for trimmers, under partitions and baths.

Floor void between joists is to be insulated with a minimum thickness of 100 mm of 10 kg/m³ proprietary sound insulation quilt; ceiling is to be minimum 15 mm plasterboard and skim.

Fix 22 mm-thick, moisture-resistant, tongue-and-grooved timber floorboards laid with joints staggered, long edge fixed across the joists and all joints positioned over joists/noggins. All boards are to be glued and screwed to floor joists, with all joints glued (using waterproof glue) and pinned, in accordance with the floorboard manufacturer's details and current BS EN standards. Allow an expansion gap around wall perimeters as manufacturer's details (typically 10–15 mm).

Floor joists are to be provided with one row of 38 × ¾ depth solid strutting at ends between joist hangers, or proprietary galvanised struts to BS EN 10327 fixed as manufacturer's details, at midspan for 2.5–4.5 m spans, and two rows at 1/3 centres for spans over 4.5 m.

### *Upgrading existing ground floors*

Thermally upgrade existing floors which have a threshold U-value worse than 0.7, to a U-value of 0.25 using guidance details for upgrading garage floors above, or alternatively take up and lay new floor as detailed below.

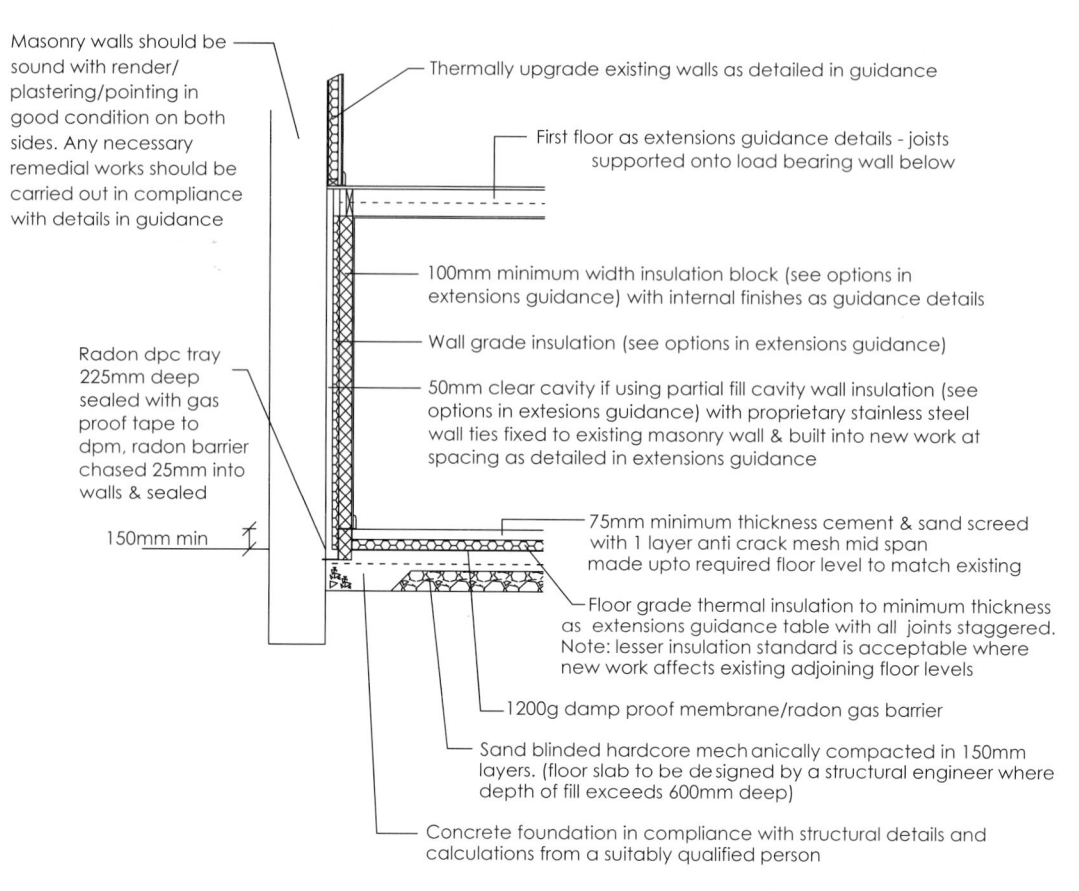

**5.46** Domestic garage and basement conversions into habitable rooms and new dwellings

**Figure 5.32:** Upper floors supported by load-bearing ground-floor walls and foundations (section detail – *not to scale*)

### *Take up existing floor and lay new concrete floor* (see Figure 5.33)

Take up existing concrete floor slab and remove off-site; lay sand-blinded hardcore, mechanically compacted in 150 mm layers (floor slab to be designed by a suitably qualified person where depth of fill exceeds 600 mm). 1200 g (300 micrometre) continuous polythene damp-proof membrane (dpm) and radon gas-proof barrier are to be laid over sand-blinded hardcore, lapped and sealed at all joints and linked to damp-proof courses (dpcs) in walls. Sealing of joints in the barrier and sealing around service penetrations are also required with radon gas-proof tape, in compliance with Part C of this guidance. Alternatively, where no dpc exists in external/internal walls that are in contact with the ground, provide a suitable British Board of Agreement (BBA or other third-party accredited) approved, chemically injected damp-proof course/tanking system to prevent passage of moisture (and passage of radon gas as necessary in radon gas-affected areas), as manufacturer's details, to all internal walls extending behind the dpm down to formation level – as agreed with building control.

Floor-grade insulation is to be laid over dpm, minimum thickness and type in accordance with domestic extension guidance table, including 25 mm-thick insulated up-stands between slab and external walls. Minimum 100 mm-thick ST2 or Gen1 concrete floor slab with a trowel-smooth surface ready for finishes is to be laid over insulation. (Note: 500 g polythene separating layer is to

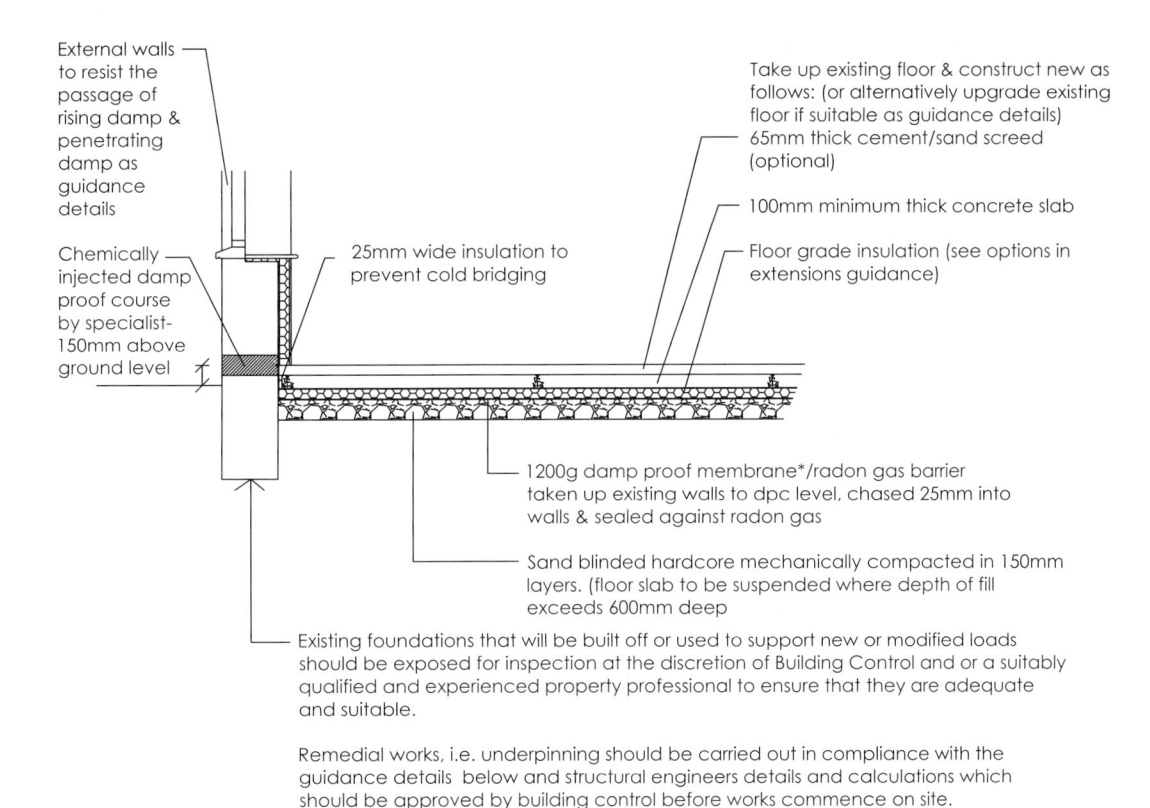

**Figure 5.33:** Take up existing floor and lay new concrete floor (section detail – *not to scale*)

Labels in figure:
- External walls to resist the passage of rising damp & penetrating damp as guidance details
- Chemically injected damp proof course by specialist-150mm above ground level
- 25mm wide insulation to prevent cold bridging
- Take up existing floor & construct new as follows: (or alternatively upgrade existing floor if suitable as guidance details) 65mm thick cement/sand screed (optional)
- 100mm minimum thick concrete slab
- Floor grade insulation (see options in extensions guidance)
- 1200g damp proof membrane*/radon gas barrier taken up existing walls to dpc level, chased 25mm into walls & sealed against radon gas
- Sand blinded hardcore mechanically compacted in 150mm layers. (floor slab to be suspended where depth of fill exceeds 600mm deep
- Existing foundations that will be built off or used to support new or modified loads should be exposed for inspection at the discretion of Building Control and or a suitably qualified and experienced property professional to ensure that they are adequate and suitable.
- Remedial works, i.e. underpinning should be carried out in compliance with the guidance details below and structural engineers details and calculations which should be approved by building control before works commence on site.

be installed between the concrete slab and insulation if using a foil-faced polyurethane/PIR-type insulation board.) Insulation is to be omitted and concrete thickness increased in areas where non-load-bearing partitions are built off the floor slab. Lay 65 mm minimum thickness cement and sand floor-grade screed bonded to floor slab, adjusted to match existing floor levels as necessary.

**Note:** Where full radon protection is required by building control, a proprietary radon gas sump should be installed below the concrete floor slab (which can be at located at the edge or side of a building) and fitted with a depressurisation pipe up-stand and signage, as detailed in Part C of this guidance.

### *Alternative options for upgrading ground floors*

Thermally upgrade existing floors that have a threshold U-value worse than 0.7 to a U-value of 0.25, using guidance details for upgrading garage floors above.

### Tanking of existing sub-structure walls with higher external ground levels (see Figure 5.34)

Higher external ground levels should be reduced at least 150 mm below internal floor levels where possible, and/or French drain/proprietary surface water drain laid. Existing walls should be underpinned as necessary in accordance with a structural engineer's details.

Where higher external ground levels cannot be reduced, the following guidance should apply.

### Existing concrete floor slab

This is to be retained if of 100 mm minimum thickness and in good condition; or alternatively, take up and lay new 150 mm-thick concrete slab with A252 steel mesh spaced mid-span, laid on hardcore mechanically compacted in 150 mm layers (floor slab is to be designed by a structural engineer where depth of fill exceeds 600 mm), as agreed with building control.

### Existing walls

They are to be structurally stable and should be able to resist the passage of rising damp and penetrating damp into the building (including radon/methane gas in affected areas), as guidance details and agreed with building control. Install a British Board of Agreement (BBA or other third-party accredited) chemically injected damp-proof course (or other damp-proofing system with accreditation approved by building control) as necessary, installed by a specialist, 150 mm above external ground level.

### Tanking systems

Tanking systems providing either barrier, structural or drained protection to the building must be assessed, designed and installed in compliance with BS 8102: 2009 Code of Practice for Protection of Below Ground Structures against Water from the Ground. The illustrated tanking section details in Figure 5.34 below are suggested details only and the actual details must be approved by building control before works commence on site. Suitable tanking systems are to have British Board of Agreement (BBA or other third-party) accreditation and be assessed by a tanking specialist as suitable for the proposed situation. Tanking systems must be designed/installed/applied by a tanking specialist in compliance with the tanking manufacturer's details and specification for a particular project, to resist the passage of water, and be vapour-permeable to prevent condensation/ interstitial condensation within the building, and where required provide additional proprietary protective application to prevent radon gas and other such ground gases and contaminates from entering the building (this may affect the vapour permeability of the tanking system). Tanking systems must be properly connected to, and made continuous with, wall damp-proof courses/radon dpc trays. Perforation of the tanking system by service entry pipes etc. should be avoided or carried out strictly in accordance with the tanking manufacturer's details.

### Floor-grade thermal insulation

This should be laid over the tanking system to minimum thickness, as domestic extension guidance tables, with all joints staggered. Note: A lesser insulation standard is acceptable where new work affects existing adjoining floor levels, as agreed with building control.

### Floor screed

75 mm sand/cement thick structural screed (mix between 1:3–1:4½ ) is to be laid over insulation with trowel-smooth finish ready for finishes, with one layer light mesh mid-span (or fibres in screed), made up to required floor level to match existing. Screed area should be limited to room sizes; floor areas exceeding 40 m$^2$ should have expansion/contraction joints as detailed in the note below. Screed

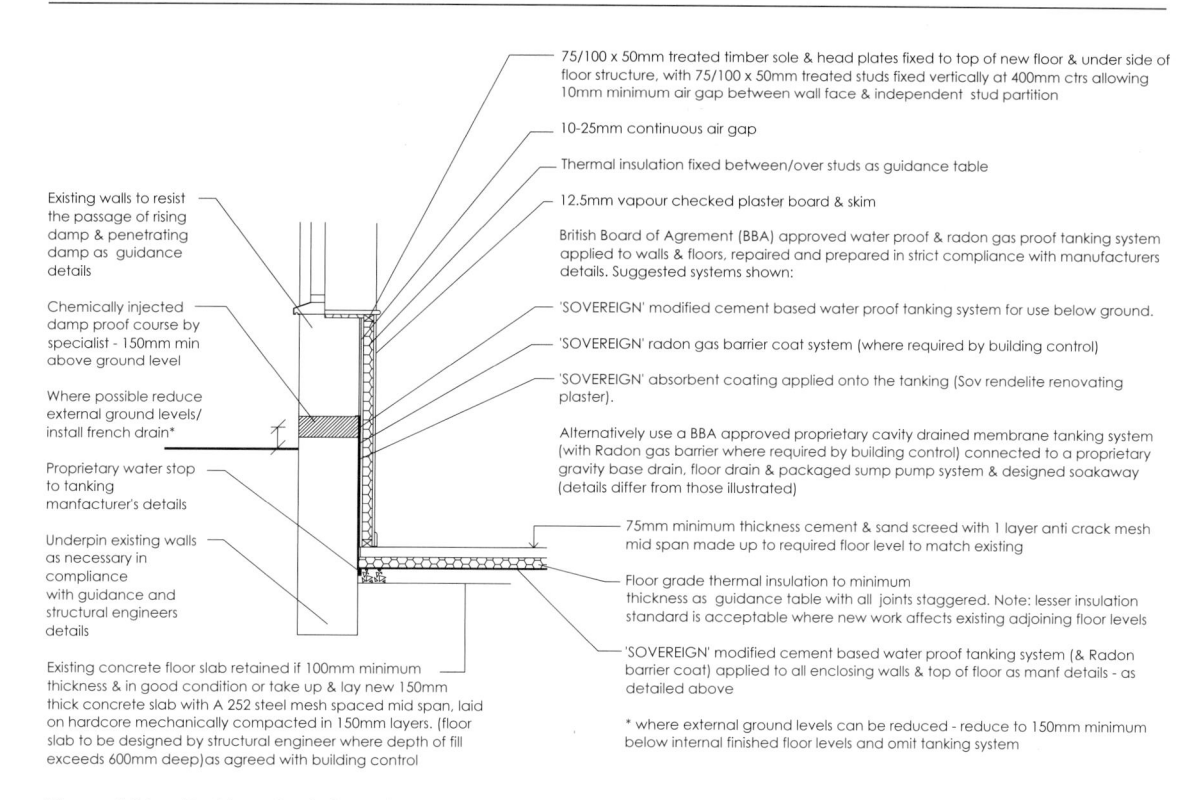

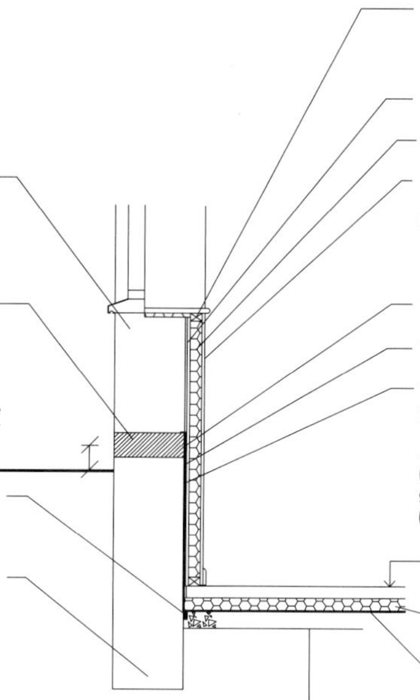
75/100 x 50mm treated timber sole & head plates fixed to top of new floor & under side of floor structure, with 75/100 x 50mm treated studs fixed vertically at 400mm ctrs allowing 10mm minimum air gap between wall face & independent stud partition
10-25mm continuous air gap
Thermal insulation fixed between/over studs as guidance table
12.5mm vapour checked plaster board & skim
British Board of Agrement (BBA) approved water proof & radon gas proof tanking system applied to walls & floors, repaired and prepared in strict compliance with manufacturers details. Suggested systems shown:
'SOVEREIGN' modified cement based water proof tanking system for use below ground.
'SOVEREIGN' radon gas barrier coat system (where required by building control)
'SOVEREIGN' absorbent coating applied onto the tanking (Sov rendelite renovating plaster).
Alternatively use a BBA approved proprietary cavity drained membrane tanking system (with Radon gas barrier where required by building control) connected to a proprietary gravity base drain, floor drain & packaged sump pump system & designed soakaway (details differ from those illustrated)
75mm minimum thickness cement & sand screed with 1 layer anti crack mesh mid span made up to required floor level to match existing
Floor grade thermal insulation to minimum thickness as guidance table with all joints staggered. Note: lesser insulation standard is acceptable where new work affects existing adjoining floor levels
'SOVEREIGN' modified cement based water proof tanking system (& Radon barrier coat) applied to all enclosing walls & top of floor as manf details - as detailed above
* where external ground levels can be reduced - reduce to 150mm minimum below internal finished floor levels and omit tanking system
Existing walls to resist the passage of rising damp & penetrating damp as guidance details
Chemically injected damp proof course by specialist - 150mm min above ground level
Where possible reduce external ground levels/ install french drain*
Proprietary water stop to tanking manfacturer's details
Underpin existing walls as necessary in compliance with guidance and structural engineers details
Existing concrete floor slab retained if 100mm minimum thickness & in good condition or take up & lay new 150mm thick concrete slab with A 252 steel mesh spaced mid span, laid on hardcore mechanically compacted in 150mm layers. (floor slab to be designed by structural engineer where depth of fill exceeds 600mm deep)as agreed with building control

**Figure 5.34:** Tanking of existing sub-structure with higher external ground levels (section detail – *not to scale*)

is to be laid over insulation with a trowel-smooth surface ready for finishes. (A 500 g polythene separating layer is to be installed between the screed and insulation if using a foil-faced polyurethane/PIR-type insulation board.) Insulation is to be omitted in areas where non-load-bearing partitions are built off the floor slab.

**Note:** Where a proprietary under-floor heating system is installed it should be fixed above insulation and under the screed layer, in compliance with the heating pipe manufacturer's/heating specialist's details. Screeds over under-floor heating should be subdivided into bays not exceeding 40 $m^2$ in area. Expansion/contraction joints in screeds should be consistent with joints in slabs.

## First-floor walls built off existing ground-floor walls (see Figures 5.35 and 5.36)

An assessment of existing walls, lintels and foundations etc. is to be made and any necessary remedial works to stabilise the wall are to be carried out in compliance with a suitably qualified person's details.

Shutter and cast a 225 mm minimum thickness in-situ concrete ring beam over the full wall width, with one layer of B503 steel mesh spaced 50 mm above top of wall, in compliance with a suitably qualified person's details.

New external wall is to be constructed as guidance details for new cavity walls, built off a continuous damp-proof course tray (allow for weep holes at 900 mm centres).

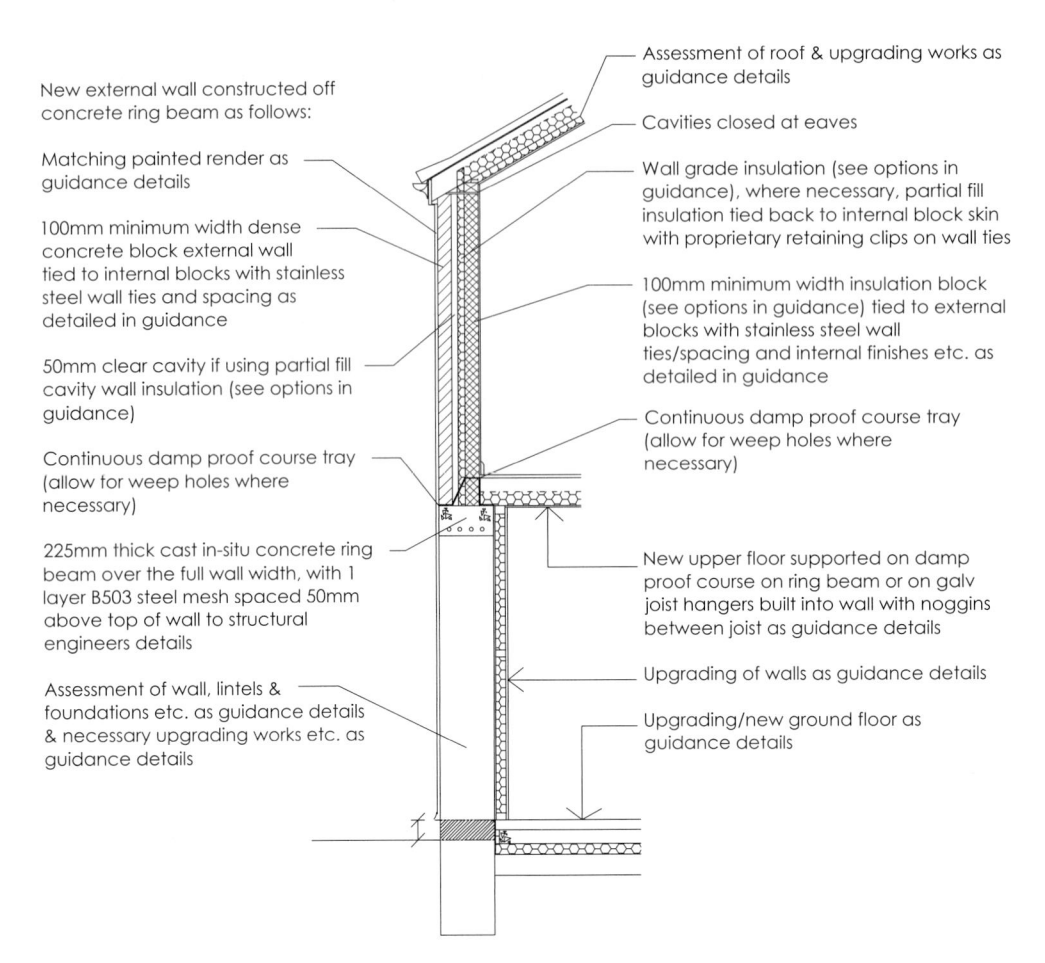

**Figure 5.35:** First-floor walls built off existing ground-floor walls (section detail – *not to scale*)

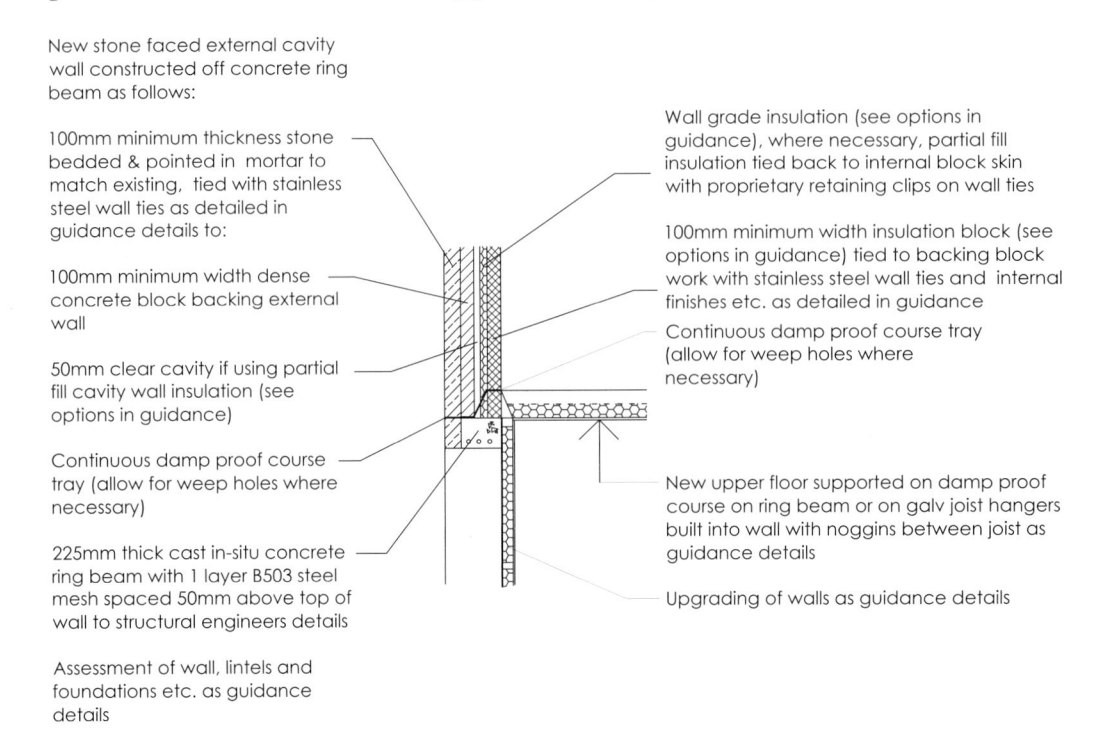

**Figure 5.36:** First-floor walls with stone facings built off existing ground-floor walls (section detail – *not to scale*)

New upper floor is to be supported on damp-proof course on ring beam or on galvanised joist hangers built into wall with noggins between joists, as guidance details.

## PART B: FIRE SAFETY AND MEANS OF ESCAPE

Follow the guidance details in Section 2 of this guidance for Domestic Extensions.

## PART C: SITE PREPARATION AND RESISTANCE TO CONTAMINANTS AND MOISTURE

Follow the guidance details in Section 2 of this guidance for Domestic Extensions.

## PART D: CAVITY WALL FILLING WITH INSULATION

Follow the guidance details in Section 2 of this guidance for Domestic Extensions.

## PART E: RESISTANCE TO THE PASSAGE OF SOUND

### Performance standards

Follow the details in Section 2 of this guidance for Domestic Extensions, with additional requirements as follows.

### Exemptions/relaxations for conversion of historic buildings into dwellings/flats

If the proposed sound insulation requirements will unacceptably alter the character or appearance of a historic/listed building, then the sound insulation standards may be exempt or improved to what is reasonably practical, or to an acceptable standard that would not prejudice the character or increase the risk of deterioration of the building fabric or fittings, in consultation with the local planning authority's conservation officer. A notice showing the actual sound test results is to be fixed in a conspicuous place inside the building, in compliance with Paragraph 0.7 of ADE.

### Party walls and floors separating conversion of buildings into new dwellings/flats

Sound insulation details for new party walls and floors are to be carried out in accordance with the relevant details contained within this guidance and Approved Document E, as follows:

- 43 dB minimum values for airborne sound insulation to walls, floors and stairs; and
- 64 dB maximum values for Impact sound insulation to floors and stairs.

### Pre-completion sound testing of party walls and floors

Robust details are to be adopted or pre-completion sound testing, to be carried out to demonstrate compliance with ADE1 as follows:

Pre-completion sound testing is to be carried out by a suitably qualified person or specialist with appropriate third-party accreditation (UKAS or ANC registration), to demonstrate compliance with ADE1, and a copy of the test results is to be sent to building control.

### *Testing rate requirements*

At least one set of recorded tests is required for every 10 new dwellings on the same site, and additional testing may be required where requested by building control.

### *Remedial works and retesting*

Remedial works and retesting will be required where the test has failed, in compliance with Section 1 of ADE.

### New internal walls and floors

Sound insulation details between new internal walls and floors separating bedrooms, or a room containing a WC and other rooms, is to be carried out in accordance with the relevant details contained within this guidance and Approved Document E.

### Existing internal walls and floors

Existing internal walls and floors may already achieve the performance standards required by Approved Document E without the need for remedial works, as agreed with building control before works commence.

## PART F: VENTILATION TO NEW DWELLINGS

Follow the details in Section 3 of this guidance for New Dwellings. Note: Air pressure testing of dwellings and flats formed by a material change of use is not required

## PART G: SANITATION, HOT-WATER SAFETY AND WATER EFFICIENCY

Follow the guidance details in Section 3 of this guidance for New Dwellings

## PART H: DRAINAGE AND WASTE DISPOSAL

Follow the guidance details in Section 2 of this guidance for Domestic Extensions.

## PART J: COMBUSTION APPLIANCES AND FUEL STORAGE SYSTEMS

Follow the guidance details in Section 2 of this guidance for Domestic Extensions.

## PART K: PROTECTION FROM FALLING, COLLISION AND IMPACT

Follow the guidance details in Section 2 of this guidance for Domestic Extensions.

## PART L: CONSERVATION OF FUEL AND POWER IN CONVERSIONS

Follow the details in Section 2 of this guidance for Domestic Extensions, with the following additional requirement.

### Energy Performance Certificate (EPC)

Energy Performance Certificates (EPC) are required for new-built dwellings (see section 3 above) and when a material change of use occurs in specified circumstances – in which a building or part of a building that was previously used for one purpose group will be used for another – for example, conversion of an agricultural barn into a dwelling. Where there is a material change of use and the conversion/modification includes the provision or extension of any fixed services for heating, hot water, air conditioning or mechanical ventilation, an EPC will be required as follows:

(i) Within 5 days after the work has been completed: the person carrying out the work shall give an energy performance certificate (EPC) for the building to the building owner and give notice to building control that the EPC has been issued, including the reference number under which the EPC has been registered, in accordance with the building regulations 2010 (as amended). The EPC must contain information in accordance with the building regulations 2010 (as amended).

Note: From April 2008 all new dwellings (and conversions into dwellings) must have an Energy Performance Certificate (EPC) to fulfil Building Regulations requirements, and these can only be produced by an accredited Energy Assessor.

## PART M: ACCESS TO AND USE OF BUILDINGS FOR DISABLED

Follow the guidance details in Section 2 of this guidance for Domestic Extensions.

## PART N: SAFETY GLAZING, OPENING AND CLEANING

Follow the guidance details in Section 2 of this guidance for Domestic Extensions.

## PART P: ELECTRICAL SAFETY

Follow the guidance details in Section 2 of this guidance for Domestic Extensions.

# Section 6  Upgrading old buildings using lime and modern applications

| | |
|---|---|
| Upgrading old buildings using lime and modern applications | 6.3 |
|     Traditional 'breathing' construction | 6.3 |
|     Inappropriate maintenance of old buildings | 6.3 |
|     Modern applications to maintain old stone/brick buildings | 6.3 |
| Re-pointing and repair of existing buildings | 6.4 |
|     Examination of the existing masonry wall/render/plaster finishes | 6.5 |
|         Analysis of the mortar/render | 6.6 |
|     Re-pointing of existing stone/brick walls | 6.6 |
|     Repair/rebuilding of existing stone/brick walls | 6.8 |
|     Repair and replacement of external render/internal plaster to walls | 6.9 |
| Types of lime mortar, lime renders/plaster and decorative finish suitable for breathable buildings | 6.10 |
|     1. Non-hydraulic lime (known as lime putty or fat lime) | 6.10 |
|         Non-hydraulic lime mortar mixes | 6.10 |

*Guide to Building Control: For Domestic Buildings*, First Edition. Anthony Gwynne.
© 2013 John Wiley & Sons, Ltd. Published 2013 by John Wiley & Sons, Ltd.

| | |
|---|---|
| 2. Hydraulic lime | 6.11 |
|     Hydraulic lime mortar mixes | 6.11 |
|     Non-hydraulic/hydraulic lime render/plaster mixes | 6.12 |
|     Type of lime binder and number of coats | 6.12 |
|     Haired lime plaster | 6.12 |
|     Pozzolanic materials for lime plaster | 6.12 |
| 3. Breathable paints | 6.12 |
|     Limewash (and shelter coat) | 6.12 |
|     Clay paint | 6.16 |
|     Plant-based/natural emulsion wall paints | 6.16 |
|     Mineral-based wall paints | 6.16 |
|     Clear protective coatings | 6.16 |
| Application of coatings/paints/systems, protection, storage and aftercare | 6.17 |

# Upgrading old buildings using lime and modern applications

Extracts in this section (including text, figures and tables) from the Ty-Mawr Lime Handbook and associated publications are reproduced by permission of Ty-Mawr Lime Ltd.

Caring appropriately for old buildings requires an understanding of how they were constructed and how they function; only then is it possible to identify the right materials and repair them. This is particularly important when dealing with heritage or similar existing buildings that require a sympathetic approach. Certain works carried out to existing buildings may be regarded as repairs and will not require Building Regulations approval; however, other works may be regarded as 'building work' and will require such approval. If in doubt, you are advised to contact building control, as the proposed works may have to comply with modern standards in accordance with the current Building Regulations, and such works should be specified and carried out by a suitably qualified and experienced conservation specialist. You are also advised to contact the local authority's conservation officer for your area, as consent may be required for any proposed repairs/works to listed buildings and buildings in conservation areas.

## Traditional 'breathing' construction

Old buildings were traditionally constructed with technologies, handed down through generations, that allowed the building to breathe naturally. The building fabric was constructed in natural materials, typically with solid walls providing good permeability and flexibility. External surfaces were designed to deflect the rain, while penetrating and rising damp were absorbed by the structure, which allowed the moisture to evaporate away naturally through the porous surfaces. Natural ventilation was provided through gaps in poorly fitting windows and doors and through chimneys, keeping the building in a state of equilibrium, as indicated by Figure 6.1 below.

## Inappropriate maintenance of old buildings

When old buildings are maintained and upgraded with inappropriate modern, hard, impervious materials, membranes and finishes, these can trap moisture and potentially lead to the deterioration of the building fabric and finishes. Additional problems can occur with condensation and mould growth caused by high levels of water vapour produced by the occupants, and lack of natural ventilation caused by the sealing up of gaps and blocking up of open flues and chimneys, as indicated by Figure 6.2 below.

## Modern applications to maintain old stone/brick buildings

Old stone/brick buildings can be maintained and upgraded using a mixture of traditional and appropriate new technologies, which allow the building to breathe naturally, typically as detailed in Figures 6.3 and 6.4 below, and such works should be specified and carried out by a suitably qualified and experienced conservation specialist.

Assessment of the building should be carried out and any moisture problems should be remedied prior to refurbishing and upgrading the building. Advice about lime, aggregates, breathable insulation and breathable paint products, condensation risk analysis, U-value calculations and product specification advice can be obtained from Ty-Mawr at: www.lime.org.uk.

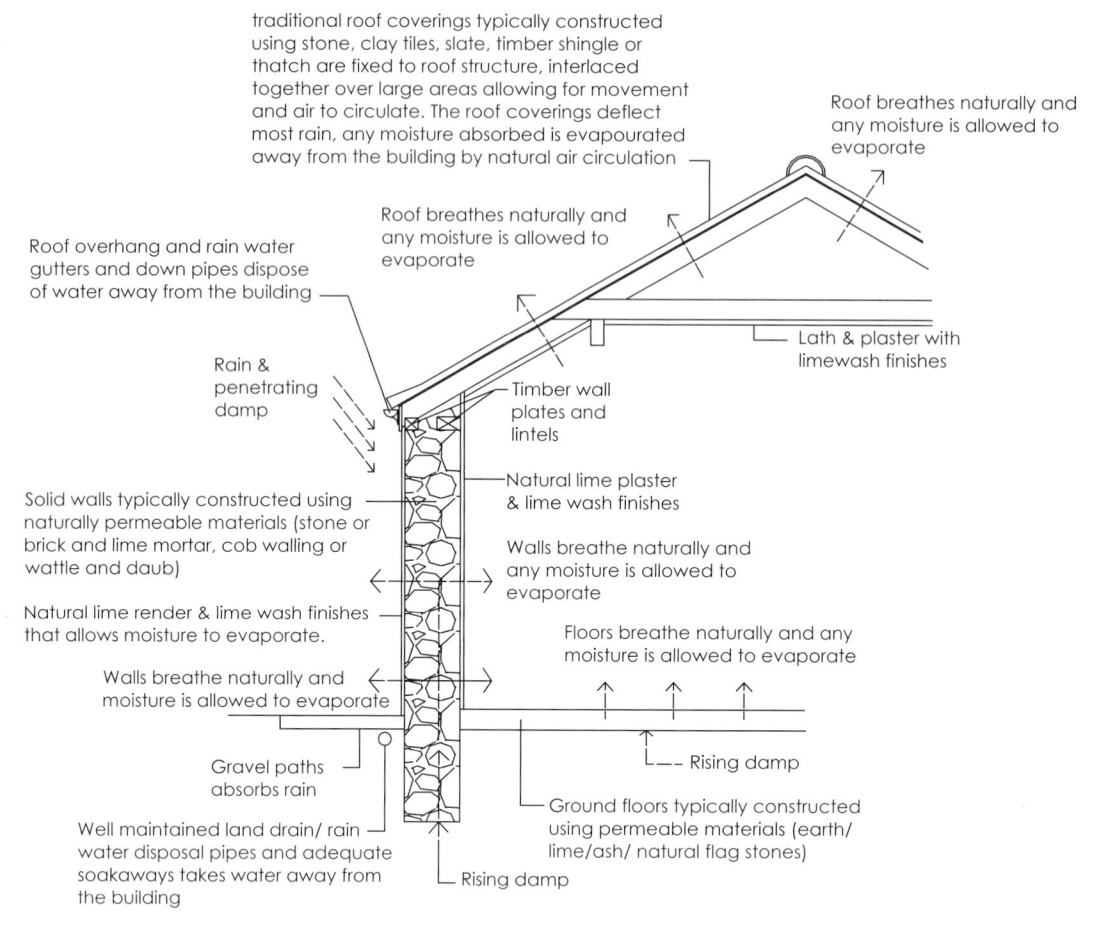

## 6.4 Upgrading old buildings using lime and modern applications

**Figure 6.1:** Typical section through a traditional 'breathing' building *(not to scale)* (Reproduced by permission of Ty-Mawr Lime Ltd.)

You are advised to contact building control before works commence, as the proposed works may require Building Regulations approval and may have to comply with modern standards, in accordance with current Building Regulations. For example, in the conversion of buildings into dwellings you will have to consider how you will prevent the passage of moisture into the building and, where necessary, provide protection against radon gas (further information is provided in Section 5 of this guidance for barn conversions). You are also advised to contact your local authority's planning officer and conservation officer before works commence, as the proposed works may require planning permission/listed building/conservation area consent.

## Re-pointing and repair of existing buildings

Cement mortars or renders and gypsum plasters are normally used in modern buildings that have cavity or solid walls supported on foundations and where strong, rapid sets are required. Cement and gypsum binders have poor permeability and flexibility, which makes them unsuitable for older, traditionally constructed breathable buildings.

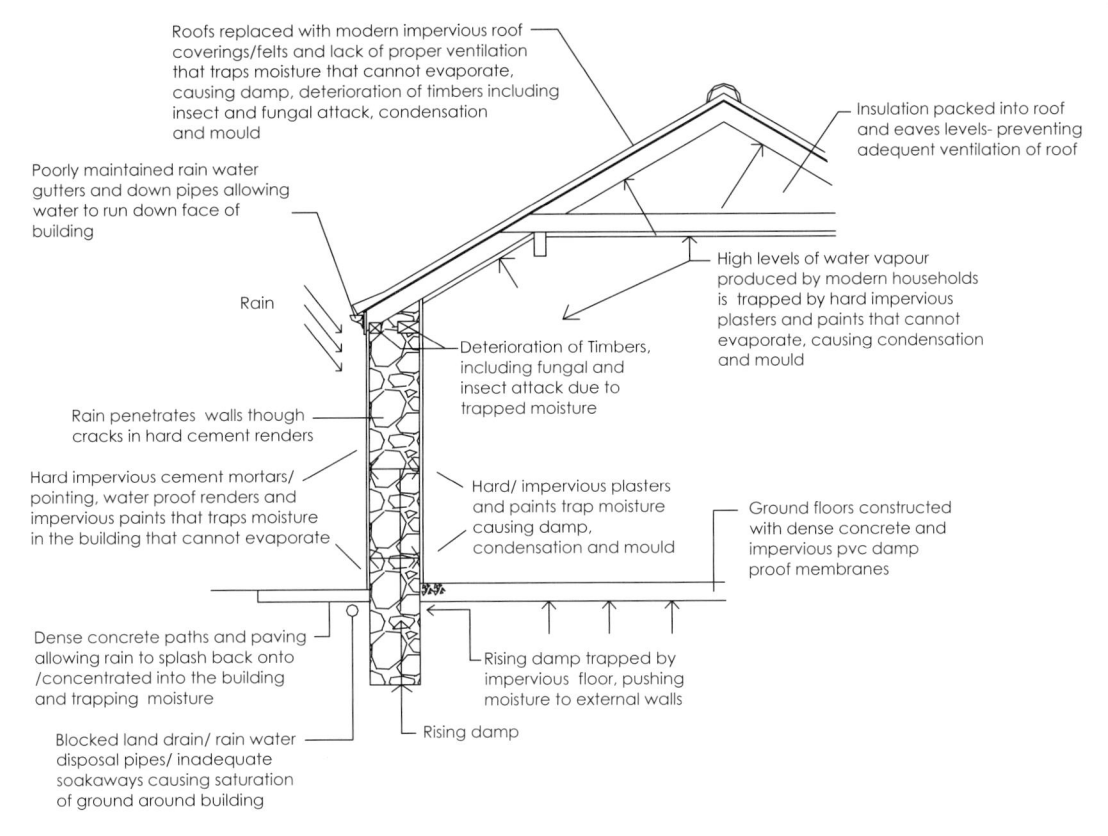

**Figure 6.2:** Inappropriate maintenance to old buildings (section detail – *not to scale*) (Reproduced by permission of Ty-Mawr Lime Ltd.)

Normally, pre-1919 buildings are constructed with solid walls and often without substantial foundations. These buildings require mortars, plasters and renders that are more flexible, and breathable lime binders should be used in mortars and renders/plasters to allow water within the walls, either from penetrating or rising damp, to be released by evaporation. This process controls damp and condensation within the building.

A suitably qualified and experienced conservation specialist should be consulted for all aspects of the proposed works, and work should be carried out by a suitably experienced conservation bricklayer/stone mason/plasterer.

## Examination of the existing masonry wall/render/plaster finishes

Assessment of the existing wall/finishes should be carried out to determine:

- Nature of the masonry units (stone/brick, soft and porous, or hard and dense)
- Type of mortar/render/plaster (lime, cement, gypsum) and type and constituents of aggregates, and reinforcement of render/plaster finishes
- Construction method, including type of bond, width and finish of joints.
- Nature of defect (mortar weathered out of joints, cracking, water penetration, etc.).

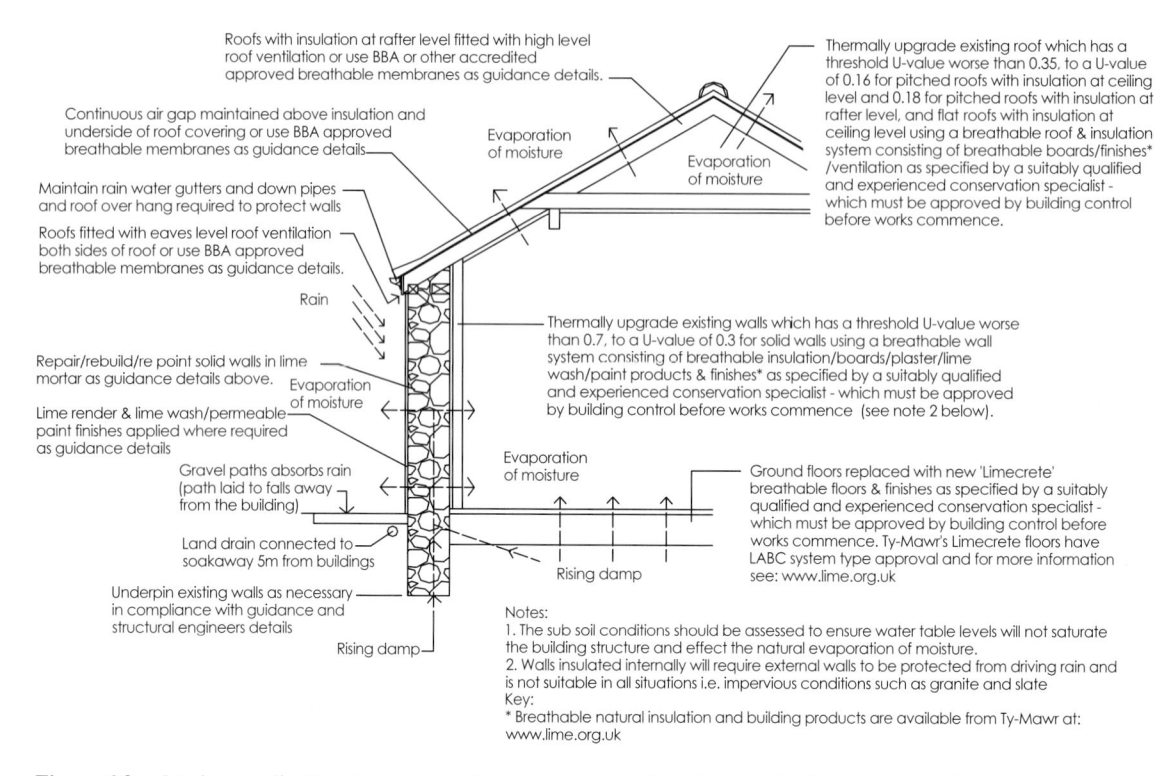

**6.6** Upgrading old buildings using lime and modern applications

**Figure 6.3:** Modern applications to maintain old stone/brick buildings (section detail – *not to scale*) (Reproduced by permission of Ty-Mawr Lime Ltd.)

### *Analysis of the mortar/render*

Analysis of the mortar for re-pointing, repair and rebuilding of masonry should be carried out by a suitably qualified and experienced conservation specialist to ensure the following:

- The binder and aggregate ratios match the existing mortar, including colour, texture and detailing.
- The mortar is softer in compressive strength than the masonry units and should be as porous as the existing mortar.

Analysis can be carried out visually or by laboratory examination (dissolution analysis), depending on the level of information required, e.g. for a listed building. The exact mix should be selected after the presentation of sample panels by the contractor. This service can be carried out by Ty-Mawr; see www.lime.org.uk.

### Re-pointing of existing stone/brick walls

Re-pointing of walls is required to refill outer parts of the joints where the original pointing has weathered out, as indicated on Figure 6.5 below.

Rake out and remove existing defective mortar as necessary to a minimum depth equal to twice the joint thickness to form a square-backed recess, using hand tools – i.e. plugging chisels or similar

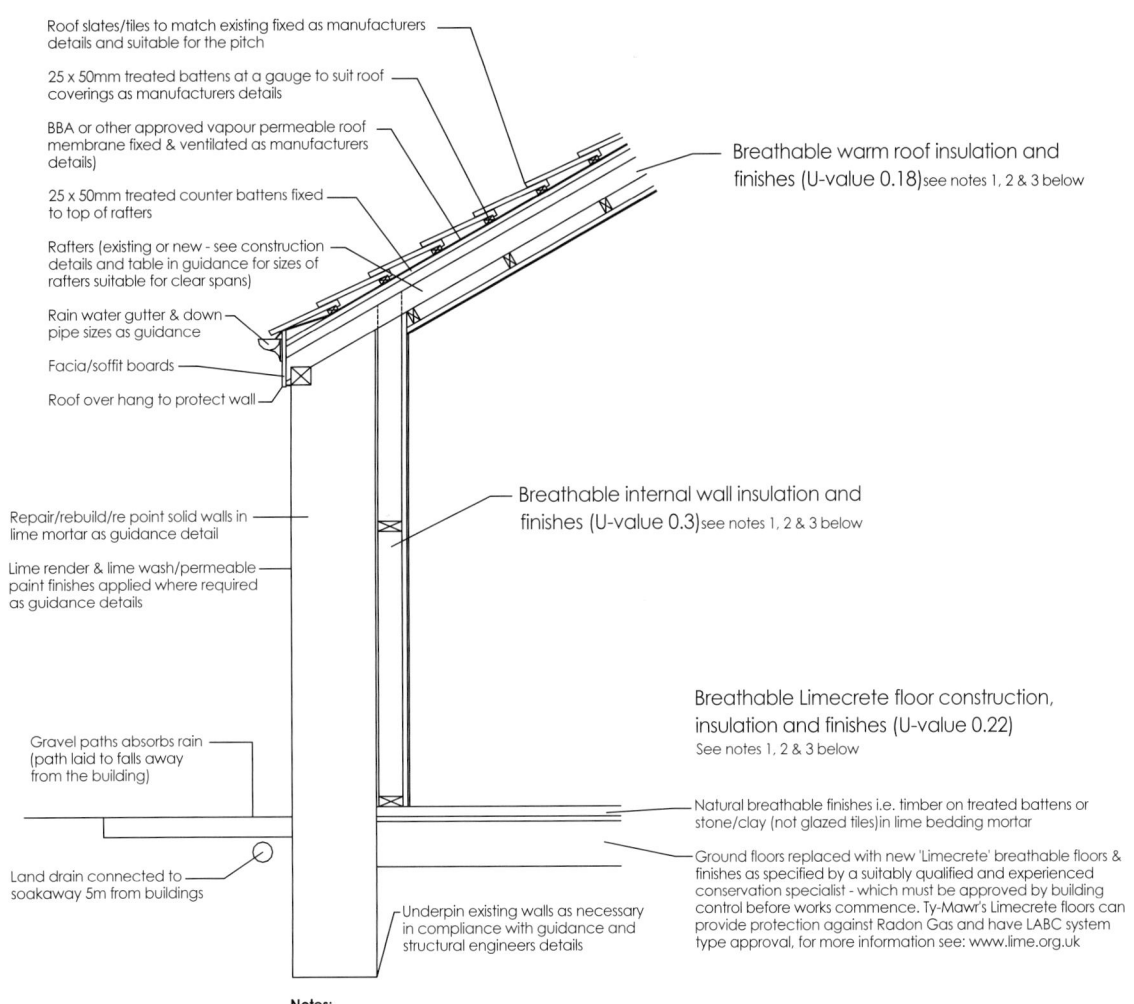

**Figure 6.4:** Upgrading thermal insulation to old stone/brick buildings using breathable insulations and finishes (section detail – *not to scale*) (Reproduced by permission of Ty-Mawr Lime Ltd.)

(power tools such as disc cutters or similar mechanical devices should not be used if there is any possibility of damage to masonry units). Brush out all joints clean of all dust and loose material, using suitable non-ferrous bristle brushes, and thoroughly flush out with clean water but not saturated.

Joints should be fully packed with mortar, pointed (slightly proud of the face) as work proceeds using a pointing trowel, and left to carbonate. Apply mortar in layers (up to 10–15 mm to allow initial set, or it will crack and fail). Deep-pack pointing is similar to ordinary rake-out and re-pointing, except joints will be up to 200 mm deep and may involve the occasional removal

**6.8**  Upgrading old buildings using lime and modern applications

*Existing stone/brick wall with weathered face & joints*

*Rake out and remove existing defective mortar as necessary to a minimum depth equal to twice the joint width*

*Brush out joints clean off all dust and loose material using suitable non ferrous bristle brushes and thoroughly flush out with clean water avoiding unnecessary saturation.*

*Joints should be fully packed with mortar, pointed slightly proud of the face as work proceeds.*

*After the initial set, brush and beat the mortar back to the weathered stone/brick surface using a natural bristle brush to close the mortar joint against the weathered surface, remove any feather edges/excess, and expose the aggregate.*

*Protect the works against direct wind, sunlight, rain and frost etc*

*Old mortar is removed to form a square backed recess using hand tools*

*Existing mortar joints*

*Existing stone/brick wall*

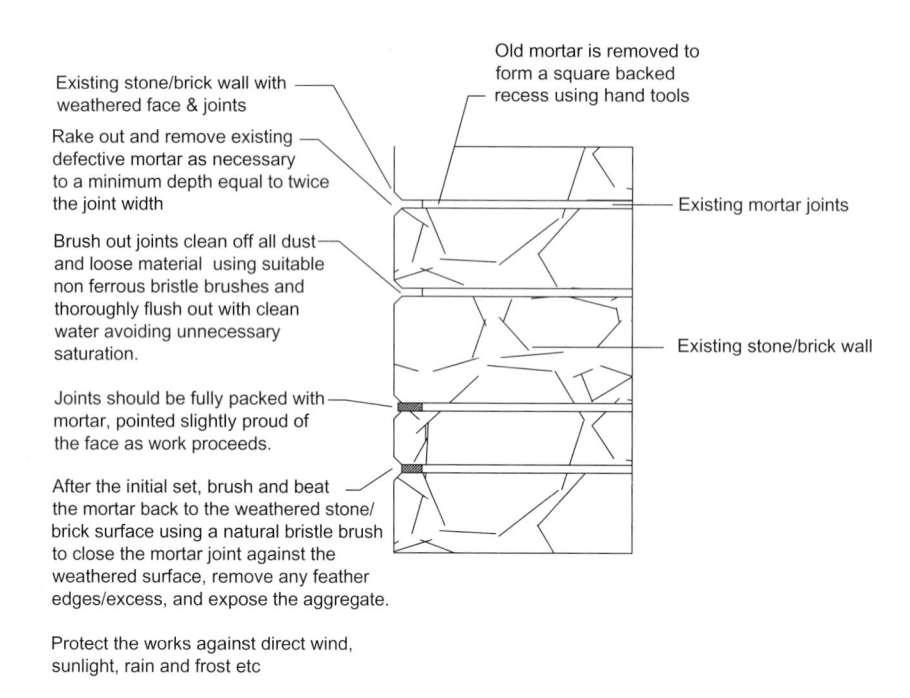

**Figure 6.5:**  Re-pointing of existing stone/brick walls (section detail – *not to scale*) (Reproduced by permission of Ty-Mawr Lime Ltd.)

and replacement of stones. Large holes should be packed out with small slivers of matching stone (termed 'galletting') or brick to prevent large joints and shrinkage. Do not overwork lime mortars.

Brush and beat the mortar joint surfaces to consolidate and close up any cracks. This should be done when the mortar can no longer be dented with the thumb, but can still be marked with the thumbnail. Protect the works as necessary from direct winds, sun, rain and frost using protective coverings, i.e. hessian, bubble wrap, polythene sheet, etc.

Do not use frozen materials, or lay on frozen surfaces. The set slows down below 10°C and stops at 5°C. If the temperature is 5°C or below on a falling thermometer, stop work.

Mortar mixes are to be in accordance with the guidance tables below and should be carefully selected to match the existing, including binder and aggregate ratios, colour, texture and detailing

### Repair/rebuilding of existing stone/brick walls

Repair and rebuilding of walls is required where the individual stone/brick units have become loose in the wall due to the mortar weathering out, or defects in the wall that have made the wall unstable. Remedial works are necessary to stabilise the wall and prevent the passage of moisture into the building.

Carefully cut out individual defective masonry units or take down defective areas of wall using hand tools (not power tools, which may damage masonry units) and set aside masonry units for reuse. Masonry must be stored clear of the ground, to avoid absorption of water and salts from the ground, and be protected from adverse weather.

Replacement stone/brick should follow the original coursing, bonding, wall line and joint profile and must be well bonded to the existing material. Second-hand materials must be sourced in a sound condition, free from cracks, fissures and defects, and must match the existing materials.

Dampen masonry before and during construction as necessary to control suction. Facing stone/brick is to commence not less than 150 mm below the finished level of external paving or soil levels. Lay stone/bricks on a full, even bed of mortar, with all joints filled and between 10 and 18 mm wide, recessed by double the width of the joint to allow re-pointing of the works when completed to match the existing.

Lay natural stones on their natural bed and evenly distribute different shapes, sizes and colours throughout the face of the wall to give a consistent overall appearance and good bond, with no long, contiguous vertical joints. For walls that are faced both sides, build in bonding stones of a length two-thirds the thickness of the wall, one to every square metre of each side of the wall and staggered. Build up the wall and point up as a separate process to ensure consistency; mortar should be left recessed to allow for re-pointing when building works have been completed. Do not overwork the mortar.

Pointing can then be undertaken as previously described. Protect the works as necessary from direct winds, sun, rain and frost using protective coverings, i.e. hessian/polythene. Wetting may also be required to ensure that the joints do not dry out too quickly and cause failures.

Do not use frozen materials, or lay on frozen surfaces. The set slows down below 10°C and stops at 5°C. If temperature is 5°C or below on a falling thermometer, stop work.

Mortar mixes are to be in accordance with the Tables below and should be carefully selected to match the existing material, including binder and aggregate ratios, colour, texture and detailing.

## Repair and replacement of external render/internal plaster to walls

Repair and rendering of existing walls are required where failure of the existing external rendered or internal plastered finishes has occurred, possibly due to; water penetration, (penetrating and rising damp), lack of maintenance, inadequate protection, inappropriate repairs, poor or incorrect materials and/or workmanship causing shrinkage cracking, loss of adhesion and surface defects, etc.

Hack off defective finishes at least 300 mm beyond the last defect or to existing joint lines; remaining sound finishes should be cut back to sound edges, undercut for good key. Remedy any structural deficiencies allowing walls to fully dry out if wet and re-point/repair existing stone/brick/lath walls as necessary as detailed in guidance; prepare walls and dub out deep hollows in maximum 8 mm-thick coats (or use hemp plaster for thicker coats internally where there are significant voids), ready to receive new render/plaster finishes. Do not use beads and stops unless specified; use proprietary manufactured stainless-steel beads and stops, fixed in accordance with the manufacturer's details where specified.

Thoroughly wet the wall to control suction of moisture before application of first render coat, form arisses and ensure correct alignment with all features as necessary. Apply external render/internal plaster coatings in an even, consistent and firm manner, to achieve good adhesion (ensure nibs are created behind timber laths where necessary). Use appropriate tools, finished to a true plane, with walls and reveals plumb and square unless otherwise specified. Provide key for next coat by combing render coats and cross-scratching plaster coats using appropriate tools, ensuring the undercoat is not penetrated. Note: Base coats should have fibre in the form of synthetic or natural hair. Finishes should be appropriate for the building in which they are being applied, e.g. a worker's cottage and a formal Georgian town house will have dramatically different finishes.

Keep each coat damp with polythene/hessian coverings or spray with water to prevent dry out; curing times are to be accordance with manufacturer's details for the type of material used. Allow each coat to dry and shrinkage to occur before applying next coat. (Rule of thumb – plaster must be hard enough not to indent with thumb, but soft enough to indent with a thumbnail.)

Render/plaster mixes are to match the existing ones in accordance with the Tables below, including binder and aggregate ratios, colour, texture and detailing.

## Types of lime mortar, lime render/plaster and decorative finish suitable for breathable buildings

There are two main types of lime binder used in mortars, renders and plasters: non hydraulic and hydraulic lime, as detailed below:

### 1. Non- hydraulic lime (known as lime putty or fat lime)

This consists of fairly pure limestone, burned in a factory process to drive off carbon dioxide; an excess of water is added to slake the resulting quicklime into a lime putty. It hardens by exposure to air in the presence of water, in order to carbonate, and over a long period of time it reverts back to limestone. Commercially produced non-hydraulic limes are available from Ty-Mawr at: www.lime.org.uk. Other producers are available nationwide.

#### *Non-hydraulic lime mortar mixes*

The lime putty is to be premixed with aggregates to match the existing mortar in the required ratio, depending on the type of stone or brick and degree of exposure and in accordance with Table 6.1 below. Turn, beat and ram the mortar as necessary to make it more plastic, without the addition of water in most cases. For walls to be rendered, leave the pointing finished 6 mm back from the stone/brick face to provide a key.

Non-hydraulic lime may be more appropriate for use on historic buildings where a slower set and softer mortar are required to provide maximum permeability and flexibility of the wall structure.

**Table 6.1:** Typical non-hydraulic lime putty mortar mixes

| Type of material in wall | Non hydraulic lime putty mortar mix (gauged) | |
|---|---|---|
| | **Sheltered application** Lime putty: Mortar | **Exposed application** Lime putty: Mortar |
| Stone/brick – poor durability | 1:3 | 1:2 |
| Stone/brick – medium durability | 1:3 (use hydraulic lime table for sandstone) | Use hydraulic lime table |
| Stone/brick – good durability | 1:3 (use hydraulic lime table for sandstone) | Use hydraulic lime table |
| Fine joints (up to 3 mm) | 1:1 | 1:1 |

**Notes:** The above mortar mixes are only suggested mixes and the actual mortar mix is to be specified by a suitably qualified and experienced conservation specialist to be suitable for the type of wall material and degree of exposure. The exact ratio will depend on the sand/aggregate used. The colour, texture and workability of the mortar are influenced by the sand/aggregate. The softer the stone/brick, the softer will be the mortar mix required.
Reproduced by permission of Ty-Mawr Lime Ltd.

Lime mortars can take from several months to a year to fully cure and should be left to weather naturally without the application of any artificial weathering, which may damage the mortar.

Only breathable paints as detailed in this guidance should be applied to breathable walls and breathable buildings.

Pozzolanic materials can be added to the non-hydraulic mortar mix to increase initial set times where specified/required, and carried out in strict consultation with an experienced conservation specialist's details. Note: Hydrated or bagged lime is normally used as a plasticiser and is added to a cement/mortar mix; it can be used as a mortar but not always with good results..

## 2. Hydraulic lime

This consists of limestone containing a natural proportion of clay in addition to calcium and magnesium carbonates; it is burned in a factory process to produce chemical compounds similar to Portland cement, which are stronger but less workable than non-hydraulic limes. It hardens by chemical reaction with water and by carbonation. The higher the percentage of natural clay and minerals in the lime, the higher the strength and the faster the initial set times will be, but the poorer the permeability and flexibility. Commercially produced hydraulic lime is available in varying compressive strengths and setting times for specific projects (which can vary between grades and producers), in strict consultation with an experienced conservation specialist's details. Natural Hydraulic Lime (NHL) should always be specified as some limes which are not natural can contain cement. Commercially produced hydraulic limes are available from Ty-Mawr; see www.lime.org.uk. Other producers are available nationwide.

### *Hydraulic lime mortar mixes*

The lime used for re-pointing and building has to be mixed with aggregates to match the existing mortar in the required ratio, depending on the type of stone or brick and degree of exposure and in accordance with Table 6.2 below.

**Table 6.2:** Typical hydraulic lime mortar mixes

| Type of material in wall | Hydraulic lime mortar mix (gauged) | |
|---|---|---|
| | **Sheltered application** **Lime: Mortar** | **Exposed application** **Lime: Mortar** |
| Stone/brick – poor durability | $1^{NHL2}$:3 or softer (use non-hydraulic lime table for limestones) | $1^{NHL2}$:3 or softer (use non-hydraulic lime table for limestones) |
| Stone/brick – medium durability | $1^{NHL3.5}$:3 | $1^{NHL3.5}$:2.5 |
| Stone/brick – good durability | $1^{NHL5}$:3 | $1^{NHL5}$:2.5 |
| Fine joints (up to 3 mm) | Use non-hydraulic lime table | Use non-hydraulic lime table |

**Key:** $^{NHL2}$ Natural hydraulic lime containing up to 12% clay (slow set); $^{NHL3.5}$ Natural hydraulic lime containing 12–18% clay (moderate set); $^{NHL5}$ Natural hydraulic lime containing up to 25% clay (faster set). All are natural hydraulic lime.
**Notes:** The above mortar mixes are only suggested mixes and the actual mortar mix is to be specified by a suitably qualified and experienced conservation specialist to be suitable for the type of wall material and degree of exposure. The exact ratio will depend on the sand/aggregate used. The colour, texture and workability of the mortar are influenced by the sand/aggregate. The softer the stone/brick, the softer will be the mortar mix required.
Reproduced by permission of Ty-Mawr Lime Ltd.

Mortar must not be allowed to dry out too quickly and surrounding masonry must be kept damp. Pointing should be kept moist for seven days – the carbonation set can complete only in the presence of moisture. Building can be carried out at the same rate as for Portland cement, depending on the hydraulic lime used and the weather conditions.

Hydraulic lime is more appropriate where a strong, rapid set is required. Lime mortars can take several months to fully cure and should be left to weather naturally. The application of artificial weathering finishes may reduce the life of the mortar.

Only breathable paints, as detailed in this guidance, should be applied to breathable walls and breathable buildings.

## Non-hydraulic/hydraulic lime render/plaster mixes

### Type of lime binder and number of coats

Non-hydraulic or hydraulic lime used for external renders and internal plasters should be suitable for the wall type and degree of exposure and mixed with aggregates (to match the existing where necessary), in accordance with the Tables below: Lime render/plasters can take several months to fully cure and should be left to weather down naturally without the application of any artificial weathering, which could damage the render/plaster. Only breathable paints as detailed in this guidance should be used on breathable renders/plasters and should be applied in accordance with the paint manufacturer's details.

### Haired lime plaster

This incorporates hair/fibre (typically goat's hair/horsehair, synthetic hair or other approved material at 1.5 kg per tonne) to provide tensile strength where necessary, cut into 50 mm lengths and added (teased) into the mix in the proportion/ratio as specified by a suitably qualified and experienced conservation specialist. Note: Synthetic hair is often used in premixed plasters, as natural hair will degrade in uncarbonated lime after a few weeks.

### Pozzolanic materials for lime plaster

Pozzolanic materials containing silica and alumina such as brick dust, pulverised fuel ash (PFA) and calcined clay can be added to non-hydraulic lime putty (also known as fat lime) where necessary to increase the setting time to be similar to that of hydraulic lime. The type and ratio of pozzolanic material is to be specified by a suitably qualified and experienced conservation specialist.

## 3. Breathable paints

All solid stone/brick walls with lime render/plaster finishes should be decorated with a breathable finish as follows.

### Limewash (and shelter coat)

Limewash is suitable for internal and external surfaces. It is made using a high-calcium (fat) lime putty and is commercially available premixed from specialist manufacturers/suppliers. Limewash is

Table 6.3: Lime render/plaster mixes (suggested mixes)

| Wall construction[o] | Internal plaster or External render | Base/levelling coat(s) | Top/finishing coat (top/finishing coat should not be harder than the base coat) | Number and thickness of base/levelling coat(s) | Number and thickness of top coat |
|---|---|---|---|---|---|
| Cob, rammed earth, straw bale[1] (haired base coats) | Internal plaster | Fat Lime Base Coat Plaster | Fat Lime Top Coat Plaster | 2 × 9 mm | 1 × 3 mm |
|  | External render | Fat Lime Base Coat Plaster or Hydraulic Lime NHL2 | Fat Lime Top Coat Plaster or Hydraulic Lime NHL2 | 2 × 9mm  2 × 9mm | 1 × 6mm  1 × 6mm |
| Reed mat, reed board (haired base coat) | Internal plaster** | Fat Lime Plaster for Boards | Fat Lime Top Coat Plaster | 2 × 9 mm | 1 × 3 mm |
|  | External render | N/A | N/A |  |  |
| Celenit** walls and ceilings | Internal** walls and ceilings | Fat Lime Plaster for Boards (un haired) or Hydraulic Lime NHL3.5/ NHL2 (with beach aggregate) | Fat Lime Top Coat Plaster or Hydraulic Lime NHL 3.5 | 2 × 6mm  1 × 6mm | 1 × 3 mm  1 × 6 mm |
| Wood Wool boards (mesh base coat) | External render | Hydraulic Lime NHL3.5 (with beach aggregate) | Hydraulic Lime NHL 3.5 | 1 × 9 mm | 1 × 9 mm |
| Wood fibre boards (mesh base coat) | Internal plaster** | Fat Lime Plaster for Boards or Hydraulic Lime NHL3.5 (with beach aggregate) | Fat Lime Top Coat Plaster  Hydraulic Lime NHL 3.5 | 2 × 6mm  1 × 6mm | 1 × 3 mm  1 × 6 mm |
|  | External render | Hydraulic Lime NHL3.5 (with beach aggregate) | Hydraulic Lime NHL 3.5 | 1 × 9 mm | 1 × 9 mm |
| Lath (internal only) or soft stone (haired base coats) | Internal plaster | Fat Lime Base Coat Plaster | Fat Lime Top Coat Plaster | 2 × 9 mm | 1 × 3 mm |
|  | External render | Hydraulic Lime NHL 3.5/ NHL2 | Fat Lime Top Coat Plaster or Hydraulic Lime NHL 3.5 / NHL2 | 2 × 9 mm | 1 × 6 mm  1 × 6 mm |

(Continued)

**Table 6.3:** *(Continued)*

| Wall construction[o] | Internal plaster or External render | Base/levelling coat(s) | Number and thickness of base/levelling coat(s) | Top/finishing coat (top/finishing coat should not be harder than the base coat) | Number and thickness of top coat |
|---|---|---|---|---|---|
| Soft brick (haired base coats) | Internal plaster | Fat Lime Base Coat Plaster | 2 × 9 mm | Fat Lime Top Coat Plaster | 1 × 3 mm |
| | External render | Fat Lime Base Coat Plaster or Hydraulic Lime NHL 3.5/ NHL2 | 2 × 9 mm 2 × 9 mm | Fat Lime Top Coat Plaster or Hydraulic Lime NHL 3.5 / NHL2 | 1 × 6 mm 1 × 6 mm |
| Hard stone (haired base coats) | Internal plaster | Hydraulic Lime NHL 2 | 2 × 9 mm | Fat Lime Top Coat Plaster | 1 × 3 mm |
| | External render | Hydraulic Lime NHL 3.5 | 2 × 9 mm | Hydraulic Lime NHL 3.5 | 1 × 6 mm |
| Hard engineering brick or dense concrete blocks (10 mm mesh or haired base coat) | Internal plaster | Hydraulic Lime NHL 3.5/ NHL2 | 2 × 9 mm | Fat Lime Top Coat Plaster or Hydraulic Lime NHL 3.5 / NHL2 | 1 × 3 mm 1 × 6 mm |
| | External render | Hydraulic Lime NHL 3.5 | 2 × 9 mm | Hydraulic Lime NHL 3.5 / NHL2 | 1 × 6 mm |
| Insulation blocks* (10 mm mesh or haired base coat) | Internal plaster | Hydraulic Lime NHL 3.5/ NHL2 | 2 × 9 mm | Fat Lime Top Coat Plaster | 1 × 3 mm |
| | External render | Hydraulic Lime NHL 3.5 | 2 × 9 mm | Hydraulic Lime NHL 3.5 / NHL2 | 1 × 6 mm |

**Key:** Fat Lime = non hydraulic lime;
NHL 1 or 2: Natural hydraulic lime containing up to 12% clay (slow set);
NHL 3.5: Natural hydraulic lime containing 12–18% clay (moderate set);
NHL 5: Natural hydraulic lime containing up to 25% clay (faster set).

[o]Dub out uneven surfaces prior to applying first coat. [1]May require more coats due to waviness of bales.
*Insulation blocks have very high suction; be careful to maintain moisture content in render/plaster mixes in accordance with manufacturer's details. **Lime Hemp plaster is preferred in these situations, applied in accordance with lime specialist's details (available from Ty-Mawr at: www.lime.org.uk).

**Notes:** The above table contains suggested mortar mixes only and the actual mortar mix, build-up ant thickness of coats is to be specified by a suitably qualified and experienced conservation specialist to be suitable for the type of wall material and degree of exposure. Exposed elevations may require additional coats.
Reproduced by permission of Ty-Mawr Lime Ltd.

**Table 6.4:** Mix ratio for lime render/plaster coats

| Application | Type of lime | Lime : aggregate mix ratio by volume | Comments |
|---|---|---|---|
| **Internal plaster** | | | |
| Base/levelling coats | As above table | 1:2.5 or 1:3 sand/aggregate | Add hair/fibre at 1.5 kg per tonne to provide tensile strength (unless using polypropylene render mesh, and it is trowelled into the first coat.) |
| Top/finishing coat | As above table | 1:2.5 or 1:3 fine sand | Use finer sand |
| **External render** | | | |
| Base/levelling coats | As above table | 1:2.5 or 1:3 sand/aggregate | |
| Top/finishing coat | As above table | 1:2.5 or 1:3 fine sand | Use finer sand |
| **Harling/roughcast finish coat** | As above table | 1:2.5 or 1:3 course sand | Apply to external render with harling trowel or Tyrolene machine |

Notes: The above are suggested render/plaster mixes only and the actual mix is to be specified by a suitably qualified and experienced conservation specialist to be suitable for the type of wall material and degree of exposure.
Reproduced by permission of Ty-Mawr Lime Ltd.

**Table 6.5:** Compressive strengths for lime

| Type of lime | Typical compressive strength (N/mm$^2$) (tested at 28 days – greater strengths achieved thereafter) |
|---|---|
| **Traditional Fat Lime**(non-hydraulic)* | 0.3–0.5 |
| **Hydraulic lime*** | |
| NHL 2 | 1.3–2.0 |
| NHL 3.5 | 2.0–4.5 |
| NHL 5 | 5.0–10.0 |
| Limecrete floors (produced by Ty-Mawr with LABC Type Approval) | 4.0 (increases to 6.5 at 56 days and 8.3 at 90 days) |

Notes: *Increased strength reduces permeability and flexibility.
Reproduced by permission of Ty-Mawr Lime Ltd.

naturally white or off-white and has a matt finish. It can be coloured using the addition of pigments, which can cause slight colour variation across the surface and a slightly blotchy appearance, which is normal. Limewash adheres by suction to lime renders and plasters, stone, brick and similar materials, but not to modern materials. It sets when exposed to carbon dioxide in the air.

Where necessary, a shelter coat consisting of lime putty and fine aggregate can be applied with a soft bristle brush over bare stone to provide a key on hard, non-porous surfaces and that allows the limewash to stick and remain on the surface.

Limewash is to be applied and protected in accordance with the limewash manufacturer's details or as specified by a suitably qualified and experienced conservation specialist. Limewash is normally applied vigorously and pushed into the surface/cracks with a stiff brush in thin layers (to the consistency of single cream – otherwise, if too thick, it will crack and crumble), applied in three coats minimum, allowing at least 12 hours between each coat for carbonation to take place before the next coat is applied. Gently water-mist surfaces between coats. Protect the works as necessary from direct winds, sun, rain and frost using protective coverings, i.e. hessian and polythene. Limewash normally requires reapplication every 4–5 years, depending on exposure and application.

### *Clay paint*

Clay paint is suitable for most internal wall surfaces. It is a solvent-free, breathable paint, helps to balance the indoor humidity of the room and is available in a range of soft and rich colours. Clay paint is to be applied and protected in accordance with the paint manufacturer's details or as specified by a suitably qualified and experienced conservation specialist.

### *Plant-based/natural emulsion wall paints*

Natural emulsion/resin wall paints are plant-based, vapour-permeable and aesthetically soft, but durable and suitable for most backgrounds internally. These paints are to be applied and protected in accordance with the paint manufacturer's details or as specified by a suitably qualified and experienced conservation specialist.

### *Mineral-based wall paints*

Silicate paint is a vapour-permeable and durable paint with a lustre similar to limewash, and can be applied over existing internal coatings with the correct preparation/primers/bonding coats, as specified by the paint manufacturer.

Silicate masonry paint (not silicone paint), is also available as an exterior paint, free of resins, solvents and biocides. It is vapour-permeable (and can be used as an alternative to limewash), is water-repellent, non-flammable, non-toxic, light-fast and mould-resistant. It has a serviceable life of 25 years when used in conjunction with a clear hydrophobing agent correctly applied as the manufacturer's details. **Note:** This paint bonds to the wall surface and should only be applied in consultation with the local authority's conservation officer if it is applied directly to stone/brick on a listed building. Mineral paints are to be applied and protected in accordance with the paint manufacturer's details, or as specified by a suitably qualified and experienced conservation specialist.

### *Clear protective coatings*

Proprietary clear protective coatings consist of highly alkali-resistant hydrophobing agents (silane–siloxane-based organosilicon substances) and can be applied to absorbent, porous stone/brick surfaces to provide a clear, long-term protection from penetrating humidity, pollution and infiltration through noxious substances in porous mineral building materials, while maintaining vapour permeability. They maintain the aesthetic appearance of the stone/brick, but provide extra protection from rain where required. Proprietary clear coatings are to be used only where specified by a suitably qualified and experienced conservation specialist and must be applied and protected in accordance with the protective coating manufacturer's details. **Note:** Clear protective coating should only be applied, in consultation with the local authority's conservation officer, if it is applied directly to stone/brick on a listed building. Further information and protective coatings can be obtained from Ty-Mawr; see www.lime.org.uk.

### Application of coatings/paints/systems, protection, storage and aftercare

These are to be in accordance with the manufacturer's details, as specified by a suitably qualified and experienced conservation specialist.

Provide all personal protective equipment (PPE) in accordance with current health and safety legislation and temporary protection to the works as necessary, and in accordance with the manufacturer's details.

# Index

accessibility   2.145, 3.26–3.32
agricultural buildings   1.9
air change rates   2.99, 2.100, 3.23–3.24
air ducts, fire safety   2.87
air permeability   3.6, 3.9–3.10
air pressure testing   3.9–3.10
air vents   2.129
   *see also* mechanical extract ventilation
air-admittance valves   2.112
alternating tread stairs   2.139–2.140
ancient monuments   2.99, 2.141
ancillary buildings   1.10
Approved Documents   1.4
approved inspectors   1.8
asbestos   1.15, 5.13
attached garages
   doorways from domestic accommodation   2.27
   electrical installations   2.149
   typical plan layout   5.17, 5.30
   *see also* integral garages

background ventilation   2.100, 3.23
barn conversions   5.40–5.49
basements   2.16–2.18
   conversions   5.38–5.40
   means of escape   2.85–2.86
   ventilation   2.100
bat protection   1.13
bathrooms
   design room temperature   2.124
   electrical installations   2.148, 2.149
   provision   2.101
   ventilation   2.100, 3.24
BBA (British Board of Agreement)   1.5
beams   *see* structural columns/beams
below ground pipes   2.104, 2.106
below ground structures   *see* basements
block paving   2.150
blockwork   *see* masonry
boarding   *see* weatherboarding
boilers, hot water   2.124–2.125

boundaries, fire protection distances   2.91
breathable buildings   5.13, 6.3–6.4
breathable paints   6.12
breather membranes
   pitched roofs   2.68
   timber-framed walls   2.44, 2.45
brickwork   *see* masonry
British Board of Agreement (BBA)   1.5
Building (Approved Inspectors etc.) Regulations 2010   1.8
Building Act 1984   1.3
building control approval   1.7–1.8
building control notices   1.9
building log book   2.145
Building Notice application   1.7
Building Regulations 2010   1.3
buttressing walls   2.28–2.29, 2.30, 2.37

carbon emission targets   3.4–3.5
carbon monoxide alarms   2.129
carports   1.10
cavity barriers   2.35, 2.43
   around windows   2.86
   fire resistance   2.90
cavity closers   2.35, 2.43, 2.49
   basement conversions   5.40
   flat roofs   2.75
   party walls   2.55
cavity spacers   2.35, 2.38
cavity walls   2.35–2.41, 2.97
   garage conversions   5.29–5.34
   inspection   5.11
   insulation   2.35, 2.37, 2.39–2.40, 2.98, 3.13
      timber-framed walls   2.44, 2.45, 2.46, 3.14
   loft conversions   4.5, 4.7
   timber-framed   2.41–2.47, 3.14
   *see also* external walls
CDM (Construction (Design and Management) Regulations) 2007   1.15
ceiling joists   2.62, 2.63
   notches/holes/cuts in   5.9
   timber sizes   2.65

*Guide to Building Control: For Domestic Buildings*, First Edition. Anthony Gwynne.
© 2013 John Wiley & Sons, Ltd. Published 2013 by John Wiley & Sons, Ltd.

ceilings  2.57
    fire resistance  2.90
    height  2.60
    surface spread of flame  2.89
cement mortars and renders  2.75–2.77
certification  1.5
cesspools  2.118–2.119
change of use  5.5–5.6
chimneys
    fire protection  2.125, 2.131
    flue block  2.130
    heights  2.132
    inspection and cleaning  2.130–2.131
    loft conversions  4.4
    masonry  2.125–2.126, 2.127
    metal  2.129, 2.130, 2.131
    providing restraint  2.32
    see also flues
cladding
    timber  2.41–2.43
    timber-framed walls  2.45–2.46, 3.15
clay heave protection  2.13, 2.14–2.15, 2.21–2.24
clay paint  6.16
clear protective coatings  6.16
Code for Sustainable Homes  3.12, 3.15, 3.18
cold water supplies  2.100–2.102, 2.143
columns  see structural columns/beams
combined sewers  2.120, 2.124
combustion air vents  2.100, 2.129, 2.132
combustion appliances  see gas heating appliances; oil heating appliances; solid fuel appliances
commissioning certificates
    combustion appliances  2.135
    fixed building services  2.103, 2.144
    gas heating appliances  2.133
    hot-water storage  2.101
    oil heating appliances  2.133
    renewable energy systems  2.135
compartment floors
    fire resistance  2.90
    openings  2.86
    separating buildings  2.91
compartment walls
    fire resistance  2.90
    lengths  2.28
    minimum thickness  2.29–2.30, 2.32
    openings  2.86
    separating buildings  2.91
Competent Person Schemes (CPS)  1.10–1.11
concrete, frost protection  1.13–1.14
concrete mixes  2.9–2.10
concrete paths, etc.  2.149–2.150
condensation risks  1.5–1.6, 2.18, 2.97
consequential improvements  2.144
conservation areas
    energy conservation  2.141
    planning consents  1.14

conservation of fuel and power  see energy performance
Conservation of Habitats and Species Regulations 2010  1.13
conservatories  1.10
Construction (Design and Management) Regulations 2007 (CDM)  1.15
consumer units
    accessible  3.28
    electrical installations  2.149
    flood protection  2.147
contaminants, resistance to  2.92–2.96
    see also radon protection
contaminated materials/soil  1.15
contravention of regulations  1.8
Control of Asbestos Regulations 2012  1.15
controls, accessible  3.28, 3.29
conversions
    barns  5.40–5.49
    basements  5.38–5.40
    feasibility  5.6–5.13
    garages  5.16–5.38
    lofts Section  4
    material change of use  5.5–5.6
    sound insulation  5.51–5.52
    underpinning for  5.14–5.16
corridor widths  3.27
covered yards/ways  1.10
CPS (Competent Person Schemes)  1.10–1.11
cracking, masonry  5.7
curtain walls, U values  3.6

damp penetration  5.11, 6.3
damp-proof course trays, etc.  2.96
damp-proof courses (dpcs)  2.96, 5.11
daylighting  3.5
demolitions  1.12
Design Emission Rate (DER)  3.6
disabled access  2.145, 3.26–3.32
discharge pipes from safety devices  2.102, 3.24–3.25
discharge stacks  2.110, 2.111, 2.112
doors
    accessibility  3.27
    fire-resisting  2.83, 4.13–4.15
    safety glazing  2.146, 2.147
    U values  2.142, 3.6
    see also external doors; internal doors
dormer roofs  2.69
    loft conversions  4.6, 4.9–4.10
    U values  2.144
double glazing, U values  2.142, 3.7
drain trenches  2.106–2.107
drainage  see foul-water drainage; rainwater drainage
drainage fields  2.115, 2.117
drainage mounds  2.115
draught-proofing  2.142, 3.8
drives  2.149–2.150
Dwelling Emission Rate (DER)  3.5

electrical installations   2.147–2.149
electrical water heating   2.102
elements of structure, fire resistance   2.89, 2.90
emergency escape   *see* means of escape
emulsion paint   6.16
energy performance
   additional requirements   1.4–1.5
   conversions   5.13, 5.53
   domestic extensions   2.141–2.145
   new buildings   3.4–3.19
   operating information   3.10
   *see also* thermal insulation
Energy Performance Certificates (EPC)   3.5, 5.53
energy-efficient lighting   2.143, 3.9
entrance doors, accessibility   3.27
escape doors   2.85
escape from building   *see* means of escape
escape stairs   2.86–2.87, 2.88
   loft conversions   4.15–4.16
   *see also* protected stairways
escape windows   2.85, 2.99, 4.12–4.13
exempt buildings and work   1.9–1.10
extensions Section   2
   cavity wall insulation   2.97–2.98
   combustion appliances   2.124–2.133
   conservation of fuel and power   2.141–2.145
   disabled access   2.145
   drainage   2.103–2.124
   electrical safety   2.147–2.150
   fire safety and means of escape   2.78–2.92
   fuel storage   2.134
   glazing safety   2.146
   intermediate upper floors   2.56–2.60
   internal partitions   2.56
   mortars, renders and gypsum plasters   2.75–2.76
   protection from falling, collision and impact   2.135–2.140
   renewable energy   2.134–2.135
   resistance to contaminants   2.92–2.96
   resistance to moisture   2.96–2.97
   roofs
      flat   2.70–2.75
      pitched   2.60–2.70
   sanitation   2.100–2.103
   separating walls and floors   2.53–2.55
   sound insulation   2.98–2.99
   sub-structure   2.9–2.27
   superstructure   2.28–2.53
   ventilation   2.99–2.100
   water supply   2.100–2.103
external boarding   *see* weatherboarding
external doors
   accessibility   3.27
   closing around   2.141–2.142, 3.8
   means of escape   2.85
   opening widths   3.27
   sealing and draught-proofing   3.8

thresholds   3.27
   U values   2.142, 3.6
external lighting   3.9
external paths, etc.   2.149–2.150
external sheathing   2.41–2.43
external stairs
   accessible   3.26–3.27
   fire exits   2.86–2.87, 2.88, 4.15–4.16
   guarding and landings   2.138–2.139
external walls   2.28–2.53
   barn conversions   5.43–5.44, 5.48, 5.49–5.51
   cavity walls   *see* cavity walls
   defective   5.7
   fire resistance   2.86–2.87, 2.88, 2.89, 2.90
   garage conversions   5.21–5.26
   insulation   2.44, 2.45–2.46, 3.13–3.15
   lateral restraint   2.28–2.29
   loft conversions   4.5, 4.7
   minimum thickness   2.29–2.30, 2.32
   sizes and proportions   2.28
   solid masonry   *see* solid masonry walls
   timber-framed   *see* timber-framed walls
   U values   2.144, 3.6
   *see also* cavity walls
extract ventilation   *see* mechanical extract ventilation

fire detection and alarm systems   2.78
fire exits   *see* means of escape
fire resistance
   cavity barriers   2.90
   ceilings   2.57, 2.90
   compartment floors   2.90
   compartment walls   2.90
   elements of structure   2.89, 2.90
   external walls   2.86–2.87, 2.88, 2.89, 2.90
   lift shafts   2.90
   load-bearing walls   2.90
   party floors   2.91
   party walls   2.53, 2.54–2.55, 2.91
   protected stairways   2.90
   roofs   2.90
   separating floors   2.53–2.54, 2.90
   separating walls   2.90
   structural columns/beams   2.90
   and thermal insulation   2.43
   upper floors   2.84, 2.90
fire separation   2.83
   between an integral garage and dwelling   2.86, 2.87
   loft conversions   4.15–4.16
fire-fighting vehicle access   3.19
fireplaces   2.125
   flues   2.130
   minimum air gap and wall thickness   2.43–2.44
fire-resisting construction   2.83–2.84, 2.86
   *see also* cavity closers; protected stairways
fire-resisting doors   2.83, 4.13–4.15
flame spread, internal linings   2.89

flat roofs
  with cold deck  2.70–2.72
    coverings  2.70–2.71
  garage conversions  5.19–5.20
  green roofs  2.74
  insulation  2.72, 2.73, 2.145
  timber sizes  2.76
  U values  2.144
  with warm deck  2.72–2.74
    coverings  2.72, 2.74
    inverted  2.74
  workmanship  2.75
floating floors  2.24–2.25
flood protection  2.97, 2.147
floor areas  1.10, 2.28
floor joists  2.56–2.58
  conversions  5.8
  notches/holes/cuts in  5.9
  strapping to walls  2.49–2.51
floor screed  2.23, 5.26–5.28, 5.35–5.36, 5.48–5.49
floorboards  2.57–2.58
floors
  sound insulation  2.98
  thermal insulation  2.21–2.27, 2.145
  U values  2.144, 3.6, 3.11–3.12
  see also ground floors; upper floors
flues
  configuration  2.127, 2.130
  connecting pipes  2.129
  heights  2.132
  inspection and cleaning  2.130–2.131
  liners  2.125–2.126
  minimum air gap and wall thickness  2.42–2.43, 2.125, 2.128
  notice plates  2.133
  repair/relining  2.132–2.133
  sizes  2.131
  see also chimneys
foul-water disposal  2.108–2.113, 5.12
foul-water drainage  2.103–2.108, 5.12
foundations  2.9–2.16
  clay heave protection  2.13–2.15
  concrete mixes  2.9–2.10
  conversions  5.7
  underpinning  5.14–5.16
frost protection  2.143
fuel storage tanks  2.134
Full Plans application  1.7

gable-end walls  2.51
galleries, means of escape  2.85
galletting  6.8
garages  1.10
  conversions
    cavity wall  5.29–5.34
    single wall  5.16–5.29
  electrical installations  2.149

external walls  2.47
fire separation  2.86
floor slope  2.86
ground-bearing concrete floors  2.27
separating walls  2.37, 2.41
sizes and proportions  2.47, 2.48
garden sheds  1.10
gardens  2.149–2.150
gas heating appliances  2.133, 3.6
Gas Safe  2.133
glazing
  double and triple  2.142, 3.7
  on escape routes  2.84
  insulation  2.141–2.142
  protected stairways  4.14, 4.15
  total area of  2.141, 3.22
  U values  2.142, 3.6, 3.7
  see also windows
gravel paths  2.150
green roofs  2.74
greenhouses  1.9
grey water harvesting  2.120
ground floors  2.19–2.27
  conversions  5.7–5.8
    barn conversions  5.45–5.51
    garage conversions  5.26–5.29, 5.34–5.38
  insulation  2.21–2.27, 3.11–3.12
ground-bearing solid concrete floors  2.20–2.21
  conversions  5.8, 5.36–5.37, 5.46–5.48
guarding, stairs, etc.  2.136–2.137, 2.138–2.139, 3.28
gullies  2.106
gypsum plasters  2.77

haired lime plaster  6.12
handrails  2.136, 2.137, 2.138, 2.140, 3.27
headroom  2.60
  loft conversions  4.4
  stairs  2.135, 2.139–2.140
Health and Safety at Work etc. Act 1974 (HSWA)  1.15
hearths
  notice plates  2.133
  open fireplaces  2.125
  solid-fuel stoves  2.127, 2.128
heat alarms  2.78
heating appliances  see hot-water systems; space heating appliances
height of building  2.28
height of ceilings  see headroom
height of storeys  2.28, 2.29
hot water supplies  2.100–2.102
hot-water safety  3.24
hot-water storage  2.101
hot-water systems  2.124–2.125
  insulation  2.125, 2.143
  thermal efficiency  3.6
hydraulic lime  6.11–6.12

insect attack  5.10
inspection chambers  2.106, 2.108, 2.109
insulation  *see* sound insulation; thermal insulation
integral garages
    fire separation  2.86, 2.87, 2.90
    radon protection  2.27
    *see also* attached garages
intermediate upper floors  *see* upper floors
internal doors
    air gap  3.22
    fire-resisting  2.83, 4.13–4.15
internal lighting  2.143, 3.9
internal load-bearing walls
    loft conversions  4.11
    minimum thickness  2.30, 2.56
internal partitions  2.56
internal stairs  2.135–2.138
    minimum width  3.27
    *see also* protected stairways
internal walls
    sound insulation  2.98, 3.21, 5.52
    surface spread of flame  2.89
intumescent paint/paper system  2.83

Japanese knotweed  1.14
joist hangers/caps  2.57

kitchens
    design room temperature  2.124
    electrical installations  2.148
    ventilation  3.24

lamps, energy-efficient  2.143, 3.9
landfill gas  2.96
landings  2.136, 2.137, 3.27
lateral restraint  2.28–2.29
    roofs to walls  2.51–2.52
    upper floors to walls  2.49–2.51
    walls at ceiling level  2.52–2.53
lead paint  1.15
lean-to roof abutment  2.62–2.63, 2.70
Lifetime Homes  3.31–3.32
lift shafts, fire resistance  2.90
light wells  2.68
lighting, energy-efficient  2.143, 3.9
lime, compressive strength  6.15
lime mortars, renders and plasters  6.5–6.8, 6.10–6.12
lime putty  6.10–6.11
lime techniques Section  6
limewash  6.12, 6.15–6.16
lintels  2.48
    cavity walls  2.35, 2.37, 2.43
    defective  5.7
listed buildings
    conversions  5.51
    energy conservation  2.141

planning consents  1.14
    sound insulation  3.20
load-bearing walls, fire resistance  2.90
local authority building control  1.7
lofts
    conversions Section  4
        feasibility  4.3–4.4
        fire safety and means of escape  4.11–4.16
        structure  4.5–4.11
    fixed ladders  2.140
    hatches, doors and light wells  2.69
    stairs to  2.139
LPG storage tanks  2.134

maintenance of old buildings Section  6
manholes  2.109
manual controls  3.28, 3.29
masonry
    chimneys  2.125–2.126, 2.127
    compressive strength  2.32–2.34
    cracking  5.7
    external walls
        external skin to timber-framed walls  2.42–2.43
        inspection  5.11
        loft conversions  4.7
    internal partitions  2.56
    maximum height of building  2.28
    maximum wall lengths  2.28
    minimum thickness  2.29–2.30, 2.32
    movement joints  2.49
    natural stone-faced cavity walls  2.35–2.37, 2.38
    party walls  2.53–2.54, 2.55
    *see also* solid masonry walls
masonry paint  6.16
material change of use  5.5–5.6
materials and workmanship  1.5
means of escape  2.79–2.86
    basements  2.85–2.86
    loft conversions  4.4, 4.11–4.16
mechanical extract ventilation  2.99–2.100, 3.22, 3.24
    interaction with heating appliances  2.131, 2.133
metal chimneys  2.129, 2.130
metallic roof trims  2.75
methane protection  2.96
micro generation systems  2.134
mineral-based wall paints  6.16
moisture penetration  5.11, 6.3
mortars
    analysis  6.6
    lime-based  6.5–6.8
movement joints
    masonry  2.49
    pipework  2.107

natural emulsion paints  6.16
natural hydraulic lime (NHL)  6.11–6.12
natural lighting  3.5

natural stone-faced cavity walls   2.35–2.37, 2.38
natural ventilation   2.99, 3.21–3.22
new dwellings   Section 3
   cavity wall insulation   3.20
   combustion appliances   3.26
   conservation of fuel and power   3.4–3.19
   disabled access   3.26–3.32
   drainage   3.25–3.26
   electrical safety   3.32
   fire safety and means of escape   3.19
   fuel storage   3.26
   glazing safety   3.32
   hot-water safety   3.24–3.25
   protection from falling, collision and impact   3.26
   resistance to contaminants   3.20
   resistance to moisture   3.20
   sanitation   3.24–3.25
   sound insulation   3.20–3.21
   ventilation   3.21–3.24
   waste disposal   3.25–3.26
   water efficiency   3.25
non-hydraulic lime   6.10–6.11
non-traditional construction   5.10
Northern Ireland
   building regulations   1.3
   Lifetime Homes   3.32

oil heating appliances   2.133
oil storage tanks   2.134
oil/fuel separators   2.121–2.122
open fireplaces   2.125
   flues   2.130
   ventilation   2.132
opening up existing structure   1.13, 4.5
openings
   cavity walls   2.37
   closing around   3.8
   fire protection   2.90
   sealing   3.8
   trimmed joists   2.58, 2.59
   ventilation   2.99, 3.22, 3.23
OSB (orientated strand board)   2.41
outbuildings   1.10

paints   6.16–6.17
   lead   1.15
party floors
   conversions   5.51–5.52
   fire resistance   2.91
   sound insulation   2.98
Party Wall Act 1996   1.15–1.17
party walls   2.53
   conversions   5.51–5.52
   fire resistance   2.91
   sound insulation   2.98, 3.20, 5.9–5.10
   U values   3.6, 3.7
passageway widths   3.27
passenger lifts, fire safety   2.86

PassivHaus standard   3.18–3.19
paths   2.149–2.150
patios   2.149–2.150
paved areas   2.150
percolation tests   2.116–2.117
photovoltaic panels   3.9–3.10
piers   2.28–2.29, 2.32
piled foundations   2.16
pipework
   drainage   2.104–2.106
   insulation   2.102–2.103, 2.143
   movement joints   2.107
   penetrating though walls   2.105, 2.142, 3.8, 3.9
   see also waste pipes
pitched roofs   2.60–2.69
   barn conversions   5.41–5.43
   coverings   2.61
   cut roof construction   2.62
   garage conversions   5.14–5.19, 5.29–5.32
   insulation   2.65–2.68, 2.145, 3.11, 3.12
   loft conversions   4.6, 4.9
   roof trusses   2.62
   structure   2.62–2.69
   timber sizes   2.62, 2.64–2.65, 2.66
   U values   2.144
   underlay and undersheeting   2.61
   valleys and lead work   2.68
   ventilation   2.66
planning permission   1.14
plasters
   lime-based   6.5–6.6, 6.9–6.10, 6.12, 6.13–6.15
   repairs   6.9–6.10
porches   1.10
pozzolanic materials   6.12
precast concrete slabs   2.150
preliminary works   1.11–1.14
pressure testing   3.9–3.10
private sewers, transfer arrangements   1.12, 2.122–2.123
private water supplies   2.101
professional advice   1.6
protected stairways   2.81–2.82, 2.83–2.84
   fire resistance   2.90
   loft conversions   4.13–4.15
protection of the works   1.13–1.14
protective coatings   6.16–6.17
public sewers   1.12
   building over or close to   2.122–2.123
   connections to   2.120, 2.123
pumped drainage   2.101, 2.112–2.113
purge ventilation   2.99, 3.21–3.22

radiator valves   2.124
radon protection   2.19, 2.20, 2.21, 2.93–2.95, 5.10
raft foundations   2.13, 2.16
rainwater disposal   2.120–2.121
rainwater down pipes   2.119
rainwater drainage   2.104, 2.119–2.122, 2.150
rainwater gutters   2.119

rainwater harvesting  2.120
ramps  2.140, 3.26, 3.29
recesses
    cavity walls  2.37
    fireplace walls  2.135
reed-bed systems  2.115–2.116
refuse storage  3.26
Regularisation certificates  1.7
renders  2.75–2.77
    analysis  6.6
    lime-based  6.5–6.6, 6.9–6.10, 6.12, 6.13–6.15
    repairs  6.9–6.10
    timber-framed walls  2.44, 2.45, 3.15
renewable energy systems  2.134
replacement windows, fire safety  2.86
repointing  6.6–6.8
retaining walls  2.16–2.17
Rights of Light  1.17–1.18
rising damp  5.11, 5.48, 6.3
roof pitch  2.63
roof trusses  4, 2.62, 4.3
roof underlay  2.61
roof valleys  2.68, 2.70
roof windows, U values  2.142, 3.6
roofs
    conversions  5.8–5.9
        garage conversions  5.16–5.20
        loft conversions  4.3–4.4, 4.5–4.10
    coverings  2.91, 2.92, 5.9
    fire resistance  2.90
    insulation  3.12, 3.16–3.17
    strapping to walls  2.51–2.52
    U values  2.144, 3.6, 3.16–3.17
    ventilation  3.12, 5.12
    *see also* dormer roofs; flat roofs; pitched roofs
room temperatures  2.124

safety glazing  2.146, 2.147
sanitary pipework  2.108, 2.111
sanitary rooms  *see* bathrooms; WCs
SAP (Standard Assessment Procedure)  3.4
Scotland, building regulations  1.3
sealing
    openings  2.142, 3.8, 3.9
    service penetrations  2.105, 2.142, 3.8, 3.9
separating floors
    fire resistance  2.53–2.54, 2.90
    garages  2.90
    sound insulation  3.20, 5.9–5.10
separating walls  2.53–2.56
    fire resistance  2.90
    garages  2.37, 2.41, 2.90
    lengths  2.28
    minimum thickness  2.29–2.30, 2.32
    *see also* party walls
septic tanks  2.113–2.114
service penetrations, sealing  2.142, 3.8, 3.9
sewage pump sets  2.113

sewage treatment systems  2.114–2.119
sewers  *see* private sewers; public sewers
sheds  1.10
silicate paint  6.16
site assessment  1.11–1.12
small detached buildings  1.10
smoke alarms  2.78
smoke testing  2.113
soak-aways  2.121
sockets, accessible  3.28, 3.29
soil types, foundations  2.11, 2.13
soil-and-vent pipes (SVP)  2.60, 2.110, 2.111
solar heat gain  3.5, 3.8
solar pv  3.9–3.10
solar water heating  2.102
solid fuel appliances  2.125–2.133
    air supply  2.129, 2.132
    free-standing stoves  2.126–2.129
    interaction with extract ventilation  2.131
solid masonry walls
    barn conversions  5.43–5.45
    inspection  5.11, 6.5
    insulation  2.40, 2.145, 3.14
    lime-based methods  6.5–6.10
    loft conversions  4.7–4.8
    repair/rebuilding  6.8–6.9
solid waste storage  2.124
sound insulation  2.98–2.99, 3.20–3.21
    floor voids  2.57
    party walls  2.53, 2.54–2.55
        conversions  5.9–5.10, 5.51–5.52
        upgrading  2.55
    timber stud partitions  2.56
sound testing  2.55, 2.98, 5.52
space heating appliances  2.124–2.133
    thermal efficiency  3.6
sprinkler systems  2.87–2.89, 3.19
    loft conversions  4.16
stages of works, notices  1.9
stairs  *see* external stairs; internal stairs
Standard Assessment Procedure (SAP)  3.4
statutory service authorities  1.12
stepped foundations  2.12
stone slabs  2.150
stone-faced cavity walls  2.35–2.37, 2.38
storey heights  2.28, 2.29
storm-water drainage  2.104
strapping  2.49–2.53
strip foundations  2.10–2.11
structural columns/beams  2.41, 2.48
    defective  5.7
    fire resistance  2.90
structural timber  1.13
    moisture content  2.75
    notches/holes/cuts in  2.41, 2.59, 5.9
    treatment  2.75
stub stacks  2.112
sub-structure walls  2.18–2.19, 5.47, 5.49

Supply (Water Fittings) Regulations 1999   2.103
surface coatings   6.16–6.17
surface finish, paths, etc.   2.150
surface spread of flame   2.89
surface water disposal   2.120–2.121, 5.12
surface water drainage   2.120, 2.150, 5.12
suspended beam-and-block ground floors   2.22–2.24, 3.12
suspended reinforced in-situ concrete ground floor slabs   2.21–2.22, 3.11
suspended timber ground floors   2.25–2.27
   garage conversions   5.37–5.38
sustainable homes   1.5, 3.12, 3.15, 3.18
swimming pool basins   2.144
switches, accessible   3.28, 3.29

tanking systems   2.18, 2.97, 5.11–5.12
   barn conversions   5.47, 5.48
   basement conversions   5.39–5.40
   garage conversions   5.25
Target Emission Rate (TER)   3.4
tarmac areas   2.150
temperature control   2.124
temporary buildings   1.10
thermal bridging   3.6, 3.8
thermal insulation
   closing around window and door openings   2.141–2.142, 3.8
   cold water supplies   2.143
   condensation risks   1.5–1.6
   and fire resistance   2.43
     floors   2.21–2.27, 2.60, 2.145, 3.11–3.12
     floating floors   2.24–2.25
     ground-bearing floor slabs   2.21
     suspended beam-and-block ground floors   2.23–2.24
     suspended reinforced in-situ concrete ground floor slabs   2.22
     upper   2.60, 3.11, 3.15
   hot-water systems   2.125, 2.143
   loft conversions   4.5–4.10
   pipework   2.102–2.103, 2.143
   roofs   3.11, 3.16–3.17
     cold deck flat roofs   2.70, 2.72, 2.145
     dormer   4.9–4.10
     pitched roofs   2.65–2.68, 3.11
     warm deck flat roofs   2.72, 2.73
   upgrading   2.144–2.145, 6.7
   walls
     cavity   2.35, 2.37, 2.39–2.40, 2.44, 2.97–2.98
     external in new buildings   3.11, 3.15
     between heated and unheated areas   2.37, 2.41, 3.14
     solid masonry   2.40, 2.145, 3.14, 4.7–4.8, 6.7
     timber-framed   2.44, 2.46, 3.14–3.15, 4.12
   water tanks/cisterns   2.143
   *see also* U values
thermal mass   3.6

third-party-accredited certification   1.5
timber cladding   2.41–2.42
timber decay   5.10
timber stud partitions   2.56
timber-framed construction   2.41, 5.10
timber-framed walls   2.41–2.47
   cladding   2.44, 2.47
   fire safety   2.43–2.44
   insulation   3.14–3.15, 4.12
   loft conversions   4.11
   party walls   2.54–2.55
   renders   2.44, 2.45
   U values   3.14–3.15, 4.12
timber-sizing tables   1.6
toilets   *see* WCs
trees, heave protection   2.12, 2.14–2.15
trench-fill foundations   2.11–2.13
trimmer/trimming joists   2.58, 2.59
triple glazing, U values   3.7
trussed roofs   2.62, 4.3
tundish   2.102, 3.24–3.25
twin-walled metal chimneys   2.129, 2.130

U values
   doors   2.142, 3.6
   floors   2.144, 3.6
     exposed upper   3.15
     garage conversions   5.45
     ground   3.11–3.12
   glazing   2.142, 3.6, 3.7
   roof lights   3.6
   roofs   2.144, 3.6, 3.16–3.17
     garage conversions   5.16, 5.19, 5.29, 5.33, 5.42–5.45
   swimming-pool basins   2.144
   walls
     cavity   3.13, 3.14, 4.5–4.6, 4.8, 5.33–5.34
     curtain walling   3.6
     external   2.144, 3.6, 3.15
     garage conversions   5.33–5.34
     between heated and unheated areas   3.14
     loft conversions   4.5, 4.7, 4.8, 4.12
     party walls   3.6, 3.7
     solid masonry   4.7, 4.8, 5.43
     timber-framed walls   3.14–3.15, 4.12
   windows   2.142, 3.6
under-floor heating systems   2.24
underground pipes   2.104, 2.106
underpinning   5.14–5.16
unvented hot-water storage   2.101
upgrading
   sound insulation   2.55
   thermal insulation   2.144–2.145, 6.7
upper floors   2.56–2.60, 2.61
   conversions   5.8, 5.43–5.45
   fire resistance   2.84, 2.90
   insulation   2.60, 3.15

lateral restraint 2.49
means of escape 2.84
sound insulation 5.52
thermal insulation 3.11
*see also* lofts, conversions
uPVC
cladding 2.45–2.46
pipework 2.86, 2.104, 2.110
soffits, fascias and barge boards, etc. 2.62
urea–formaldehyde (UF) 2.98
utility companies 1.12

vapour-control barrier 2.75
vapour-permeable membranes *see* breather membranes
vented hot-water storage 2.101
ventilation 2.99–2.100, 3.21–3.24
garage conversions 5.12, 5.42–5.43
to heating appliances 2.100, 2.129, 2.132
loft conversions 4.4
roofs 3.11
ventilation ducts, fire safety 2.87
ventilation rates 2.100, 3.23–3.24
vertical lateral restraint 2.28–2.29

Wales
building regulations 1.3
conservation of fuel and power 1.5
Lifetime Homes 3.32
planning requirements 3.15
sprinkler systems 3.19
wall plates 2.52
wall ties 2.35, 2.36, 2.43, 2.49
wall tiling 2.45
walls
abutments 2.47
defective 5.7
between heated and unheated areas 2.37, 2.41, 3.14
heights and lengths 2.28, 2.31, 2.32
lateral restraint 2.28–2.29

U values 3.6
*see also* external walls; internal walls; party walls
waste pipes 2.108–2.112
waste storage 3.25–3.26
waste traps 2.108, 2.110
waste water treatment 2.113
water efficiency 3.25
water heating *see* hot-water systems
water supply service pipes 2.103
water tanks/cisterns 2.101
insulation 2.103, 2.143
WCs 2.101
accessible 3.29–3.30
design room temperature 2.124
ground floor 3.29
ventilation 3.22
weatherboarding 2.41, 2.45, 2.46
*see also* cladding
weep holes 2.48
wet rooms
ventilation 2.100, 3.22, 3.23
*see also* bathrooms; kitchens; WCs
wetland drainage treatment 2.115–2.116
wheelchair access 3.27–3.30
wholesome water supply 2.100–2.101
Wildlife and Countryside Act 1981 1.13
windows
closing around 2.141–2.142, 3.8
energy conservation 2.141–2.142
guarding 2.139
maximum area of 2.141, 3.22
means of escape 2.85, 2.99, 3.22, 4.12–4.13
sealing and draught-proofing 2.142, 3.8
U values 2.142, 3.6
ventilation 2.99, 3.21–3.22
*see also* glazing
work stages, notices 1.9
workmanship *see* materials and workmanship
workshops 1.10

# WILEY-BLACKWELL

## Other Books Available from Wiley-Blackwell

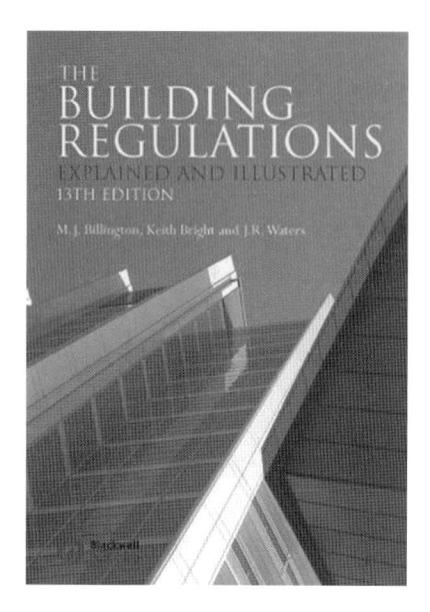

The Building Regulations:
Explained and Illustrated
13<sup>th</sup> Edition
Billington, et al.
978-1-4051-5922-7

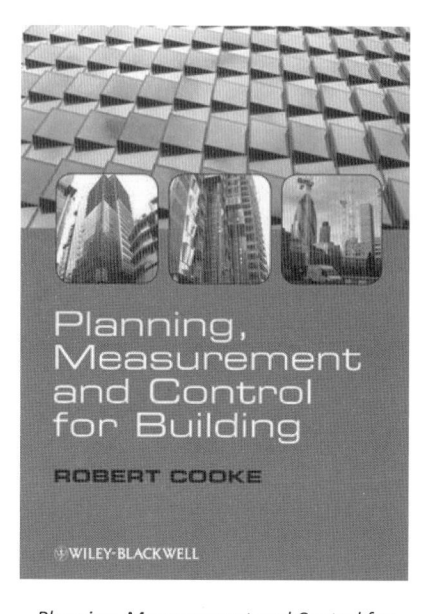

Planning, Measurement and Control for Building
Robert Cooke
978-1-4051-9139-5

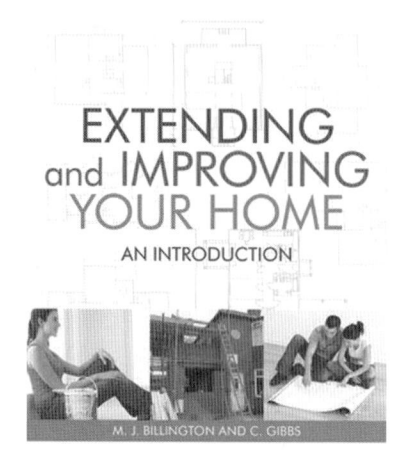

Extending and Improving Your Home
Billington & Gibbs
978-1-4051-9811-0

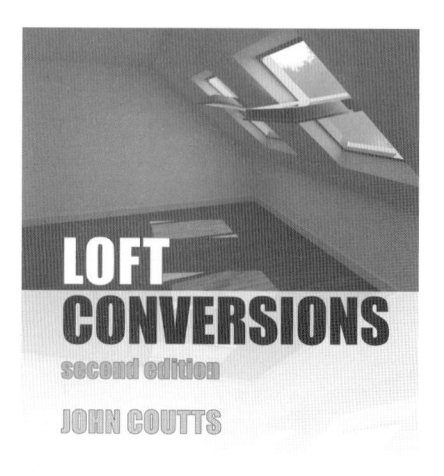

Loft Conversions
2<sup>nd</sup> edition
Coutts
978-1-1184-0004-3

www.wiley.com/go/construction